Magnetic Nanom
Analytical Chemis

Magnetic Nanomaterials in Analytical Chemistry

Edited by

Mazaher Ahmadi
Faculty of Chemistry, Bu-Ali Sina University, Hamedan, Iran

Abbas Afkhami
Faculty of Chemistry, Bu-Ali Sina University, Hamedan, Iran

Tayyebeh Madrakian
Faculty of Chemistry, Bu-Ali Sina University, Hamedan, Iran

ELSEVIER

Elsevier
Radarweg 29, PO Box 211, 1000 AE Amsterdam, Netherlands
The Boulevard, Langford Lane, Kidlington, Oxford OX5 1GB, United Kingdom
50 Hampshire Street, 5th Floor, Cambridge, MA 02139, United States

British Library Cataloguing-in-Publication Data
A catalogue record for this book is available from the British Library

Library of Congress Cataloging-in-Publication Data
A catalog record for this book is available from the Library of Congress

ISBN: 978-0-12-822131-0

For Information on all Elsevier publications
visit our website at https://www.elsevier.com/books-and-journals

Publisher: Susan Dennis
Acquisitions Editor: Kathryn Eryilmaz
Editorial Project Manager: Cole Newman
Production Project Manager: Debasish Ghosh
Cover Designer: Christian J. Bilbow

Typeset by MPS Limited, Chennai, India

Contents

List of contributors

Mahnaz Adeli Knowledge Development & University Relationship Department, Iran Khodro Company, Tehran, Iran

Abbas Afkhami Faculty of Chemistry, Bu-Ali Sina University, Hamedan, Iran

Roya Afsharipour Department of Chemistry, Faculty of Science, Yazd University, Yazd, Iran

Sajjad Aftabi Department of Human Anatomy and Cell Science, Max Rady College of Medicine, Rady Faculty of Health Sciences, University of Manitoba, Winnipeg, MB, Canada

Samira Aghajani Hamedan University of Technology, Hamedan, Iran

Miguel Ángel Aguirre Department of Analytical Chemistry and Food Science and University Institute of Materials, Faculty of Science, University of Alicante, Alicante, Spain

Mazaher Ahmadi Faculty of Chemistry, Bu-Ali Sina University, Hamedan, Iran

Yasser Al-Eryani Department of Electrical and Computer Engineering, University of Manitoba, Winnipeg, MB, Canada

Zahra Amouzegar Faculty of Chemistry, Bu-Ali Sina University, Hamedan, Iran

Kholoud Arafa Center for Materials Science, Zewail City of Science and Technology, Giza, Egypt

Hasan Bagheri Chemical Injuries Research Center, Systems Biology and Poisonings Institute, Baqiyatallah University of Medical Sciences, Tehran, Iran

Mohammad Reza Bayatloo Department of Analytical Chemistry and Pollutants, Shahid Beheshti University, Tehran, Iran

Hadi Beitollahi Environment Department, Institute of Science and High Technology and Environmental Sciences, Graduate University of Advanced Technology, Kerman, Iran

Antonio Canals Department of Analytical Chemistry and Food Science and University Institute of Materials, Faculty of Science, University of Alicante, Alicante, Spain

Soheila Chalavi Department of Analytical Chemistry and Pollutants, Shahid Beheshti University, Tehran, Iran

Maryam Dadashi Department of Oral Biology, Dr. Gerald Niznick College of Dentistry, Rady Faculty of Health Sciences, University of Manitoba, Winnipeg, MB, Canada

Shayessteh Dadfarnia Department of Chemistry, Faculty of Science, Yazd University, Yazd, Iran

Seyyed Mosayeb Daryanavard Department of Chemistry, Faculty of Science, University of Hormozgan, Bandar Abbas, Iran

Zahra Dourandish Environment Department, Institute of Science and High Technology and Environmental Sciences, Graduate University of Advanced Technology, Kerman, Iran

Ibrahim M. El-Sherbiny Center for Materials Science, Zewail City of Science and Technology, Giza, Egypt

Mohammad Faraji Research Group of Food, Halal and Agricultural Products, Research Department of Food Technology and Agricultural Products, Standard Research Institute (SRI), Karaj, Iran

Farnaz Fazlali Research and Development Department, Farin Behbood Tashkhis LTD, Tehran, Iran

Mostafa Fytory Center for Materials Science, Zewail City of Science and Technology, Giza, Egypt

Ghazal Ghavami Faculty of Science, University of Waterloo, Waterloo, ON, Canada

Saeid Ghavami Department of Human Anatomy and Cell Science, Max Rady College of Medicine, Rady Faculty of Health Sciences, University of Manitoba, Winnipeg, MB, Canada; Research Institute in Oncology and Hematology, CancerCare Manitoba, University of Manitoba, Winnipeg, MN, Canada; Autophagy Research Center, Shiraz University of Medical Sciences, Shiraz, Iran; Faculty of Medicine, Katowice School of Technology, Katowice, Poland

Arash Ghoorchian Faculty of Chemistry, Bu-Ali Sina University, Hamedan, Iran

Ali Hajian Institute of Sensor and Actuator Systems, TU Wien, Vienna, Austria

Pegah Hashemi Research and Development Department, Farin Behbood Tashkhis LTD, Tehran, Iran

Peyman Mohammadzadeh Jahani Department of Basic Sciences, School of Medicine, Bam University of Medical Sciences, Bam, Iran

Mahdie Kamalabadi Faculty of Chemistry, Bu-Ali Sina University, Hamedan, Iran

Hosein Khoshsafar Research and Development Department, Farin Behbood Tashkhis LTD, Tehran, Iran

Tayyebeh Madrakian Faculty of Chemistry, Bu-Ali Sina University, Hamedan, Iran

Parisa Mahdavi Department of Analytical Chemistry and Pollutants, Shahid Beheshti University, Tehran, Iran

Fariba Garkani Nejad Environment Department, Institute of Science and High Technology and Environmental Sciences, Graduate University of Advanced Technology, Kerman, Iran

Tien Anh Nguyen Department of Physics, Le Quy Don Technical University, Ha Noi, Vietnam

Saeed Nojavan Department of Analytical Chemistry and Pollutants, Shahid Beheshti University, Tehran, Iran

Farshid Noormohammadi Chemical Engineering Division, College of Environment, UoE, Karaj, Iran

Zachary William Oldham Faculty of Engineering, University of Waterloo, Waterloo, ON, Canada

Narges Salehi Department of Chemistry, Varamine (Pishva) Branch Islamic Azad University, Varamin, Iran

Hedieh Sattarifard Department of Human Anatomy and Cell Science, Max Rady College of Medicine, Rady Faculty of Health Sciences, University of Manitoba, Winnipeg, MB, Canada

Ali Mohammad Haji Shabani Department of Chemistry, Faculty of Science, Yazd University, Yazd, Iran

Somayeh Tajik Research Center for Tropical and Infectious Diseases, Kerman University of Medical Sciences, Kerman, Iran

Lorena Vidal Department of Analytical Chemistry and Food Science and University Institute of Materials, Faculty of Science, University of Alicante, Alicante, Spain

Yadollah Yamini Department of Chemistry, Faculty of Sciences, Tarbiat Modares University, Tehran, Iran

Preface

Nanotechnology has become the primary choice of researchers to enhance the efficacy of various systems due to the different chemical and physical properties of nanoscale materials compared to bulk materials. The electrical, optical, and magnetic properties of nanomaterials are significantly dissimilar from bulk materials. Taking advantage of these novel properties and favorable dimensions of nanomaterials is particularly meaningful in analytical techniques. In this realm, higher sensitivity, lower analysis times, higher extraction efficiencies, and small environmental impacts are considered desired goals. In Analytical Chemistry, various nanosystems have been utilized to develop new methodologies. Magnetic nanomaterials with the ability to be controlled using external magnetic fields have opened new doors to develop new sample preparation methods. Also, magnetic nanomaterials have become one of the most interesting materials in developing sensors and biosensors. Magnetic nanomaterials have attracted close attention in the development of novel magnetic sorbents for food analysis, bioanalysis, environmental monitoring, and forensic chemistry. These components have a wide application usage in different fields such as sensing materials in the development of colorimetric sensors, resonance light scattering sensors, plasmonic sensors, electrochemical sensors and biosensors, and magnetic field sensors.

The journey of this book started in 2009 when we first used maghemite nanoparticles as an adsorbent for wastewater treatment in our laboratory. Although this was not the first report, it was a new path for us. Those particles were easy to synthesize and functionalize, and we could easily control them using a simple handheld magnet without any high-speed centrifugation. Since then, various applications for these magnetic nanoparticles were derived at our lab, and around 120 research papers result from this continuous effort.

The editors have taught different analytical methods at the undergraduate and doctoral levels. They consider that there is a lack of a textbook dedicated to the application of magnetic nanomaterials in this context. Thus the main purpose of this book is to provide the readers with a comprehensive review of the application of magnetic nanomaterials in Analytical Chemistry. This book has also covered the fundamental information of magnetic nanomaterials, including their classification, synthesis, functionalization, and characterization methods, separation and isolation techniques, and toxicity, fate, and safe disposal of magnetic nanomaterials.

Finally, we are grateful to our distinguished group of international contributors who have made this project truly unique and relevant. Further, we acknowledge the support of the Faculty of Chemistry at Bu-Ali Sina University, Iran. We thank our families for their patience and understanding of the time commitments necessary to conduct this project. Particularly, Dr. Ahmadi welcomes his beloved newborn daughter, Avin, who made every moment of the last 9 months special and memorable. Finally, we thank the editing and production staff at Elsevier for their valued assistance.

Mazaher Ahmadi, Abbas Afkhami and Tayyebeh Madrakian

October 2020

Section I

Fundamentals

Physical properties, classification, synthesis, and functionalization of magnetic nanomaterials

1

Ibrahim M. El-Sherbiny, Kholoud Arafa and Mostafa Fytory
Center for Materials Science, Zewail City of Science and Technology, Giza, Egypt

1.1 Physical properties of MNMs

1.1.1 *Zeta potential and colloidal stability*

Colloidal stability is vital for the applicability of the MNMs. Given the lack of applied magnetics forces, the stability of the particles is influenced by the effect of repulsive forces (mainly caused by steric clashes or electrostatic interactions), which are counteracted by attractive forces (mainly presented by van der Waals interactions) [1]. While the surface charge of MNMs relies on their synthesis components, however, the zeta potential (ZP, ζ) is mainly controlled by the charge of the formed particles as well as the ionic content and strength of the solution [2]. Conclusively, the surface charge dictates the ZP subsequently affecting the solution stability of the MNMs [3,4].

Practically, to improve the stability of bare magnetic nanoparticles (MNPs) in suspension, carboxylates, inorganic compounds, and polymeric compounds are usually used as stabilizers. These stabilizers act either via enhancing interparticulate hydrophilic interactions or contributing as steric or electrostatic barriers to overcome the undesired aggregation due to van der Waals forces [5].

The MNMs surface charges are measured by ZP, which is considered a function that correlates surface charge density with the shear plane location along with the surface structure. The knowledge of ZP is crucial as it controls the stability of MNMs both in vitro and in vivo. Thus different approaches are developed to assess ZP experimentally including streaming current, electrophoretic mobility, and electric conductivity techniques [6]. Afterward, ZP can be inferred indirectly through theoretical calculations of the obtained empirical results. In addition to the occasionally employed light-based methods, the ultrasound technique is alternatively deemed recently more attractive as it can assess concentrated samples without prior dilution [7].

1.1.2 *Size and shape*

The particle size can be regulated by varying several reaction parameters including precursors' concentration, type of the surfactant, temperature, and time all of which

Magnetic Nanomaterials in Analytical Chemistry. DOI: https://doi.org/10.1016/B978-0-12-822131-0.00015-7

play an important role in nucleation rate, growth, and nanoparticles (NPs) stability [8]. Several studies reported that employing organic-phase methods with capping properties such as oleylamine to synthesize MNMs efficiently fine-tuned their particle size as well as shape (e.g., octahedral MNMs and nanoprisms) [9,10]. Likewise, the surfactant molar ratio can regulate MNMs' growth in specific directions affecting its final morphology [11].

Importantly varying the MNMs' size not only affects their magnetic behavior but also controls their colloidal stability as well. The latter effect can be assessed through evaluating the change in the hydrodynamic particle size over a certain time course using the dynamic light scattering technique upon varying the ionic strength and/or pH of the surrounding media [12]. Turbidity measurements are also used to investigate the stability of MNMs via evaluating the aggregation kinetics [13].

The size and the shape of MNMs can be characterized using different techniques including high-resolution transmission electron microscopy and field-emission scanning electron microscopy with subatomic resolution revealing the atomic arrangement in the nanosized particles. Additionally, photon correlation spectroscopy, quasi-elastic light scattering, and X-ray diffraction techniques are employed for the determination of mean particle size and size distributions [14].

1.1.3 Coating

One of the problems faced with the development of MNMs is that they easily oxidize in atmospheric air. Thus coating their surface is frequently done to protect the magnetic content [15]. Coatings act as a shell guarding against the direct contact of the magnetic core with additional agents conjugated to the shell-material surface, thus hindering any undesired interactions. Additionally, coatings provide modifiable surfaces for the immobilization of catalysts through chemical covalent binding. Moreover, branched coatings help increase the MNMs surface area [16]. Also, coated MNMs are considered thermally stable, a property that promotes homogeneous dispersion of particles [4]. Some of the most frequently used coatings are briefly mentioned in the following sections along with an essence of their potential applications, depicted in Fig. 1.1.

1.1.3.1 Polymeric coating

There are plentiful choices of polymers either natural or synthetic that can be used to coat MNMs for various applications including polyethylene glycol (PEG), poly (vinyl alcohol), polyethylenimine, chitosan, dextran, and alginate. Surface coating of the MNMs with polymeric materials is employed to enhance the water dispersibility along with the physical and chemical colloidal stability [14]. One of which is chitosan, a naturally ubiquitous polysaccharide, frequently used to coat MNMs. Chitosan-coated MNMs are vastly employed in the biomedical field [17]. Recently various studies developed protein-imprinted polymers-coated MNPs that acted as sorbents for target proteins. Magnetic molecularly imprinted polymers were found to be both selective and robust in recognition and separation of template proteins [18]. A surface-imprinted

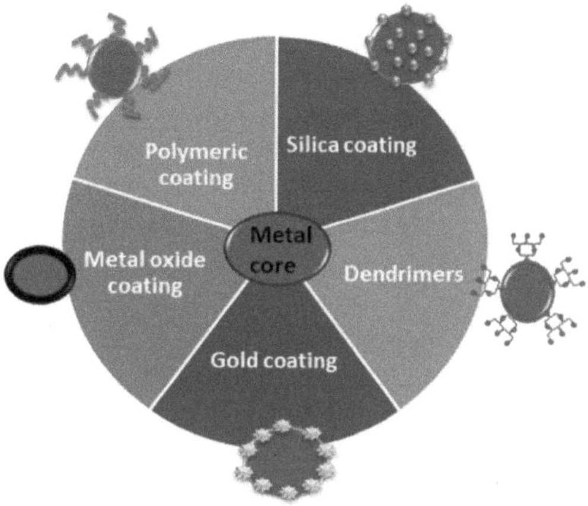

Figure 1.1 Schematic representation for coating routes of magnetic nanomaterials.

biosensor with a signal amplifier made of chitosan-coated MNPs-modified multiwalled carbon nanotubes (MNPs/CS-MWCNTs) was reported to successfully detect bovine serum albumin (BSA). The detection of BSA exhibited a linear range (10^2 to 10^{-4} μg mL^{-1}) with a low detection limit (2.8×10^{-5} μg mL^{-1}) [19].

1.1.3.2 Silica (SiO_2) coating

Silica is a naturally occurring mineral, and it is also synthesized for specific applications. Silica-coated MNMs proved to be instrumental in catalysis as well as biomedical applications. Moreover, silica coating imparts that MNMs enhanced stability in aqueous media as well as versatility in surface functionalization potential [15]. Ahmadi et al. reported the fabrication of tryptophan-modified magnetite nanospheres coated by silica. The developed MNMs have proven to be efficient in enantioseparation of model chiral molecules [20]. Similarly, hemoglobin-immobilized silica-coated MNPs were also employed as a chiral magnetic nanoselector for L-tryptophan [21].

1.1.3.3 Carbon coating

Carbon coating is preferred over silica and polymeric coatings as it is more chemically and thermally stable. Additionally, it is impermeable to most chemicals thus shielding the inner magnetic core from possible oxidation and acid erosion. From the synthetic approach, carbon coating can be applied directly to nanoscale metal core yielding MNMs with a higher magnetic moment [15]. The electrochemical properties of the glass carbon electrodes modified by carbon-coated nickel MNPs (C–Ni/GCE) were studied to develop a sensitive electrochemical sensor for the accurate detection of acetaminophen in pharmaceutical dosage forms. Results showed that detection was linear in a concentration range of 7.8×10^{-3} to 1.1×10^{-1} mmol L^{-1} [22].

1.1.3.4 Metal oxides and hydroxide coating

Inorganic oxides, such as titania, alumina, and ceria, are usually used as catalyst supporting platform and are lately utilized in a diverse array of applications after combining with MNMs. A recent study reported the use of Aβ antibody-conjugated multifunctional magnetite/ceria nanoassemblies to extracorporeal to cleanse the blood from the Aβ peptides. Aβ peptides are misfolded proteins responsible for neurodegeneration in Alzheimer's disease due to the formation of amyloid plaques that are toxic to neurons [23]. Thus the nanoassemblies are composed of magnetite NPs core that helps the magnetic isolation of the captured Aβ peptides. Additionally the nanoassembly shell mimics superoxide dismutase (SOD) acitivity via scavenging the reactive oxygen species produced by the immune system [24].

1.2 Magnetic properties

Generally, bulky magnetic materials are defined as multimagnetic domain structures. However, as they reduce in size reaching a nanoscale, MNMs become single magnetic domain structures with paramagnetic properties. Upon exposure to an external magnetic field, paramagnetic materials lack any magnetic behavior macroscopically, only showing a very weak magnetic property [25]. As the size of the MNMs reaches a critical size equivalent to or less than 20 nm, their magnetic spin gets disordered acting as a "single super spin" and turns superparamagnetic [25,26]. The superparamagnetic phenomenon implies that MNMs only exhibit high magnetic susceptibility rendering them rapidly responsive to external magnetic stimuli till reaching the saturation magnetization [26]. Upon removal of the external magnetic field, magnetic response ceases instantly with negligible remanence and coercivity [27].

1.3 Classification of MNMs

1.3.1 Monocomponent MNMs

1.3.1.1 Metals (Fe, Ni, Co, and Au)

Iron, cobalt, and nickel are ubiquitously available ferromagnetic elements that constitute the building blocks for various MNMs. Studying the characteristics of ferromagnetic elements at the nanoscale will serve to understand the underpinning properties of MNMs opening avenues of utilization in the fields of ferrofluids, biomedicine, and catalysis [28,29].

Iron NPs are considered the most abundant type of nanostructures. Initially, the synthesis of Fe MNPs involved the decomposition of the metastable organometallic compound $Fe(CO)_5$ in the presence of an organic solvent. This manufacturing process successfully yielded a monodisperse-sized NPs yet remained inefficient due to the low stability of developed NPs under air [30]. To overcome the sensitivity of Fe MNMs toward oxygen, coating the metal surface of NPs with poly(*N*-vinylpyrrolidone) resulted in its passivation against oxidation [31]. Additionally, Fe NPs size can be regulated through the addition of the bulky halides that have a strong binding affinity to the NPs surface in turn slowing their growth kinetics [32].

Preparation methods of Fe NPs can be extrapolated to develop Co MNPs, where small Co MNPs (2 − 6 nm) were prepared via the insertion of a bulky trialkylphosphine that hindered their growth effectively [33]. Moreover, incorporating surfactants in the synthesis steps enhanced the production of small monodisperse MNPs. Where monopopulated Ni MNPs (3.7 nm) were synthesized by the reduction of its acetylacetonate complex $Ni(acac)_3$ in the organic medium [34]. From the application perspective, monocomponent MNMs have a wide array of applications. Lately, MNPs are heavily used in biochemical assays for the detection and quantification of DNA samples. Nie et al. have relied on magnetic separation and fluorescence detection sandwich assay for the separation of the target DNA sequence. The recognition element of this DNA sensor is made of iron oxide NPs functionalized with an oligonucleotide sequence complementary to the target DNA sequence. The transducer element composed of Au NPs is functionalized with two different probes one of which is complementary to a nonoverlapping sequence of the target DNA while the other probe acts as a barcode DNA. This barcode DNA can hybridize with fluorescently tagged oligonucleotide probes for signal transduction [35].

1.3.1.2 Metal alloy nanostructures

Metal alloys are made by mixing metals together or with nonmetals. These combination materials attain privileged characteristics including enhanced hardness, improved tensile strength, and lowered melting points. Also, casting metals as an alloy imparts it with inertness toward oxidation and corrosion.

Metal alloys can form via either atom exchange or the interstitial mechanism. The type produced is dictated by the relative size of each element in the mixture. Mixing similar-sized atoms yields to the formation of a substitution alloy, in which partial atomic exchange occurs between some of the new atoms with that composing the parent metallic crystals. Alternatively, the interstitial alloys are formed when substitution is not possible thus the smaller atoms become trapped in the interstitial space of the base crystal matrix [36].

FePd and FePt are the most abundantly used metal alloys due to their favored chemical stability and high magnetocrystalline anisotropy [37,38]. Monodisperse FePd MNPs were successfully prepared employing adamantine carboxylic acid and tributylphosphine as stabilizers [39]. Alternatively, face-centered tetragonal FePt/MgO MNPs were prepared from face-centered cubic FePt-Fe_3O_4@MgO precursors via thermal annealing. Functionally, MgO protected the MNPs against aggregation and was removed by etching away afterward [40].

A diverse array of metal alloy nanostructures can also be produced via manipulating the synthesis steps of FePt NPs. Facile formation of monodisperse metal alloys especially CoPt via the coreduction of their acetylacetonate metal salts with $Pt(acac)_2$ in oleylamine was reported. Oleylamine (OA) acted as the surfactant, solvent, and reducing agent and this method led to the formation of face-centered cubic-CO NPs with tunable superparamagnetic properties [41].

1.3.1.3 Metallic oxides

MFe_2O_4 MNPs (M = Fe, Co, Mn, etc.) possess unique magnetic properties and chemical stability making them of great practicability in the research field. This might be attributed to the facile NPs manufacturing process relying on Fe complexes that make it more appealing [42]. Among other metal oxides, magnetite specifically exhibits high applicability in the fields of magnetic separation and biomedicine given its favored structured packing as well as semimetallic performance. A recent study reported the fabrication of Fe_3O_4 @Pt MNP and multiwalled carbon nanotube modified carbon paste electrode for the electrochemical determination of the chemotherapeutic agent doxorubicin (DOX). The developed biosensor was sensitive to DOX with a linear calibration curve in the concentration range of 5×10^{-5} to 7×10^{-2} mmol L^{-1} and the obtained LOD was 1 nmol L^{-1} [43].

1.3.1.4 Metal carbides

Carbides can be categorized based on the periodic classification of the metal constituents resulting in different types, mainly interstitial carbides, intermediate transition metal carbides (TMCs), and acetylides [44].

Compared to iron oxides, iron carbides such as Fe_5C_2, Fe_3C, and Fe_2C are known for their enhanced magnetic saturation and stability making them more suitable for use in the biomedical applications. Recently, Fe_5C_2 MNMs were synthesized via the decomposition of $Fe(CO)_5$ in octadecylamine in the presence of bromide ions [45]. Furthermore, various iron carbide MNMs based on Fe@Fe_3O_4 nanostructures can be formed by introducing halide ions that control the phase transformation thus tuning the atomic bonding between Fe and C during the synthesis [46].

Levy and Boudart were the first to reveal that early TMCs are catalytically active. TMCs are synthesized via the addition of carbon to elemental transition metals, which resulted in novel materials with the properties of precious metals. This opened avenues for research that is focused on the generation of various TMCs and exploring their potential applications [47,48].

Two-dimensional TMCs (MXenes) possess both high metallic conductivity as well as hydrophilic nature due to the presence of either hydroxyl or oxygen terminal groups on their surfaces. A recent study used ($Ti_3C_2T_x$ MXenes) was deposited on a glassy carbon electrode and coated by the Nafion network for the selective determination of dopamine. The developed sensor accomplished a wide detection range (15×10^{-3} to 10 μmol L^{-1}) and high selectivity compared to graphene oxide sensors [49].

1.3.2 Multicomponent MNMs

1.3.2.1 Heterostructures

Magnetic heterostructures are mainly made of a magnetic part along with additional functional components resulting in multifunctional composites with distinct tailored properties. The newly acquired physical properties are tunneling magnetoresistance, giant magnetoresistance, proximity effects, interlayer exchange coupling, exchange bias, spin-injection, and spin-transport. A seed-mediated manufacturing process is using monocomponent MNPs as the first nucleus on which heteroepitaxial growth or chemical deposition of other components takes place.

Core–shell heterostructures are considered as prominent platforms of MNMs as they enhance particle protection, offer vast possibilities of shell modifications, and enable synergistic effects from both components. Exploiting the aforementioned advantages, a research team designed a novel photoelectrochemical sensor relying on Fe_3O_4@SiO_2@TiO_2 magnetic–optical bifunctional beacon to recognize and select progesterone in complex biosamples. This approach ensures the effective removal of complex interfering species. Results showed that the linear progesterone concentration window is 1×10^{-3} to 6 nmol L^{-1} with a LOD of 0.3 pmol L^{-1} [50].

FePt-based heterostructures have been intensively studied as they exhibit unique magnetic properties. Also, Au imparted FePt-based heterostructures with wider utilization in the fields of catalysis and biomedical applications including MRI applications, drug delivery systems, hyperthermia treatment platforms, and theranostics. Kostevšek et al. fabricated multimodal FePt/SiO_2/Au hybrid NPs to treat cancer. The developed heterostructures attained a combination of photothermal and magnetic properties, which can be utilized for local hypothermia treatment. The advantage of such a structure is attributed to superparamagnetic properties necessary for biomedical purposes originating from its magnetic cores of FePt, while the gold nanoshells absorb light in the near-infrared range implicated in the photothermal activity [51].

1.3.2.2 Exchange-coupled MNMs

Exchange-coupled MNMs are composed of hard and soft phases uniting both the high coercivity characteristics of the hard phase with the potent magnetization of the soft phase. This approach has been highly adopted as it did not only yield permanent magnets with enhanced strength but also produced magnets with increased capacity to store energy, a parameter which can be estimated by the energy product $(BH)_{max}$ [52].

Coating hard magnetic cores with soft magnetic shells is the most successful approach to obtain an exchange-coupling effect. fct-FePt@Co MNPs can be produced through coating fct-FePt MNPs with $Co(acac)_2$ via a reductive reaction. This synthesis route led to increasing the saturation magnetization significantly [40]. In addition, a single-domain nanomagnet can be done via preparing $Sm[Co(CN)_6] \cdot 4H_2O$@ graphene oxide. Subsequently, reductive annealing with calcium granules formed the core–shell structured $SmCo_5$@Co and $SmCo_5$@Sm_2Co_{17} MNMs with larger coercivity and high saturation magnetization [53].

1.4 Synthesis of MNMs

MNMs synthesis pathways can be classified into three main categories namely biomineralization, physical methods, and chemical methods. The following section will briefly summarize those main approaches and their advantages.

1.4.1 Biomineralization

The biomineralization approach relies on the preparation of MNPs via employing living microorganisms [54]. Among those living microorganisms are the magnetotactic bacteria that are recently considered as good model organisms for the implementation of magnetite biomineralization [55]. Thus magnetotactic bacteria magnetosomes produced by this approach are MNMs surrounded by biofilms and organized in a chain-like structure. The yielded MNPs are known for their regular shape, ordered particle size, and good crystallinity. The phospholipids and fatty acids present in the membrane of the magnetosome serve as a nanoreactor that controls the accurate preparation of MNPs. Moreover, proteins found in the magnetosome membrane play an important role in the biomineralization process by controlling iron transport, nucleation, redox reaction, nucleation, and MNPs growth. An example of these MNPs that are prepared by bacteria is the magnetosome that is protein-coated by iron oxide. Given the fact that the microorganisms naturally utilize the MNPs as a compass to get their favorite habitation in the anaerobic environment at the deepwater [56]. Accordingly mimicking the nature MNPs were successfully synthesized at the laboratory in the range of 20–50 nm under anaerobic conditions that are similar to their native environment [57–59]. Additionally, the magnetosomes were found to exhibit excellent magnetic behavior that can be used for biomedical application (particularly hyperthermia) [60,61]. However, the medical utilization of magnetosomes is still restricted until now due to their bacterial protein coating that can elicit undesired immunological reactions. Current research efforts on magnetosomes concentrate on clarification and optimization of the procedure of biomineralization [62,63] to improve wet chemical synthesis procedures that mimic the biologic environment, thus providing MNPs with the same magnetic properties. Till now, the biomineralization process of magnetosome is still not understood and the large-scale production of magnetosome still needs more progress [64].

1.4.2 Physical methods

Top-down and bottom-up are the two main routes for the physical synthesis of MNMs. In the top-down approach, MNMs are synthesized through the reduction of bulk magnetic materials to very fine species at the nanoscale by means of milling. The main disadvantage of these approaches is the difficulty of controlling the desired particle shape and size [65]. Also, the milling methods cause lattice defects that lead to deviations in the magnetic characteristics compared to representative particles of the same size [66]. Alternatively, in the bottom-up approaches MNPs are synthesized by condensation through either gaseous or liquid phase. Interestingly, laser evaporation is considered one of the most promising bottom-up approaches for the preparation of the MNPs. In this method, powder of rough metal oxides in the range of few micrometer sizes evaporates by laser beam [67,68]. The particle size resulted from evaporation methods ranges from 20 to 50 nm [69].

1.4.3 Chemical methods

The chemical methods serve as alternative synthesis routes for MNMs by bottom-up approaches. We will discuss shortly the most prominent chemical methods.

1.4.3.1 Coprecipitation

Coprecipitation is a very simple method for the preparation of MNMs. The aqueous media are the most commonly used in precipitation approach. Commonly, magnetic iron oxides are synthesized through the coprecipitation of an aqueous solution of ferric, ferrous, and iron salt, then the base is added. The particle size and magnetic phase of the NPs can be adjusted through variation of iron salts ratio, pH, temperature, and the kind of base used. Khalafalla and Massart have been reported pioneering work on this synthesis approach [70,71]. Vasić et al. prepared MNPs via this method [72]. The particle size can be increased up to 40 nm by changing reaction conditions. Over recent years, many modifications have been applied to this method. Single crystals of superparamagnetic with size in the range from 10 to 15 nm can be prepared using high pressure during the precipitation process [73] or through slowing the reaction conditions [74], which exhibit very promising characteristics for medical application [75,76]. Besides, upon applying high temperature during the reaction the size of MNPs can be adjusted.

1.4.3.2 Hydrothermal

Hydrothermal synthesis is carried out in aqueous solution at elevated temperatures reaching above 200ºC in autoclaves under high pressures above 2000 psi. This approach shows water capability in hydrolyzing and dehydrating metal salts at raised temperatures. Because of the poor solubility of the found metal oxide particles in water at such temperatures [77,78], the precipitation occurs and through changing of temperature, concentration, and autoclaving time, morphology and particle size can be adjusted [79,80]. By reducing the duration of autoclaving, monodisperse particles with sizes ranging between 10 and 50 nm were successfully obtained. [80].

1.4.3.3 Microemulsion

The microemulsion approach is based on a double phase system aimed at the creation of closely monodisperse MNMs. For that, a water-in-oil microemulsion system is established by the dropping of nanosized water droplets onto the oil phase, and surfactants are used as a stabilizer at the water/oil interface [81]. The MNMs obtained from this system have a very uniform morphology due to the limitation in both nucleation and growth of NPs [82]. Owing to their small size distribution, MNMs obtained from the microemulsion system act as a promising tool for various medical applications [83].

1.4.3.4 Thermal decomposition

Thermal decomposition is based on the preparation of MNMs from nonmagnetic precursors by boiling organometallic compounds in an organic solvent. The obtained particles exhibit a narrow size distribution. Very often, iron acetylacetonates or iron carbonyls serve as nonmagnetic precursors and fatty acids are used as surfactant molecules. The morphology and the size of synthesized MNMs can be conceivably manipulated through changing the ratio of precursors, using different solvent system and adjusting surfactants concentration. The thermal decomposition of nonmagnetic precursors results in a very pure iron metal. Subsequently, in an additional step, the resulted metal particle can be oxidized to iron oxide in oxidative environments at high temperatures. A one-step approach for synthesizing MNMs is introduced by the thermal decomposition of precursors with cationic iron centers as Tris(acetylacetonate) iron (III). MNPs synthesized by thermal decomposition have been reported by Hyeon et al. [84]. Furthermore, the synthesis of MNMs through thermal decomposition with favorable outcomes was achieved by Park et al. [85], on which many modifications have been applied by numerous research groups resulting in MNMs with size up to 30 nm and almost monodisperse size distribution.

1.4.3.5 Polyol method

The polyol is the process in which MNMs are prepared in high-boiling and multivalent alcohols. This method is based on the hydrolysis of the oxidative base of iron salts in a polyol mixture such as PEG and *N*-methyl diethanolamine. The size and structure of the obtained MNMs can be modified by either the conditions of reaction or the applied solvents. However, the obtained particles are of poor monodispersity in size. MNMs with flower shape can be prepared through this method under specific reaction parameters [86], which display outstanding heating action for hyperthermia [87]. Similar to coprecipitated clusters, these particles show a multicore structure and are made of single cores in the range between 8 and 10 nm, which can form a cluster with a size up to 30 nm. These particles show interesting characteristics for hyperthermia treatment as mentioned earlier [75]. There are other approaches available for MNMs synthesis such as laser pyrolysis, microwave-assisted synthesis [88], glass crystallization, and sol–gel method [89]. However, these methodologies are not included in this chapter as the resulting particles are rarely used for analytical purposes [90].

1.5 Functionalization of MNMs

The functionalization of MNMs is central for wide range of applications such as biolabeling, bioseparation, and catalysis, where it provides the systems with the advantages of high reactivity, high dispersibility, and easy separation. Without coating the surface of MNMs, iron oxide NPs possess hydrophobic surfaces with a high ratio between surface area and volume. The surface coating of the MNMs is essential for physical and chemical stability.

Overall, coating processes for MNMs may be classified into adsorptive and covalent methods. Covalent techniques could be further subclassified into (1) grafting-to, (2) grafting-from, and (3) grafting-through techniques as summarized in Fig. 1.2.

1.5.1 Adsorptive functionalization

Adsorptive surface functionalization can be introduced through two possible strategies, one by adsorption of end-modified polyelectrolyte or by polyelectrolyte chain adsorption while the other can be achieved using either the polyelectrolyte anionic groups or other groups as copolymer. Besides, the functional group of the polyelectrolyte, appropriate copolymers that are known as anchors such as catechol derivatives (arsenic acid, phosphate [91], or dopamine [92]) can be used as surface modifiers. Carboxylic groups are most commonly used among all other groups for immobilization processes. Generally, numerous carboxylic groups per polymer chain are used for the functionalization to consciously decrease the possibility of polymeric shell detachment at low concentrations. The binding techniques for each

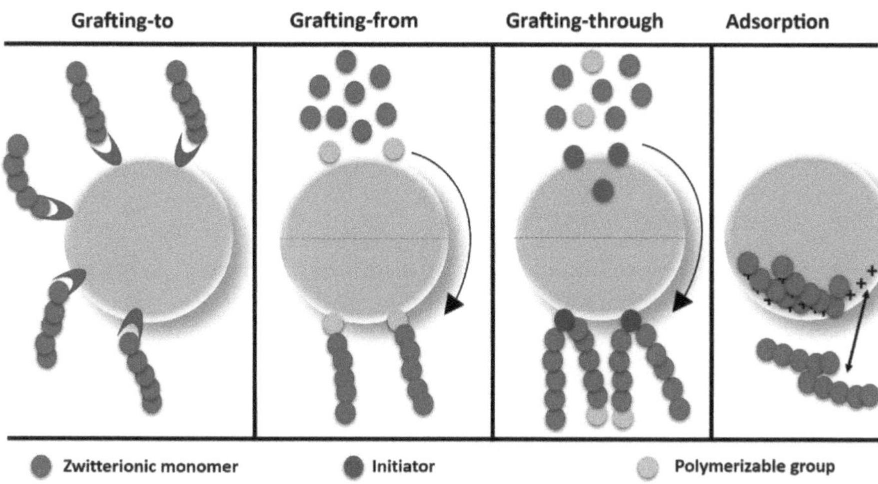

Figure 1.2 Schematic design showing different grafting approaches of polymers to nanoparticle surfaces.

carboxylate can be bidentate bridging (1), bidentate chelation (2), or monodentate chelation (3) depending on the solution conditions in addition to carboxylic acid substituent (R) [93]. Additional distinguished anchoring groups are catechols that mimic natural marine muscle anchoring mechanisms, where they use dopamine groups in the adhesive of mussel foot proteins. Due to the robust binding affinity of catechols to surfaces (in specific to iron oxide), a single catechol group per polymer provides sufficient anchoring at the surface of nanomaterials. Catechol derivatives groups can also powerfully affect the binding affinity on the surface of iron oxide. Overall, catechol derivatives are advantageous due to the presence of electron pulling substituents that enhance their binding affinity and consequently increase the resulting composite materials stability [94]. Less commonly used as an anchoring group for the functionalization of polymers at MNMs surfaces is the phosphate anchoring group.

1.5.2 Covalent functionalization

Covalent functionalization can be accomplished via coating the surface of MNMs by a polymeric shell prior to modifications implementation. The most frequent approach is coating the surface of MNMs with a thin shell of SiO_2 that can be synthesized by the Stöber process, using silane as a precursor. The obtained SiO_2 modified surface exhibits more functional groups as thiol or amine [95–97]. These groups can be used later on in the grafting process of polyelectrolytes.

1.5.2.1 Grafting-to approach

In the grafting-to method, the individual polyelectrolyte is functionalized by a suitable end group able to react with the modified surface of nanomaterials. For this approach, the surface of nanomaterials is first modified with an initiator, followed by polymerization of the initiated surface. Covalent grafting-to can be realized by gathering polyelectrolytes with triethoxysilanes, which can be attached to the functionalized nanomaterials surface (e.g., silica functionalization).

1.5.2.2 Grafting-from approach

This method can be achieved by modifying the surface of nanomaterials with initiators that are used in the atom transfer radical polymerization [98].

1.5.2.3 Grafting-through approach

The most employed methods to functionalize polyelectrolyte to the surface of MNMs are either physisorption or chemisorption through electrostatic interaction between the nanomaterials and the polymer, surface complexation with iron ions, or by manipulating hydrophobic interactions. A common example in the chemisorption approach is the immobilization of polymers with carboxylic moieties on pristine MNMs as reported by von der Lühe et al. [99]. Very recent work has been shown by Biehl who reported MNMs modification using thiol-based grafting technique [100].

Another work was achieved by Pombo Garcia et al. who functionalized amphiphilic zwitterionic polymers on the hydrophobic surface of MNMs through hydrophobic interactions [101]. Other ways that have been used are to add the polyelectrolyte during the preparation of the MNMs as that one reported by Mincheva et al. [102].

1.6 Conclusion

The increasing utilization of magnetic nanomaterials (MNMs) in the analytical chemistry arena, functionalization of MNMs as well as the recent advances in synthesis approaches of MNMs were the focus of this chapter. Despite of the leaps of progress in developing MNMs yet several challenges still need further optimization: (i) The particle shape and size distribution of MNMs is hard to efficiently control; (ii) MNMs have a propensity to oxidize and agglomerate; (iii) The toxicity and degradation mechanisms of MNMs still needs thorough investigation to avoid contamination of tested analytes and ensure reproducible results. By overcoming such limitations, MNMs still offer materials with distinctive properties owing to their superparamagnetic behavior along with their unique dimensions. Through the application of external magnetic field, the manipulation of the interparticle interaction at a nanoscale precision was facilitated. Which in turn led to advancement in analyte sensing, species purification and robust quantitative analysis. Benefiting from their unique properties, MNMs should become increasingly integrated in industrial, environmental, and biomedical applications.

References

[1] J. Valle-Delgado, J. Molina-Bolivar, F. Galisteo-Gonzalez, M. Gálvez-Ruiz, Study of the colloidal stability of an amphoteric latex, Colloid Polym. Sci. 281 (2003) 708–715.
[2] F. Dilnawaz, A. Singh, C. Mohanty, S.K. Sahoo, Dual drug loaded superparamagnetic iron oxide nanoparticles for targeted cancer therapy, Biomaterials 31 (2010) 3694–3706.
[3] A.A. Belanova, N. Gavalas, Y.M. Makarenko, M.M. Belousova, A.V. Soldatov, P.V. Zolotukhin, Physicochemical properties of magnetic nanoparticles: implications for biomedical applications in vitro and in vivo, Oncol. Res. Treat. 41 (2018) 139–143.
[4] V. Ayala, A.P. Herrera, M. Latorre-Esteves, M. Torres-Lugo, C. Rinaldi, Effect of surface charge on the colloidal stability and in vitro uptake of carboxymethyl dextran-coated iron oxide nanoparticles, J. Nanopart. Res. 15 (2013) 1874.
[5] K. Thode, R.H. Müller, M. Kresse, Two-time window and multiangle photon correlation spectroscopy size and zeta potential analysis—highly sensitive rapid assay for dispersion stability, J. Pharm. Sci. 89 (2000) 1317–1324.
[6] O. El-Gholabzouri, M.Á. Cabrerizo-Vílchez, R. Hidalgo-Álvarez, Zeta-potential of polystyrene latex determined using different electrokinetic techniques in binary liquid mixtures, Colloids Surf. A: Physicochem. Eng. Asp. 291 (2006) 30–37.

[7] A.S. Dukhin, V. Shilov, Y. Borkovskaya, Dynamic electrophoretic mobility in concentrated dispersed systems. Cell model, Langmuir 15 (1999) 3452–3457.
[8] U. Jeong, X. Teng, Y. Wang, H. Yang, Y. Xia, Superparamagnetic colloids: controlled synthesis and niche applications, Adv. Mater. 19 (2007) 33–60.
[9] L. Zhang, J. Wu, H. Liao, Y. Hou, S. Gao, Octahedral Fe_3O_4 nanoparticles and their assembled structures, Chem. Commun. (2009) 4378–4380.
[10] Y. Zeng, R. Hao, B. Xing, Y. Hou, Z. Xu, One-pot synthesis of Fe_3O_4 nanoprisms with controlled electrochemical properties, Chem. Commun. 46 (2010) 3920–3922.
[11] Y. Hou, Z. Xu, S. Sun, Controlled synthesis and chemical conversions of FeO nanoparticles, Angew. Chem. Int. Ed. 46 (2007) 6329–6332.
[12] M. Schudel, S. Behrens, H. Holthoff, R. Kretzschmar, M. Borkovec, Absolute aggregation rate constants of hematite particles in aqueous suspensions: a comparison of two different surface morphologies, J. Colloid Interface Sci. 196 (1997) 241–253.
[13] J. Viota, J. De Vicente, J. Duran, A. Delgado, Stabilization of magnetorheological suspensions by polyacrylic acid polymers, J. Colloid Interface Sci. 284 (2005) 527–541.
[14] L.H. Reddy, J.L. Arias, J. Nicolas, P. Couvreur, Magnetic nanoparticles: design and characterization, toxicity and biocompatibility, pharmaceutical and biomedical applications, Chem. Rev. 112 (2012) 5818–5878.
[15] L.M. Rossi, N.J. Costa, F.P. Silva, R. Wojcieszak, Magnetic nanomaterials in catalysis: advanced catalysts for magnetic separation and beyond, Green. Chem. 16 (2014) 2906–2933.
[16] Y. Deng, D. Qi, C. Deng, X. Zhang, D. Zhao, Superparamagnetic high-magnetization microspheres with an Fe_3O_4@ SiO_2 core and perpendicularly aligned mesoporous SiO_2 shell for removal of microcystins, J. Am. Chem. Soc. 130 (2008) 28–29.
[17] P.I. Soares, A.I. Sousa, I.M. Ferreira, C.M. Novo, J.P. Borges, Towards the development of multifunctional chitosan-based iron oxide nanoparticles: optimization and modelling of doxorubicin release, Carbohydr. Polym. 153 (2016) 212–221.
[18] R.I. Boysen, Advances in the development of molecularly imprinted polymers for the separation and analysis of proteins with liquid chromatography, J. Sep. Sci. 42 (2019) 51–71.
[19] H.-J. Chen, Z.-H. Zhang, L.-J. Luo, S.-Z. Yao, Surface-imprinted chitosan-coated magnetic nanoparticles modified multi-walled carbon nanotubes biosensor for detection of bovine serum albumin, Sens. Actuators B: Chem. 163 (2012) 76–83.
[20] M. Ahmadi, T. Madrakian, A. Afkhami, Chiral magnetic nanospheres resonance light scattering properties studies for selective determination of naproxen and phenylglycine enantiomers, Sens. Actuators B: Chem. 210 (2015) 439–445.
[21] H. Bagheri, M. Ahmadi, T. Madrakian, A. Afkhami, Preconcentration and spectrofluorometric determination of L-tryptophan in the presence of D-tryptophan using a chiral magnetic nanoselector, Sens. Actuators B: Chem. 221 (2015) 681–687.
[22] S.-F. Wang, F. Xie, R.-F. Hu, Carbon-coated nickel magnetic nanoparticles modified electrodes as a sensor for determination of acetaminophen, Sens. Actuators B: Chem. 123 (2007) 495–500.
[23] C. Haass, D.J. Selkoe, Soluble protein oligomers in neurodegeneration: lessons from the Alzheimer's amyloid β-peptide, Nat. Rev. Mol. Cell Biol. 8 (2007) 101–112.
[24] D. Kim, H.J. Kwon, T. Hyeon, Magnetite/ceria nanoparticle assemblies for extracorporeal cleansing of amyloid-β in Alzheimer's disease, Adv. Mater. 31 (2019) 1807965.
[25] S. Blundell, Magnetism in Condensed Matter, American Association of Physics Teachers, 2003.

[26] O.A.A. Aziz, K. Arafa, A.S.A. Dena, I.M. El-Sherbiny, Superparamagnetic iron oxide nanoparticles (SPIONs): preparation and recent applications, J. Nanotechnol. Adv. Mater. 8 (2020) 21–29.
[27] S.A. Wahajuddin, Superparamagnetic iron oxide nanoparticles: magnetic nanoplatforms as drug carriers, Int. J. Nanomed. 7 (2012) 3445.
[28] N.H. Hai, R. Lemoine, S. Remboldt, M. Strand, J.E. Shield, D. Schmitter, et al., Iron and Cobalt-based magnetic fluids produced by inert gas condensation, J. Magn. Magn. Mater. 293 (2005) 75–79.
[29] D.L. Huber, Synthesis, properties, and applications of iron nanoparticles, Small 1 (2005) 482–501.
[30] K.S. Suslick, M. Fang, T. Hyeon, Sonochemical synthesis of iron colloids, J. Am. Chem. Soc. 118 (1996) 11960–11961.
[31] Y.-L. Hou, S. Gao, Solvothermal reduction synthesis and magnetic properties of polymer protected iron and nickel nanocrystals, J. Alloy. Compd. 365 (2004) 112–116.
[32] S. Zhang, G. Jiang, G.T. Filsinger, L. Wu, H. Zhu, J. Lee, et al., Halide ion-mediated growth of single crystalline Fe nanoparticles, Nanoscale 6 (2014) 4852–4856.
[33] S. Sun, C. Murray, Synthesis of monodisperse cobalt nanocrystals and their assembly into magnetic superlattices, J. Appl. Phys. 85 (1999) 4325–4330.
[34] Y. Hou, S. Gao, Monodisperse nickel nanoparticles prepared from a monosurfactant system and their magnetic properties, J. Mater. Chem. 13 (2003) 1510–1512.
[35] L.-b Nie, X.-L. Wang, S. Li, H. Chen, Amplification of fluorescence detection of DNA based on magnetic separation, Anal. Sci. 25 (2009) 1327–1333.
[36] J.L. Dossett, H.E. Boyer, Practical Heat Treating, ASM International, 2006.
[37] S. Sun, Recent advances in chemical synthesis, self-assembly, and applications of FePt nanoparticles, Adv. Mater. 18 (2006) 393–403.
[38] O. Gutfleisch, J. Lyubina, K.H. Müller, L. Schultz, FePt hard magnets, Adv. Eng. Mater. 7 (2005) 208–212.
[39] Y. Hou, H. Kondoh, T. Kogure, T. Ohta, Preparation and characterization of monodisperse FePd nanoparticles, Chem. Mater. 16 (2004) 5149–5152.
[40] F. Liu, J. Zhu, W. Yang, Y. Dong, Y. Hou, C. Zhang, et al., Building nanocomposite magnets by coating a hard magnetic core with a soft magnetic shell, Angew. Chem. Int. Ed. 53 (2014) 2176–2180.
[41] Y. Yu, W. Yang, X. Sun, W. Zhu, X.-Z. Li, D.J. Sellmyer, et al., Monodisperse MPt (M = Fe, Co, Ni, Cu, Zn) nanoparticles prepared from a facile oleylamine reduction of metal salts, Nano Lett. 14 (2014) 2778–2782.
[42] Y. Hou, J. Yu, S. Gao, Solvothermal reduction synthesis and characterization of superparamagnetic magnetite nanoparticles, J. Mater. Chem. 13 (2003) 1983–1987.
[43] T. Madrakian, K.D. Asl, M. Ahmadi, A. Afkhami, Fe_3O_4@ Pt/MWCNT/carbon paste electrode for determination of a doxorubicin anticancer drug in a human urine sample, RSC Adv. 6 (2016) 72803–72809.
[44] N.N. Greenwood, A. Earnshaw, Chemistry of the Elements, Elsevier, 2012.
[45] C. Yang, H. Zhao, Y. Hou, D. Ma, Fe_5C_2 nanoparticles: a facile bromide-induced synthesis and as an active phase for Fischer–Tropsch synthesis, J. Am. Chem. Soc. 134 (2012) 15814–15821.
[46] Z. Yang, T. Zhao, X. Huang, X. Chu, T. Tang, Y. Ju, et al., Modulating the phases of iron carbide nanoparticles: from a perspective of interfering with the carbon penetration of Fe@Fe_3O_4 by selectively adsorbed halide ions, Chem. Sci. 8 (2017) 473–481.

[47] M.G. Quesne, A. Roldan, N.H. de Leeuw, C.R.A. Catlow, Bulk and surface properties of metal carbides: implications for catalysis, Phys. Chem. Chem. Phys. 20 (2018) 6905–6916.
[48] R. Levy, M. Boudart, Platinum-like behavior of tungsten carbide in surface catalysis, Science 181 (1973) 547–549.
[49] F. Shahzad, A. Iqbal, S.A. Zaidi, S.-W. Hwang, C.M. Koo, Nafion-stabilized two-dimensional transition metal carbide ($Ti_3C_2T_x$ MXene) as a high-performance electrochemical sensor for neurotransmitter, J. Ind. Eng. Chem. 79 (2019) 338–344.
[50] H. Li, Y. Li, J. Li, F. Yang, L. Xu, W. Wang, et al., Magnetic–optical core–shell nanostructures for highly selective photoelectrochemical aptasensing, Anal. Chem. 92 (2020) 4094–4100.
[51] N. Kostevšek, K. Žužek Rožman, M.S. Arshad, M. Spreitzer, S. Kobe, S. Šturm, Multimodal hybrid FePt/SiO_2/Au nanoparticles for nanomedical applications: combining photothermal stimulation and manipulation with an external magnetic field, J. Phys. Chem. C 119 (2015) 16374–16382.
[52] H. Kirchmayr, Permanent magnets and hard magnetic materials, J. Phys. D: Appl. Phys. 29 (1996) 2763.
[53] C. Yang, L. Jia, S. Wang, C. Gao, D. Shi, Y. Hou, et al., Single domain $SmCo_5$@Co exchange-coupled magnets prepared from core/shell $Sm[Co(CN)_6]\cdot 4H_2O$@GO particles: a novel chemical approach, Sci. Rep. 3 (2013) 3542.
[54] K.M. Krishnan, Fundamentals and Applications of Magnetic Materials, Oxford University Press, 2016.
[55] J. Baumgartner, G. Morin, N. Menguy, T.P. Gonzalez, M. Widdrat, J. Cosmidis, et al., Magnetotactic bacteria form magnetite from a phosphate-rich ferric hydroxide via nanometric ferric (oxyhydr)oxide intermediates, Proc. Natl. Acad. Sci. 110 (2013) 14883–14888.
[56] D.A. Bazylinski, A.J. Garratt-Reed, R.B. Frankel, Electron microscopic studies of magnetosomes in magnetotactic bacteria, Microsc. Res. Tech. 27 (1994) 389–401.
[57] D. Faivre, D.J. Schuler, Magnetotactic bacteria and magnetosomes, Chem. Rev. 108 (2008) 4875–4898.
[58] D. Schüler, R.B. Frankel, Bacterial magnetosomes: microbiology, biomineralization and biotechnological applications, Appl. Microbiol. Biotechnol. 52 (1999) 464–473.
[59] M. Timko, M. Molcan, A. Hashim, A. Skumiel, M. Muller, H. Gojzewski, et al., Hyperthermic effect in suspension of magnetosomes prepared by various methods, IEEE Trans. Magnet. 49 (2012) 250–254.
[60] R. Hergt, R. Hiergeist, M. Zeisberger, D. Schüler, U. Heyen, I. Hilger, et al., Magnetic properties of bacterial magnetosomes as potential diagnostic and therapeutic tools, J. Magn. Magn. Mater. 293 (2005) 80–86.
[61] M. Molcan, H. Gojzewski, A. Skumiel, S. Dutz, J. Kovac, M. Kubovcikova, et al., Energy losses in mechanically modified bacterial magnetosomes, J. Phys. D Appl. Phys. 49 (2016) 365002.
[62] J. Baumgartner, M. Antonietta Carillo, K.M. Eckes, P. Werner, D. Faivre, Biomimetic magnetite formation: from biocombinatorial approaches to mineralization effects, Langmuir 30 (2014) 2129–2136.
[63] A. Scheffel, M. Gruska, D. Faivre, A. Linaroudis, J.M. Plitzko, D. Schüler, An acidic protein aligns magnetosomes along a filamentous structure in magnetotactic bacteria, Nature 440 (2006) 110–114.
[64] R. Contreras-Montoya, Y. Jabalera, V. Blanco, J.M. Cuerva, C. Jimenez-Lopez, L. Alvarez de Cienfuegos, Lysine as size-control additive in a bioinspired synthesis of

pure superparamagnetic magnetite nanoparticles, Cryst. Growth Des. 20 (2020) 533–542.
[65] C.M. Castro, B.S. Mitchell., Nanoparticles from mechanical attrition. In:, Synthesis, Functionalization and Surface Treatment of Nanoparticles. American Scientific Publishers (2002).
[66] S. Dutz, R. Hergt, J. Mürbe, R. Müller, M. Zeisberger, W. Andrä, et al., Hysteresis losses of magnetic nanoparticle powders in the single domain size range, J. Magn. Magn. Mater. 308 (2007) 305–312.
[67] H.-D. Kurland, J. Grabow, G. Staupendahl, W. Andrä, S. Dutz, M.E. Bellemann, Magnetic iron oxide nanopowders produced by CO_2 laser evaporation, J. Magn. Magn. Mater. 311 (2007) 73–77.
[68] H.-D. Kurland, J. Grabow, G. Staupendahl, F.A. Müller, E. Müller, S. Dutz, et al., Magnetic iron oxide nanopowders produced by CO_2 laser evaporation—'in situ' coating and particle embedding in a ceramic matrix, J. Magn. Magn. Mater. 321 (2009) 1381–1385.
[69] C. Stötzel, H.-D. Kurland, J. Grabow, S. Dutz, E. Müller, M. Sierka, et al., Control of the crystal phase composition of Fe_xO_y nanopowders prepared by CO_2 laser vaporization, Cryst. Growth Des. 13 (2013) 4868–4876.
[70] S. Khalafalla, G. Reimers, Preparation of dilution-stable aqueous magnetic fluids, IEEE Trans. Magnet. 16 (1980) 178–183.
[71] R. Massart, Preparation of aqueous ferrofluids without using surfactant-behavior as a function of the pH and the counterions, C.R. Acad. Sci. Paris Ser. C 291 (1980) 1–3.
[72] K. Vasić, Ž. Knez, E.A. Konstantinova, A.I. Kokorin, S. Gyergyek, M. Leitgeb, Structural and magnetic characteristics of carboxymethyl dextran coated magnetic nanoparticles: from characterization to immobilization application, React. Funct. Polym. 148 (2020) 104481.
[73] C. Grüttner, K. Müller, J. Teller, F. Westphal, A. Foreman, R. Ivkov, Synthesis and antibody conjugation of magnetic nanoparticles with improved specific power absorption rates for alternating magnetic field cancer therapy, J. Magn. Magn. Mater. 311 (2007) 181–186.
[74] S. Dutz, W. Andrä, R. Hergt, R. Müller, C. Oestreich, C. Schmidt, et al., Influence of dextran coating on the magnetic behaviour of iron oxide nanoparticles, J. Magn. Magn. Mater. 311 (2007) 51–54.
[75] S. Dutz, M. Kettering, I. Hilger, R. Müller, M. Zeisberger, Magnetic multicore nanoparticles for hyperthermia—influence of particle immobilization in tumour tissue on magnetic properties, Nanotechnology 22 (2011) 265102.
[76] S. Dutz, J.H. Clement, D. Eberbeck, T. Gelbrich, R. Hergt, R. Müller, et al., Ferrofluids of magnetic multicore nanoparticles for biomedical applications, J. Magn. Magn. Mater. 321 (2009) 1501–1504.
[77] Y. Hao, A.S. Teja, Continuous hydrothermal crystallization of α-Fe_2O_3 and Co_3O_4 nanoparticles, J. Mater. Res. 18 (2003) 415–422.
[78] Y. Lv, H. Wang, X. Wang, J. Bai, Synthesis, characterization and growing mechanism of monodisperse Fe_3O_4 microspheres, J. Cryst. Growth 311 (2009) 3445–3450.
[79] D. Chen, R. Xu, Hydrothermal synthesis and characterization of nanocrystalline Fe_3O_4 powders, Mater. Res. Bull. 33 (1998) 1015–1021.
[80] C. Xu, A.S. Teja, Continuous hydrothermal synthesis of iron oxide and PVA-protected iron oxide nanoparticles, J. Supercrit. Fluids 44 (2008) 85–91.
[81] Z. Zhou, J. Wang, X. Liu, H.S.O. Chan, Synthesis of Fe_3O_4 nanoparticles from emulsions, J. Mater. Chem. 11 (2001) 1704–1709.

[82] C. Okoli, M. Sanchez-Dominguez, M. Boutonnet, S. Jaras, C. Civera, C. Solans, et al., Comparison and functionalization study of microemulsion-prepared magnetic iron oxide nanoparticles, Langmuir 28 (2012) 8479–8485.
[83] Q. Zeng, I. Baker, J. Loudis, Y. Liao, P. Hoopes, J.B. Weaver, Fe/Fe oxide nanocomposite particles with large specific absorption rate for hyperthermia, Appl. Phys. Lett. 90 (2007) 233112.
[84] T. Hyeon, S.S. Lee, J. Park, Y. Chung, H.B. Na, Synthesis of highly crystalline and monodisperse maghemite nanocrystallites without a size-selection process, J. Am. Chem. Soc. 123 (2001) 12798–12801.
[85] J. Park, E. Lee, N.M. Hwang, M. Kang, S.C. Kim, Y. Hwang, et al., One-nanometer-scale size-controlled synthesis of monodisperse magnetic iron oxide nanoparticles, Angew. Chem. Int. Ed. 117 (2005) 2932–2937.
[86] P. Hugounenq, M. Levy, D. Alloyeau, L. Lartigue, E. Dubois, V. Cabuil, et al., Iron oxide monocrystalline nanoflowers for highly efficient magnetic hyperthermia, J. Phys. Chem. C 116 (2012) 15702–15712.
[87] L. Lartigue, P. Hugounenq, D. Alloyeau, S.P. Clarke, M. Levy, J.-C. Bacri, et al., Cooperative organization in iron oxide multi-core nanoparticles potentiates their efficiency as heating mediators and MRI contrast agents, ACS Nano 6 (2012) 10935–10949.
[88] M. Faraji, Y. Yamini, M. Rezaee, Magnetic nanoparticles: synthesis, stabilization, functionalization, characterization, and applications, J. Iran. Chem. Soc. 7 (2010) 1–37.
[89] J. Xu, H. Yang, W. Fu, K. Du, Y. Sui, J. Chen, et al., Preparation and magnetic properties of magnetite nanoparticles by sol–gel method, J. Magn. Magn. Mater. 309 (2007) 307–311.
[90] R. Müller, R. Hergt, S. Dutz, M. Zeisberger, W. Gawalek, Nanocrystalline iron oxide and Ba ferrite particles in the superparamagnetism–ferromagnetism transition range with ferrofluid applications, J. Phys. Condens. Matter. 18 (2006) S2527.
[91] C. Monteil, N. Bar, A. Bee, D. Villemin, An efficient recyclable magnetic material for the selective removal of organic pollutants, Beilstein J. Nanotechnol. 7 (2016) 1447–1453.
[92] L. Zhang, H. Xue, C. Gao, L. Carr, J. Wang, B. Chu, et al., Imaging and cell targeting characteristics of magnetic nanoparticles modified by a functionalizable zwitterionic polymer with adhesive 3, 4-dihydroxyphenyl-l-alanine linkages, Biomaterials 31 (2010) 6582–6588.
[93] K.V. Korpany, D.D. Majewski, C.T. Chiu, S.N. Cross, A.S. Blum, Iron oxide surface chemistry: effect of chemical structure on binding in benzoic acid and catechol derivatives, Langmuir 33 (2017) 3000–3013.
[94] P. Biehl, M. Von der Lühe, S. Dutz, F.H. Schacher, Synthesis, characterization, and applications of magnetic nanoparticles featuring polyzwitterionic coatings, Polymers (Basel) 10 (2018) 91.
[95] W. Stöber, A. Fink, E. Bohn, Controlled growth of monodisperse silica spheres in the micron size range, J. Colloid Interface Sci. 26 (1968) 62–69.
[96] L.G. Bach, M.R. Islam, J.T. Kim, S. Seo, K.T. Lim, Encapsulation of Fe_3O_4 magnetic nanoparticles with poly (methyl methacrylate) via surface functionalized thiol-lactam initiated radical polymerization, Appl. Surf. Sci. 258 (2012) 2959–2966.
[97] E.E. Ureña-Benavides, E.L. Lin, E.L. Foster, Z. Xue, M.R. Ortiz, Y. Fei, et al., Low adsorption of magnetite nanoparticles with uniform polyelectrolyte coatings in concentrated brine on model silica and sandstone, Ind. Eng. Chem. Res., 55, 2016, pp. 1522–1532.

[98] Xa Zhang, W. Lin, S. Chen, H. Xu, H. Gu, Development of a stable dual functional coating with low non-specific protein adsorption and high sensitivity for new superparamagnetic nanospheres, Langmuir 27 (2011) 13669–13674.
[99] M. von der Lühe, U. Günther, A. Weidner, C. Gräfe, J.H. Clement, S. Dutz, et al., SPION@ polydehydroalanine hybrid particles, RSC Adv. 5 (2015) 31920–31929.
[100] P. Biehl, F.H. Schacher, Surface functionalization of magnetic nanoparticles using a thiol-based grafting-through approach, Surfaces 3 (2020) 116–131.
[101] K. Pombo-García, S. Weiss, K. Zarschler, C.S. Ang, R. Hübner, J. Pufe, et al., Zwitterionic polymer-coated ultrasmall superparamagnetic iron oxide nanoparticles with low protein interaction and high biocompatibility, ChemNanoMat 2 (2016) 959–971.
[102] R. Mincheva, O. Stoilova, H. Penchev, T. Ruskov, I. Spirov, N. Manolova, et al., Synthesis of polymer-stabilized magnetic nanoparticles and fabrication of nanocomposite fibers thereof using electrospinning, Eur. Polym. J. 44 (2008) 615–627.

Separation and isolation of magnetic nanomaterials

2

Farnaz Fazlali[1], Ali Hajian[2], Seyyed Mosayeb Daryanavard[3] and Hasan Bagheri[4]

[1]Research and Development Department, Farin Behbood Tashkhis LTD, Tehran, Iran, [2]Institute of Sensor and Actuator Systems, TU Wien, Vienna, Austria, [3]Department of Chemistry, Faculty of Science, University of Hormozgan, Bandar Abbas, Iran, [4]Chemical Injuries Research Center, Systems Biology and Poisonings Institute, Baqiyatallah University of Medical Sciences, Tehran, Iran

2.1 Introduction

Magnetic nanoparticles (MNPs) are a category of nanoparticles (NPs) that can be operated by magnetic fields [1,2]. MNPs recommend the golden chance to generate supplies with physical features that are basically distinct from those detected in magnetic supplies established for bigger particles, their small dimensionality can produce mixed size-belonging alterations of their magnetic, optical, and electrical features, like other NPs. MNPs are the crucial topic of serious study due to their high potential in various usages from the data storage and drug delivery field. The elimination of particles from a mixture with magnetic fields can be extra selective, effective, and frequently much quicker than traditional methods such as centrifugation or filtration systems [3–5]. This chapter is proposed to offer selected viewpoint of the necessity as well as the existing research field for MNP isolations: techniques for the isolation and extraction of NPs with magnetic properties. There is a huge number of research manipulating the fabrication of MNPs for many aims. The quantity of literature on isolation methods used for the examination of MNPs is noticeably lesser, compared to this great yield, but it's pleasant anyway [6].

The isolation (separation) and examination of MNPs are crucially imperative to achieve data on their physical and chemical features such as size, form, and also surface chemical properties, allowing their factual exploit for numerous usages. In other words, there is a crucial requirement for the improvement of effective methods for the extraction of MNPs for evaluating their purity. Stephens et al. [7] studied a large number of researches explaining the separation of MNPs via practical magnetic fields and field gradients for developed isolation and investigation, in 2011. Alves et al. [6] also researched for the separation of MNPs from 2012 to 2018. The chief target of this chapter is to review isolation and separation methods of MNPs, by engaging various kinds of field-flow fractionation (FFF) [cyclical electrical field-flow fractionation (CyElFFF), microchip magnetophoresis, and magnetic field-flow fractionation

Magnetic Nanomaterials in Analytical Chemistry. DOI: https://doi.org/10.1016/B978-0-12-822131-0.00009-1

(MgFFF)], macroscale magnetophoresis [high-/low-gradient magnetic separation (HGMS/LGMS)], and asymmetrical flow field-flow fractionation (AF4) [8].

2.2 Conventional methods for MNPs separation

2.2.1 *Field-flow fractionation*

FFF techniques are flow-based isolation methods for purification of a variety of supplies comprising particles (nano and micro). FFF is a chromatography-based isolation method that can be noteworthy in the medical and biomedical areas. FFF exploits an exterior field practical to the route of flow to accomplish impressive isolation and producing differential movement of MNPs through the mixture of hydrodynamics power, Brownian motion, and exterior power field [9]. As the flow profile is laminar in a FFF passage, therefore particles that react extra powerfully by the field are found nearer to the passage walls. Consequently, they will transfer extra gently because of slower flow currents. Isolation is attained via the equilibrium among the diffusion motion of the analyte and the exterior field. Analytical separation depends entirely on the kind of usage of external field and consequently depending on the category of field applied, the isolation is done by various mechanisms. FFF has been employed to isolate macromolecules in extensive varieties of molecular masses and particle sizes as well as organized structures such as cells and microorganisms [(10^3 to 10^{16} Da) and (10^{-2} to 10^2 mm), respectively]. The latest improvements in this field have allowed enhanced trapping productivity of smaller particles. In addition to isolation, identification of samples by direct evaluation of physical and chemical factors is one of the crucial properties of FFF. Physical properties of materials can be acquired immediately from the measured time of retention, for instance, coefficient of diffusion and size distribution. Isolation by FFF has numerous exceptional benefits; it is a very choosy isolation device and therefore has great resolving potency. Furthermore, FFF is an extremely flexible method. Investigational settings may be extensively varied to adjust the rapidity, variety, and resolve of the isolation. A huge quantity of various fields, in theory, may be used in FFF for isolation. Finally, as a consequence of the parabolic rapidity profile, the NPs migrate faster than bigger particles. The selection of the field completely belongs to the character of the materials to be isolated. For instance, if the material is a combination of species by various electrical charges, undoubtedly an electrical field would be a reasonable nominee [10]. Four types of fields have conquered FFF work. These fields provide four noticeable FFF subtechniques: thermal field-flow fractionation, electrical field-flow fractionation (ElFFF), cross-flow of solvent field-flow fractionation, and MgFFF (see Table 2.1) [6].

The FFF method has a great throughput, extraordinary isolation proficiency, and has been verified to isolate particles (numerous nanometers to 100 μm). Nonetheless, wide adjustment is needed as each FFF techniques require special exterior field magnitude, material nature, solvents for performing effective isolation [8].

Table 2.1 Subtechniques of field-flow fractionation (FFF).

Subtechnique of FFF	Abbreviation	Exterior field
Magnetic field-flow fractionation	MgFFF	Magnetic field
Sedimentation field-flow fractionation	SdFFF	Centrifugal force) Gravity)
Flow field-flow fractionation	FlFFF	Cross-flow
Thermal field-flow fractionation	ThFFF	Thermal gradient
Electrical field-flow fractionation	ElFFF	Electrical field

2.2.1.1 Cyclical electrical field-flow fractionation

CyElFFF is a helpful and practical method for describing, extracting, and isolating NPs with magnetic properties established on their dimension and charge exploiting cyclical electric fields [11]. The great diffusion percentage of MNPs has banned CyElFFF from existence enforceable to particles smaller than 200 nm. ElFFF is a subtechnique of FFF in which the electric extraction and isolation area are created by implementing a probable modification over the head and down walls of the isolation channel (electrodes) [7]. CyElFFF contains a wavering square factual voltage among head and down electrodes inward the channel. Consequently, MNPs transfer back and forward among the electrodes with their dimensions and electrophoretic volubility in agreement. MNPs with great electrophoretic volubility transfer more to the middle of the channel. MNPs employ additional movement at the quicker fluid areas, therefore eluting sooner than the inferior volubility MNPs. One of the most important restrictions of this method is the band expansion of the subsequent ultraviolet (UV) fractograms and small resolution. In this technique, MNPs are organized according to the connection among their dimensions and electrical features such as electrophoretic mobility. ElFFF has been used to determine diversity of MNPs, but only limited consequences have been attained for small MNPs [12]. As the maintenance in ElFFF is susceptible to surface charge, coatings, and particle size, ElFFF suggests the capability to extra thoroughly describe electrical and surface features of MNPs relevant to other FFF techniques and MNP identification methods. Furthermore, ElFFF channels can be synthesized in a relatively simpler way compared with other FFF tools. Tasci and coworkers described to solve the diffusion matter about the isolation of MNPs by CyElFFF performing square wave voltages using upper duty cycles instead of direct current offset voltages [13]. Therefore NPs diffusion was repressed, which permitted isolation of MNPs with a mean size of around 50 until 100 nm. This research revealed the ability of CyElFFF for dimension and also the investigation of electrophoretic polystyrene sulfonate and lipid-coated MNPs. CyElFFF is described to separate and isolate MNPs based on their immigration amounts that relate to transport coefficients. In CyElFFF, the orientation of the electric field is exchanged conforming to a periodic pattern through the development of isolation. In recent times, CyElFFF has been advanced. The informed description for CyElFFF was that the irregular current small circuits the huge capacitance connected by the parallel plate channel electrodes by way of the frequency growths. Further, the great voltages of more than 1.7 V enforce water electrolysis and also bubble creation that is problematic in ElFFF. Nonetheless, the greater frequencies of CyElFFF produce fewer, qualifying isolations with greater applied voltages. Lespes and colleagues considered various functioning factors in CyElFFF to estimate their impression on the maintenance mechanisms and also the influence of fractionation for iron oxide (Fe_3O_4) super-MNPs in an extensive series of size around 15−400 nm and electrophoretic volubility [11]. The maximum isolation proficiency was examined as a role of voltage, rate of flow, stop flow effective, and frequency. Dragiciu and coworkers organized the fluid flow of the aquatic mixture through various MNPs such as Fe_3O_4, gold (Au), Au/Fe_3O_4, and Ag/Fe_3O_4 via a collection of Cr/Au electrodes. The crystals of MNPs exploited in this research were 10 nm. The electrodes were exploited to perform an appropriate and considered voltage across the mixture (around 1.5−5 V). The NPs immigration started at 2.5 V and Au/Fe_3O_4 replied most powerfully with the practical voltage. Future research for isolating the MNPs by the method of electric fields demonstrations unbelievable capacity [7].

2.2.1.2 Asymmetrical flow field-flow fractionation

AF4 was then established on the foundation of flow FFF [14]. In AF4, the bottom wall is substituted by a semipermissive membrane through a special cutoff dimension, permitting the diffusion of the mixture and minor molecules to lower the cutoff dimension but retaining material ingredients bigger than the cutoff dimension. The cross-flow, vertical to the forward channel flow, is the motivating power for sample agglomeration at the bottom wall and the membrane [15]. The isolation of materials is merely defined by the changes in their characteristic diffusion coefficients. The tunable cross-flow creates AF4 a formidable fractionating method with huge flexibility to imbed materials with great dimensions. Furthermore, due to its adaptability with organic and also aqueous buffers, AF4 has been extensively exploited to describe biological and nonbiological samples, such as proteins, lipoproteins, viruses, polysaccharides, plasmids, and different polymers. The appropriateness of AF4 has been exposed for the fractionation of magnetic polyplexes with an average diameter of 50 nm and carboxydextran-coated maghemite distributions with an average diameter of 6 until 60 nm by linking the AF4 device to UV and also multiangle laser light scattering detectors. AF4 can be a modest, speedy, and trustworthy instrument for modality control in commercial generation of MNPs when providing supplementary data associated with MNPs sample ingredients [16].

2.2.1.3 Magnetic field-flow fractionation

MgFFF is a useful and beneficial technique for isolating MNPs by developing variances in magnetic susceptibility. This method was first proposed by Giddings. MgFFF has been revealed to remain an operative and valuable technique for the isolation of MNPs when along a flowing channel an exterior magnetic field is applied. The greatest effective plans for purification and isolation of MNPs exploit functional magnetic fields to operate transportation. MgFFF has been displayed to be valuable for the isolation of MNPs by the use of an exterior magnetic field, which reasons NP magnetization. In summary, magnetic power performances on NPs are driving it to the accumulation wall. This power is different by diffusion back into the flow current so that a permanent condition NPs spreading of special height is shaped. Hydrodynamic powers then do vertical to the factual field and isolate the ingredients. Stephens and his coworker revealed isolation of various kinds of MNPs in a barefaced tube-shaped delicate column by the usage of MgFFF. Due to the magnetic power belonging to the MNPs moment, this method isolates MNPs conforming to dimension and abscess combination. The fundamental setting of the investigation is revealed in Fig. 2.1 [17]. Materials entered the MgFFF by injecting interact with the exterior magnetic field that forces the NPs to upper field gradient beside the accumulation wall. The sample that connects powerfully with the field is

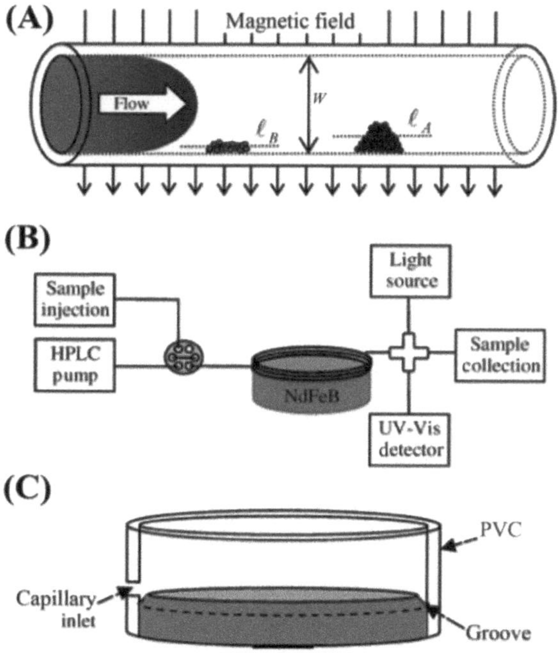

Figure 2.1 (A) Illustration of the MgFFF isolation mechanism for MNPs in a capillary channel with a practical magnetic field. (B) Illustration of the MgFFF instrument. Flow is organized by an HPLC pump and material injected by a six-port injection valve onto a 250 mm capillary. The capillary is wound about an NdFeB magnet and elution observed by UV−visible absorbance. (C) The magnet is kept strong with a 1.5-in.-long section of 3.81 cm PVC pipe, by a hole in the wall to supplement the capillary. The capillary is wound about the magnet in the groove made by the beveled edge of the magnet and the PVC interior wall. *Source*: A.H. Latham, R.S. Freitas, P. Schiffer, M.E. Williams, Capillary magnetic field flow fractionation and analysis of magnetic nanoparticles, Anal. Chem. 77 (15) (2005) 5055−5062, reprinted by permission of the American Chemical Society.

consequently limited to gentler flow currents close to the wall of the channel when matter reacts weakly with a magnetic field that abides on the road to the middle of the channel.

In MgFFF, the dimensionless maintenance factor, λ, is connected to the features of the NPs by the following equation:

$$\lambda = \left(\frac{4d}{w}\right)\left(\frac{k_bT}{\mu\Delta H}\right)^2 \tag{2.1}$$

where k_b (in J K^{-1}) is Boltzmann constant, ΔH (in T) is the drop in magnetic field power through the capillary, T (in K) is the definitive temperature, d (in m) is the NP size, w (in m) is the diameter of the capillary, and μ (in J T^{-1}) is the magnetic moment [17]. This factor λ is connected to R (the analytically specified retention ratio), can be affirmed as follows (Eq. 2.2) [17]:

$$\lambda = \frac{R}{6(1-R)^{1/3}} \tag{2.2}$$

where R is the ratio of the elution periods of retained and also unretained ingredients. Consequently, R (the ratio) can be straightly connected to the magnetic moment (see Eq. 2.1); meanwhile all other factors are recognized [18].

In exciting research, Rheinlander et al. [19] compared the proficiency of MgFFF and nonmagnetic size exclusion chromatography (SEC) to Fe_3O_4 MNPs. Various nanoparticles with magnetic feature (MNPs) were examined: Fe_3O_4 with 5–6 nm diameter coated by dextran and another iron oxide NPs (8 nm diameter) by polyethylene glycol altered surfaces. The examples were distributed over a column filled with iron spheres with 0.3 mm diameter and then the field with the magnetic property was changed to impress the maintenance of the Fe_3O_4 MNPs. The instruction of elution was established on the dependence of dimension factors. In this way, the small particles of iron oxide MNPs eluted first in the first stage and the next step the polyethylene glycol with 8 nm diameter NPs. Compared to SEC, the average NP dimension of generated fractions reduced with elution time. Investigation of the fractions isolated by SEC and MgFFF determines similar dimension distributions. Nonetheless, MgFFF is beneficial as it is quicker and has a greater volume and capacity.

In another interesting research, Beveridge et al. [7] have advanced an exciting method named differential magnetic catch and release (DMCR) to isolate MNPs with a wide range of different sizes. DMCR is almost identical to hydrodynamic chromatography on a microcapillary scale, in that it does not exploit a motionless phase, however, like MgFFF relies on an exterior magnetic field to isolate the target sample effectively. This method uses an electromagnet to create a flexible magnetic flow in tubular capillaries. They revealed the nature of magnetic flux density and also solvent flow speed on the maintenance of cobalt ferrite ($CoFe_2O_4$) MNPs. Exploiting combinations of Au and $CoFe_2O_4$ NPs as pattern systems, the column loading capacity, resolution, capture produce, and effects of mobile phase viscosity on isolation in DMCR were studied. Peak resolution in DMCR is superficially controlled by selectively discharging the NPs at an arranged time from the column by eliminating the magnet flux density. However, higher times of capture rises peak resolution at the expenditure of the capture yield. Besides, the MNPs diameter, the rapidity of the mobile phase, and also viscosity, and the practical magnetic field affect the NP capture produces. Exploiting optimized factors, a combination of cobalt ferrite MNPs whose diameters were about 8 and 12 nm was isolated into monoscatter fractions.

2.2.1.4 Magnetic quadrupole field-flow fractionation

The continuous isolation procedure suggests benefits in the capability to develop huge volumes of the cell, exploit of online observing of the isolation procedure, and capability to stage and recirculate organized fractions. The experiments have accurate control over forces captive in replacement of the labeled cell population from the label-free population to a degree ordered with the great resolution demanded among the organized fractions. The communication of the magnetic label by the exterior magnetic field is commonly an extremely complex purpose of three-dimensional coordinates that considerably imposes the number of magnetic field geometries appropriate for magnetic flow organization. A magnetic quadrupole field-flow fractionation was nominated due to the effective utilization of the obtainable exterior magnetic field energy, and also due to its exceedingly regular association of magnetic power on the situation. Magnetic quadrupole fields have been experienced for the uninterrupted dehydrated organization of coals nonetheless were never verified for continuous isolation of samples in additional viscose media.

The first example of a magnetic quadrupole field-flow fractionation (MqFFF) was investigated and estimated by Zborowski and coworkers for continuous isolation of human peripheral lymphocytes categorized with magnetic colloids [20].

In the following, this procedure is completely described by Carpino et al. [21]. The consequences exhibited that the isolation procedure was near to the foretold manner of a perfect quadrupole magnetic field.

Williams et al. [22] investigated a changed MqFFF technique in a quadrupole magnet field by engaging the isolation channel. The research works considered a commercially obtainable material of dextran-covered magnetite with an inconspicuous size range. The quadrupole magnet device exposed more than 90% of the NPs into the sample (by an average diameter of 178 nm).

In the next step, Orita et al. [23] established a modest on/off-field magnetic quadrupole field-flow fractionation to isolate and calculate two various submicrometer MNPs commercially around 90 and 200 nm at the particular magnetic field and flow situations. Moore et al. [12] used the MqFFF for the centrifugal isolation of red blood cell (RBC)by an average diameter of 8 mm. A quadrupole field was considered having an extreme field of 1.68 T at the magnet pole kinds, and an approximately constant radial field gradient inside an open cylinder-shaped. A scattering of light indicator downstream of the magnet light attenuation is affected by the eluting of the cells from the magnet as a purpose of time. The samples of the cell were collected from high RBCs, which attained through chemical alteration of hemoglobin to methemoglobin, small spin RBC, and combinations of deoxy RBCs and also white blood cells (WBCs). Cell following velocimetry was exploited to calculate the magnetophoretic movement of the RBC. Additionally, the consequences presented that the movement is related to the existence of high-spin hemoglobin. Merely high-spin RBC was fascinated with the magnet, although low-spin RBC proved magnetic capability comparable to WBCs. Consequently, it was also understood that RBCs did not wash during 15 minutes from the channel with specifications such as at flow rate 0.05 mL min^{-1} nonetheless as would be predictable quickly washed at a greater flow rate of 2.0 mL min^{-1}. These consequences agreed with previous studies about the magnetic features of hemoglobin exploiting other methods [24]. The fractionation investigates RBC and also RBCs/WBCs combination presented that a $5x10^7$ per mL suspension of cell pumped over a magnetic field. The 5% of remaining pollution was comparable to that usually realized in WBCs attained with blood centrifugation. One benefit about blood centrifugation is that magnetic quadrupole field-flow fractionation RBC isolation can be scaled to microliter instruments for RBC debulking that can be transportable and worked directly after dedication with negligible human work.

2.2.2 Isolation of MNPs through porous media

Magnetic isolation is a separation procedure where the chosen nanomagnetic samples are being isolated consuming an exterior magnetic field. Nanomagnetic isolation is done by HGMS or LGMS method.

2.2.2.1 High-gradient magnetic separation

Furthermore, scientists have concentrated on improving techniques for extra efficient isolation of MNPs by applying column flow methods. HGMS is an analytical method used to separate nanomagnetic particles from nonmagnetic particles [18]. HGMS is usually performed in typically manufacturing practice to isolate magnetic samples from nonmagnetic solutions, for instance, wastewater behavior of bacteria and also solids. High-gradient magnetic separator is an effective item that contains a column filled by magnetizable matrix, such as stainless steel wool, through an exterior magnetic field. Usually, HGMS is used to isolate microsize or greater particles, or microsize accumulation of NPs, or encapsulation of NPs in superior beads of the polymer. But, the application of HGMS to suspensions of independently scattered MNPs has been so poor discovered, so far HGMS most usually uses a column filled by micron-sized magnetic wires enclosed through an electromagnet [25]. HGMS methods commonly comprise of a column filled with magnetically vulnerable wires located inside an electromagnet. By application of a magnetic field through the column, the wires dehomogenize the magnetic field inside the column creating high field gradients around the wires to permit in obtaining MNPs onto their surfaces. The fascination of MNPs is related to the magnetic field gradients produced particle size and magnetic features.

2.2.2.2 Low-gradient magnetic separation

Though magnetic isolation by HGMS has been extensively applied in the chemical engineering manufacturing with encouraging consequences, this equipment is not without limitations. The chief disadvantage or drawback of the HGMS is ascribed to the existence of numerous underlying uncertainties [26]. First of all, the inhomogeneous magnetic gradients established inside the column are scarcely demonstrated. This consequence restricts the probable optimization of the entire isolation procedure. Additionally, MNPs might powerfully associate with the matrix and scarcely recovered. Furthermore, the engagement of great conducting electromagnets for the creation of a high-gradient magnetic field is overpriced and consuming power. For all these reasons, recent consideration has been assigned to the possibility of LGMS. Since almost all research works in the field of MNPs have been devoted to the treatment of spherical NPs, and as a result, little attention has been paid to rod-shaped MNPs. Yeap et al. [27] compared the magnetophoretic performance of spherical- and rod-shaped iron oxide (Fe_3O_4) MNPs under HGMS. Impressions of NP concentration and also a magnetic field gradient on the isolation kinetics were studied. It was revealed that when the particle concentration is low, the magnetophoresis of MNPs at a low magnetic gradient is meaningfully improved by particle anisotropy (nonspherical form), and with rod-shape MNPs taking considerably fewer time than spherical shape of MNPs to be isolated. Based on the results of the research, it can be concluded that new methods for separating MNPs according to their shape, size, dimensions, range, coverage, and also according to their other unique magnetic properties create golden opportunities for the use of MNPs.

2.2.3 Isolation of MNPs through microchannels

Conventional methods to isolate MNPs have numerous restrictions that decrease the whole efficiency of the isolation [8]. The old traditional techniques of NP and MNPs organization such as chromatography and also electrophoresis need very long hours and a huge amount of material for isolation. Microfluidics affords continuous isolation for materials with various dimensions and permits for least aggregation and material loss in the the isolation process.

Microfluidics also suggests advantage beyond the isolation procedure as it is capable of providing actual-time control, for instance, size verification via modulating the investigation factors such as buffer solution and exterior field. Additionally, microfluidics also prepares an inexpensive solution for MNPs and biomolecules isolation that could be simply combined with other methods such as mixing, calculation, exploration, and examination. On the other side, nanofluidics has been a developing field for MNP research containing MNP isolation with very great isolation proficiency and resolution [28]. Nevertheless, the synthesis of nanofluidics is more exclusive and complicated as it requires smaller and more particular fabrication to generate submicron channels with a wide range of different sizes.

Lab on a chip based on microfluidics is a massive study part of instruments due to their great potential for the speedy investigation of infinitely small amounts of reagents by great resolution in methods that should demonstrate to be both inexpensive and transportable. The research area of microfluidics is frequently evolving as reduced platforms afford rapid examination with the great resolution at a cost in law, foster transportability, and create exploit of remarkably minute quantities of reagents. The overall purpose of micro total analysis systems (mTAS) is to improve the capability to accomplish complex mixtures of analytical and also chemical functions on a single disposable chip, comprising isolation and discovery of materials [29]. The majority of study exploiting magnetic fields and also microchannels has concentrated on isolating and discovering biomolecules and cells in a ferrofluid. Even though less deeply researched in mTAS applications, NP-established technologies have the capability to perform a remarkable role. Notably, the superparamagnetic treatment of nanosize particles at ambient temperature creates them, distinguished nominees, for consolidation in mTAS instruments.

Zhou and coworkers investigated the isolation of spherical MNPs by an average diameter of 7 μm and elliptical forms in modest and operative behavior [30]. The microfluidic chip is comprised of two various inlets and outlets (two inlets and one outlet). Inlet number one was inserted by the aqueous solution of glycerol that performed as flow with buffer property when inlet number two was injected through material NPs suspended in the aqueous solution of glycerol. The microfluidic instrument was located in the center of an even magnetic field and mounted on a reversed microscope to register the routes of the MNPs. A pressure-driven flow was incorporated by the magnetic field practically perpendicular to the flow route. It was concluded that the asymmetrical revolution of the ellipsoidal MNPs, composed with the particle-wall hydrodynamic interplay of the particle wall, caused a net lift power toward the channel midpoint. Otherwise, MNPs in spherical shape stayed nearer to the wall of the channel. This even magnetic field method can be practical to multiple microfluidic channels enabling great performance parallelization for biological and biomedical applications that need the isolation of shaped MNPs.

Zhang et al. [31] investigated a microchip found on magnetophoresis by differential interference contrast discovery (see Fig. 2.2A and B). The actual time of moving routes and speeds of the MNPs at diverse magnetic field powers that are relevant on the space to the constant magnet were calculated and grounded on consecutive difference interference contrast pictures. The consequences showed that magnet with smaller spaces afforded greater magnetophoretic speeds of magnetic particles, and improved magnetophoretic speed variances among magnetic particles of different sizes. This research permitted the effective isolation and discovery of a polydisperse combination of nanomagnetic particles with 500 and 160 nm dimensions at a single stage of the particle in just around 15 seconds.

The operational application of NPs with magnetic properties extremely depends on suitable cleaning after the fabrication stage or before their usage to eliminate solvents, extra surfactants, impurities, and unwanted byproducts. As manual washing is wasting much time and unproductive, Cardoso et al. [32] considered, synthesized, and experienced a microfluidic structure for the continuous washing and isolation of MNPs with a mean diameter of 10 nm fabricated by coprecipitation, consuming ammonium hydroxide (NH_4OH) as the catalyst. Basically, research in the theoretical dimension was achieved to adjust the formation of the microfluidic

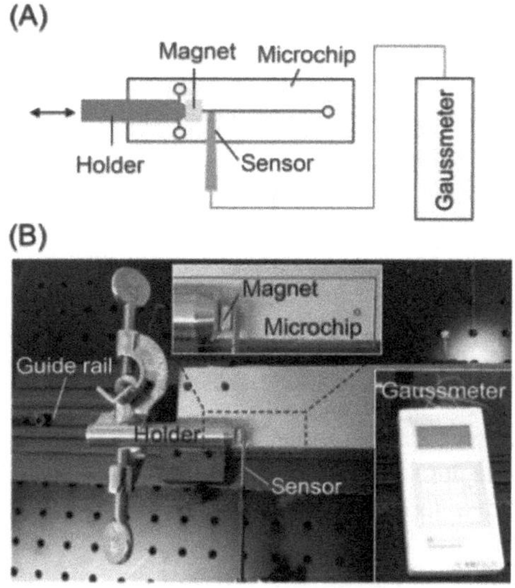

Figure 2.2 Schematic diagram (A) and photography (B) of the experimental system for magnetic field strength evaluation in the isolation region in the microchip.
Source: P. Zhang, S. Park, S.H. Kang, Size-dependent magnetophoresis of native single super-paramagnetic nanoparticles in a microchip, Chem. Commun. 49 (66) (2013) 7298–7300, reprinted by permission of the Royal Society of Chemistry.

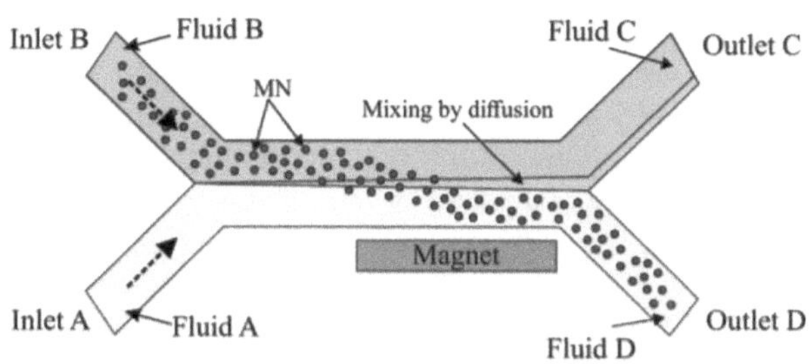

Figure 2.3 Schematic illustration of continuous washing and isolation of magnetic nanoparticles in a microfluidic method. Fluid A: washing solution; Fluid B: fabrication solution by magnetic nanoparticles; Fluid C: waste; Fluid D: washed magnetic nanoparticles.
Source: V.F. Cardoso, D. Miranda, G. Botelho, G. Minas, S. Lanceros-Méndez, Highly effective clean-up of magnetic nanoparticles using microfluidic technology, Sens. Actuators B: Chem. 255 (2018) 2384–2391, reprinted by permission of the Elsevier].

instrument in the geometrical area and the trial situations. The adjusted microfluidic organization was composed of various inlets and outlets (two inlets and two outlets) (Fig. 2.3). The washing solution containing water (fluid A) was familiarized through the inlet A and the fabrication mixture with MNPs (fluid B) over the inlet B. The leftover fluid (fluid C) departed over the outlet C, while the washed MNPs mixture (fluid D) departed over the outlet D. A constant magnet was placed close to the diffusion channel to divert the MNPs by magnetic powers from the fabrication mixture to the washing mixture. Gas chromatography was accomplished to indirectly evaluate the washing proficiency by calculating the reduction of the peak region of ammonium hydroxide. The consequences proved a washing proficiency of about more than 99.7% by monitoring the fluid flows in the microfluidic method, while manual washing attained a measure of about more than 94.3% after washing in six successive intervals. In other words, the microfluidic method suggests inconsiderable loss of MNPs, and the procedure is accomplished autonomously.

2.2.4 MNPs isolation in electric fields

Capillary electromigration techniques are of growing usage for isolations in the analytical area and description of NPs with magnetic properties. These techniques fractionate scatterings related to variations in the range of different particle sizes or calculate the electrophoretic mobilities. Electromigration is able to afford data to describe the concentration of charge and surface, dimension, and also presentation of extra feature of their chemical compound. For instance, electrophoretic mobilities are changed to zeta potentials and also retardation coefficient amounts. Capillary electrophoresis (CE) has been exploited for the isolation of inorganic colloidal NPs comprising various NPs such as silica, silver, gold/silver, and gold. In other words, CE is a strong isolation technique that has the benefits of negligible demand of material and volumes of buffer, and the absence of the creation of biological remaining. Additionally, the usage of narrow capillaries in CE by great electrical persistence permits the great arenas of electrical by negligible therm creation. The usage of immense arenas of electrical composed by the typical flow of plug-kind from electrically driven methods consequences in examination with short times and also great proficiency and resolution for approximately any kind of ionic samples. However, an outstanding restriction of bare iron oxide MNPs isolations by CE has been earlier described. Recently, attention has moved to combinations of metal oxide comprising iron oxide. Lately, d'Orlye et al. [33] described a society of maghemite NPs with 6 until 10 nm diameter, hung in acidic or basic aqueous media. The capillary walls were completely covered to prevent the absorption of cationic NPs, and the time of migration and forms of the peak were revealed to the capillary surface chemistry. The optimum covering was assigned to be di-dodecyl dimethyl-ammonium bromide that produced a reduction in NP electrophoretic movement and provided Gaussian-formed peaks. Nevertheless, CE can be analytically challenging and needs specified material due to the high electric fields that are desired, and the necessity to adjust both the phase mobility, for instance, pH and ionic strength and also capillary surface chemistry. These can be mainly challenging for NP isolation; meanwhile alterations in both solvent and chemistry of external surface can be able to conduce to accumulation, capillary sorption, or shear degradation. It should be noted that some of the selected flow characteristics such as the pH of a buffer in CE can change the charge of the particle as well as the scattering rate.

References

[1] A. Afkhami, M. Saber-Tehrani, H. Bagheri, Modified maghemite nanoparticles as an efficient adsorbent for removing some cationic dyes from aqueous solution, Desalination 263 (1–3) (2010) 240–248. Available from: https://doi.org/10.1016/j.desal.2010.06.065.

[2] H. Bagheri, A. Afkhami, M. Saber-Tehrani, H. Khoshsafar, Preparation and characterization of magnetic nanocomposite of Schiff base/silica/magnetite as a preconcentration phase for the trace determination of heavy metal ions in water, food and biological samples using atomic absorption spectrometry, Talanta 97 (2012) 87–95. Available from: https://doi.org/10.1016/j.talanta.2012.03.066.

[3] T. Madrakian, A. Afkhami, M. Ahmadi, H. Bagheri, Removal of some cationic dyes from aqueous solutions using magnetic-modified multi-walled carbon nanotubes, J. Hazard. Mater. 196 (2011) 109–114. Available from: https://doi.org/10.1016/j.jhazmat.2011.08.078.

[4] K. Molaei, H. Bagheri, A.A. Asgharinezhad, H. Ebrahimzadeh, M. Shamsipur, SiO_2-coated magnetic graphene oxide modified with polypyrrole–polythiophene: a novel and efficient nanocomposite for solid phase extraction of trace amounts of heavy metals, Talanta 167 (2017) 607–616. Available from: https://doi.org/10.1016/j.talanta.2017.02.066.

[5] M.R. Shishehbore, A. Afkhami, H. Bagheri, Salicylic acid functionalized silica-coated magnetite nanoparticles for solid phase extraction and preconcentration of some heavy metal ions from various real samples, Chem. Cent. J. 5 (1) (2011) 41. Available from: https://doi.org/10.1186/1752-153X-5-41.

[6] M.N. Alves, M. Miró, M.C. Breadmore, M. Macka, Trends in analytical separations of magnetic (nano)particles, TrAC - Trends Anal. Chem. 114 (2019) 89–97. Available from: https://doi.org/10.1016/j.trac.2019.02.026.

[7] J.R. Stephens, J.S. Beveridge, M.E. Williams, Analytical methods for separating and isolating magnetic nanoparticles, Phys. Chem. Chem. Phys. 14 (10) (2012) 3280–3289. Available from: https://doi.org/10.1039/c2cp22982j.

[8] X. Zhang, Y. Li, S. Shen, S. Lee, H. Dou, Field-flow fractionation: a gentle separation and characterization technique in biomedicine, TrAC - Trends Anal. Chem. 108 (2018) 231–238. Available from: https://doi.org/10.1016/j.trac.2018.09.005.

[9] T. Jamshaid, E.T.T. Neto, M.M. Eissa, N. Zine, M.H. Kunita, A.E. El-Salhi, et al., Magnetic particles: from preparation to lab-on-a-chip, biosensors, microsystems and microfluidics applications, TrAC - Trends Anal. Chem. 79 (2016) 344–362. Available from: https://doi.org/10.1016/j.trac.2015.10.022.

[10] F.A. Messaud, R.D. Sanderson, J.R. Runyon, T. Otte, H. Pasch, S.K.R. Williams, An overview on field-flow fractionation techniques and their applications in the separation and characterization of polymers, Prog. Polym. Sci. (Oxf.) 34 (4) (2009) 351–368. Available from: https://doi.org/10.1016/j.progpolymsci.2008.11.001.

[11] J. Gigault, B.K. Gale, I. Le Hecho, G. Lespes, Nanoparticle characterization by cyclical electrical field-flow fractionation, Anal. Chem. 83 (17) (2011) 6565–6572. Available from: https://doi.org/10.1021/ac2008948.

[12] L.R. Moore, P.S. Williams, F. Nehl, K. Abe, J.J. Chalmers, M. Zborowski, Feasibility study of red blood cell debulking by magnetic field-flow fractionation with step-programmed flow Field-Flow Fractionation, Anal. Bioanal. Chem. 406 (6) (2014) 1661–1670. Available from: https://doi.org/10.1007/s00216-013-7394-z.

[13] T.O. Tasci, W.P. Johnson, D.P. Fernandez, E. Manangon, B.K. Gale, Biased cyclical electrical field flow fractionation for separation of sub 50 nm particles, Anal. Chem. 85 (23) (2013) 11225–11232. Available from: https://doi.org/10.1021/ac401331z.

[14] K.G. Wahlund, J.C. Giddings, Properties of an asymmetrical flow field-flow fractionation channel having one permeable wall, Anal. Chem. 59 (9) (1987) 1332–1339. Available from: https://doi.org/10.1021/ac00136a016.

[15] H. Zhang, D. Lyden, Asymmetric-flow field-flow fractionation technology for exomere and small extracellular vesicle separation and characterization, Nat. Protoc. 14 (4) (2019) 1027–1053. Available from: https://doi.org/10.1038/s41596-019-0126-x.

[16] A. Litzén, K.G. Wahlund, Improved separation speed and efficiency for proteins, nucleic acids and viruses in asymmetrical flow field flow fractionation, J. Chromatogr. A 476 (C) (1989) 413–421. Available from: https://doi.org/10.1016/S0021-9673(01)93885-3.

[17] A.H. Latham, R.S. Freitas, P. Schiffer, M.E. Williams, Capillary magnetic field flow fractionation and analysis of magnetic nanoparticles, Anal. Chem. 77 (15) (2005) 5055–5062. Available from: https://doi.org/10.1021/ac050611f.

[18] P.S. Williams, F. Carpino, L.R. Moore, M. Zborowski, Magnetic field programming in quadrupole magnetic field-flow fractionation, Phys. Procedia 9 (2) (2010) 91–95. Available from: https://doi.org/10.1016/j.phpro.2010.11.022.

[19] T. Rheinländer, R. Kötitz, W. Weitschies, W. Semmler, Different methods for the fractionation of magnetic fluids, Colloid Polym. Sci. 278 (3) (2000) 259–263. Available from: https://doi.org/10.1007/s003960050041.

[20] M. Zborowski, L. Sun, L.R. Moore, P. Stephen Williams, J.J. Chalmers, Continuous cell separation using novel magnetic quadrupole flow sorter, J. Magn. Magn. Mater. 194 (1) (1999) 224–230. Available from: https://doi.org/10.1016/S0304-8853(98)00581-2.

[21] F. Carpino, L.R. Moore, M. Zborowski, J.J. Chalmers, P.S. Williams, Analysis of magnetic nanoparticles using quadrupole magnetic field-flow fractionation, J. Magn. Magn. Mater. 293 (1) (2005) 546–552. Available from: https://doi.org/10.1016/j.jmmm.2005.01.071.

[22] P.S. Williams, F. Carpino, L.R. Moore, M. Zborowski, Magnetic field programming in quadrupole magnetic field-flow fractionation, Phys. Procedia 9 (2010) 91–95. Available from: https://doi.org/10.1016/j.phpro.2010.11.022.

[23] T. Orita, L.R. Moore, P. Joshi, M. Tomita, T. Horiuchi, M. Zborowski, A quantitative determination of magnetic nanoparticle separation using on-off field operation of quadrupole magnetic field-flow fractionation (QMgFFF), Anal. Sci. 29 (7) (2013) 761–764. Available from: https://doi.org/10.2116/analsci.29.761.

[24] X. Jin, S. Abbot, X. Zhang, L. Kang, V. Voskinarian-Berse, R. Zhao, et al., Erythrocyte enrichment in hematopoietic progenitor cell cultures based on magnetic susceptibility of the hemoglobin, PLoS One 7 (8) (2012). Available from: https://doi.org/10.1371/journal.pone.0039491.

[25] S. Mirshahghassemi, A.D. Ebner, B. Cai, J.R. Lead, Application of high gradient magnetic separation for oil remediation using polymer-coated magnetic nanoparticles, Sep. Purif. Technol. 179 (2017) 328–334. Available from: https://doi.org/10.1016/j.seppur.2017.01.067.

[26] P.Y. Toh, B.W. Ng, A.L. Ahmad, D.C.J. Chieh, J.K. Lim, Magnetophoretic separation of *Chlorella* sp.: role of cationic polymer binder, Process. Saf. Environ. Prot. 92 (6) (2014) 515–521. Available from: https://doi.org/10.1016/j.psep.2014.03.010.

[27] S.P. Yeap, J.K. Lim, B.S. Ooi, A.L. Ahmad, Agglomeration, colloidal stability, and magnetic separation of magnetic nanoparticles: collective influences on environmental engineering applications, J. Nanopart. Res. 19 (11) (2017). Available from: https://doi.org/10.1007/s11051-017-4065-6.

[28] S. Yaman, M. Anil-Inevi, E. Ozcivici, H.C. Tekin, Magnetic force-based microfluidic techniques for cellular and tissue bioengineering, Front. Bioeng. Biotechnol. 6 (2018). Available from: https://doi.org/10.3389/fbioe.2018.00192.
[29] M. Unni, J. Zhang, T.J. George, M.S. Segal, Z.H. Fan, C. Rinaldi, Engineering magnetic nanoparticles and their integration with microfluidics for cell isolation, J. Colloid Interface Sci. 564 (2020) 204–215. Available from: https://doi.org/10.1016/j.jcis.2019.12.092.
[30] R. Zhou, F. Bai, C. Wang, Magnetic separation of microparticles by shape, Lab Chip 17 (2017) 401–406. Available from: https://doi.org/10.1039/C6LC01382A.
[31] P. Zhang, S. Park, S.H. Kang, Size-dependent magnetophoresis of native single superparamagnetic nanoparticles in a microchip, Chem. Commun. 49 (66) (2013) 7298–7300. Available from: https://doi.org/10.1039/c3cc43602k.
[32] V.F. Cardoso, D. Miranda, G. Botelho, G. Minas, S. Lanceros-Méndez, Highly effective clean-up of magnetic nanoparticles using microfluidic technology, Sens. Actuators B: Chem. 255 (2018) 2384–2391. Available from: https://doi.org/10.1016/j.snb.2017.08.095.
[33] F. d'Orlyé, A. Varenne, P. Gareil, Size-based characterization of nanometric cationic maghemite particles using capillary zone electrophoresis, Electrophoresis 29 (18) (2008) 3768–3778. Available from: https://doi.org/10.1002/elps.200800123.

Characterization of magnetic nanomaterials

3

Mohammad Faraji[1], *Yadollah Yamini*[2] *and Narges Salehi*[3]
[1]Research Group of Food, Halal and Agricultural Products, Research Department of Food Technology and Agricultural Products, Standard Research Institute (SRI), Karaj, Iran,
[2]Department of Chemistry, Faculty of Sciences, Tarbiat Modares University, Tehran, Iran,
[3]Department of Chemistry, Varamine (Pishva) Branch Islamic Azad University, Varamin, Iran

3.1 Introduction

Nanoscale materials usually exhibit many different properties than their bulk counterparts. Therefore the study of their properties and applications in various fields of science has attracted the attention of scientists. Nowadays, magnetic nanomaterials (MNMs) are very diverse than they were a decade ago. Also, many types of MNMs are produced more than a decade ago, and at higher values than before, therefore we need to develop more accurate and reliable techniques to characterize them. Due to the lack of access to all characterization techniques and the impossibility of interpreting the data obtained from these techniques in a research group, as well as the inherent problems of nanomaterials (NMs) compared to bulk materials, sometimes characterizations are not complete.

To this end, it is often necessary to use extended techniques and specialists to accurately characterize MNMs to make sure that the characterization process is as comprehensive as possible. In this regard, it is necessary to know the limitations and strengths of each of the techniques, so that we can choose the best technique to provide reliable information when studying a particular parameter and/or what additional techniques are needed to ensure sufficiency of characterization. For example, determining the particle size can be done with different techniques, and we need to know when to use transition electron microscopy (TEM) and when to use scanning electron microscopy (SEM).

Valid and robust characterization techniques for MNMs have a major impact on the applications of these materials in analytical chemistry. Typically, size and morphology, surface charge and surface chemistry, crystal structure, and chemical composition and magnetic property are some of the features that are investigated after MNMs synthesis for analytical applications [1–5].

In this chapter, we will present various techniques that are widely used in the characterization of MNMs. In the following sections, we will investigate and discuss the principles of the test, the information obtained, and the strengths and weaknesses of each technique. These techniques are classified into several groups based

Magnetic Nanomaterials in Analytical Chemistry. DOI: https://doi.org/10.1016/B978-0-12-822131-0.00014-5

on the information they provide about investigated properties of MNMs. These categories include size and morphology, surface properties, chemical and structural properties, magnetic properties, and other common physicochemical properties. Table 3.1 summarizes common techniques for the characterization of MNMs based on the mentioned categories.

3.2 Classification of characterization techniques

Efficiency and magnetic property of the MNMs in different applications highly depend on surface functional groups, size, structure, and morphology of them. For this purpose, different characterization techniques have been applied [1–5].

3.2.1 Size and morphology

3.2.1.1 Scanning electron microscopy

SEM is widely applied to determine the morphology and size distribution of synthesized micro- to nanoscale materials. The SEM technique reports total particle size, so it is not suitable for the characterization of core–shell nanoparticles (NPs; Fig. 3.1A and B).

Another limitation of SEM is that it is not suitable for the detection of NMs with a size of less than 20 nm, and this is due to its lower resolution than the TEM technique [2]. For instance, Zhang et al. applied the SEM technique for characterization of the synthesized fluorinate magnetic covalent organic frameworks (COFs; Fe_3O_4@TpPa-F_4) [6]. In their study, the SEM images of theFe_3O_4 and Fe_3O_4@TpPa-F_4 showed good dispersibility and regular shape. Also, analyses of SEM images have shown that compared with the Fe_3O_4 (Fig. 3.1A), the Fe_3O_4@TpPa-F_4 has a relatively rough surface (Fig. 3.1B) that indicates successful coating of COF layer on the Fe_3O_4 surface.

3.2.1.2 Transmission electron microscopy

Principals of TEM techniques are comprehensively mentioned in the literature [9]. TEM is the most recommended technique for analyzing shape, size, and homogeneity of synthesized MNMs. However, when using this technique, there are some limitations such as misleading images as a result of orientation effects and problems in quantifying large numbers of particles.

Since, in the case of very uniform samples, other techniques such as X-ray diffraction (XRD) [10] and small-angle scattering of X-rays [11] for spherical and larger NPs could provide more reliable results due to their capacity to analyze larger amounts of NPs. To get reliable results from TEM images, the sample preparation protocols are very important and should not lead to a change in the sample. For example, improper drying leads to extra aggregation in colloid suspensions [12]. Therefore the TEM technique is usually combined with other techniques such as dynamic light scattering (DLS) and UV–Vis that require less preparation and also can measure more particles [13,14]. TEM can also be applied for the

Table 3.1 Summary of common and dominant properties of magnetic nanoparticles needed to be determined and the corresponding characterization techniques along with information obtained from each of these techniques.

Entity characterized	Technique	Main information obtained	Comment
Size and morphology	SEM	Determination of morphology and size distribution of prepared particles in the scales of micro to the nanorange.	• It is not a good technique for characterization of core/shell NPs • It is not efficient for NPs with particles size lower than 20 nm
	TEM	NP size, size monodispersity, shape, aggregation state, detect and localize/quantify NPs in matrices, study growth kinetics.	• Difficulty in quantifying a large number of particles • Misleading images due to orientation effects
	HRTEM	All information by conventional TEM but also on the crystal structure of single particles. Distinguish monocrystalline, polycrystalline, and amorphous NPs. Study defects.	• The most common technique to characterize the internal structure of NPs
	DLS	Hydrodynamic size, detection of agglomerates.	• A limitation of DLS is the necessary conditions for the particles to be in suspension and undergoing Brownian motion • It lacks the resolution to detect small aggregates • Its resolution for polydisperse, heterogeneous samples is rather low
Surface properties	FTIR	Surface composition, ligand binding.	• Confirm attachment of different functional group in each step of functionalization
	TGA	Ligand binding/composition/density/arrangement/mass, surface composition, estimate the binding efficiency, mass, and composition of stabilizers.	• The thermal stability of MNMs could be obtained

(*Continued*)

Table 3.1 (Continued)

Entity characterized	Technique	Main information obtained	Comment
	Zeta potential	Estimate properties under experimental conditions, surface charge, agglomeration state.	• It is influenced by the concentration of the suspension and composition of the solvent and other additives
	BET	Surface area, specific surface area.	• Pore volume and mesoporosity are also obtained by this technique
	EPM	Evaluate the surface charge.	• The analysis could be influenced by NP concentration, pH, and natural organic matter
Chemical and structural properties	SEM-EDX	Morphology, quick examination–elemental composition.	• It is carried out on the suspension
	XRD	Crystal structure, composition, crystalline grain size.	• It is not suitable for amorphous materials and the XRD peaks are too broad for particles with a size below 3 nm
	XPS	Electronic structure, elemental composition, oxidation states.	• It can also analyze the ligand exchange interactions and surface functionalization of NPs as well as core/shell structures • Additional techniques for the characterization of the structure, composition and other main NP properties are required
	XAS	X-ray absorption coefficient (element-specific)—chemical state of species.	• The highly sensitive technique, EXAFS is a convenient way to identify the chemical state of species which may occur even in very low concentrations.

(*Continued*)

Table 3.1 (Continued)

Entity characterized	Technique	Main information obtained	Comment
Magnetic properties	SQUID	Magnetization saturation, magnetization remanence, blocking temperature.	• It is not element specific but it rather measures the whole magnetization • Nano-SQUIDs offer the advantage of direct measurement of magnetization changes in small spin systems
	VSM	Similar to SQUID through M−H plots and ZFC−FC curves.	• It is not element specific but it rather measures the whole magnetization • Saturation magnetization value ≥ 16.3 emu g^{-1} is required for magnetic separation from solution with a magnet
	Mössbauer	Oxidation state, symmetry, surface spins, magnetic ordering of Fe atoms, magnetic anisotropy energy, thermal unblocking, distinguish between iron oxides.	• Identify the magnetic phases in a sample

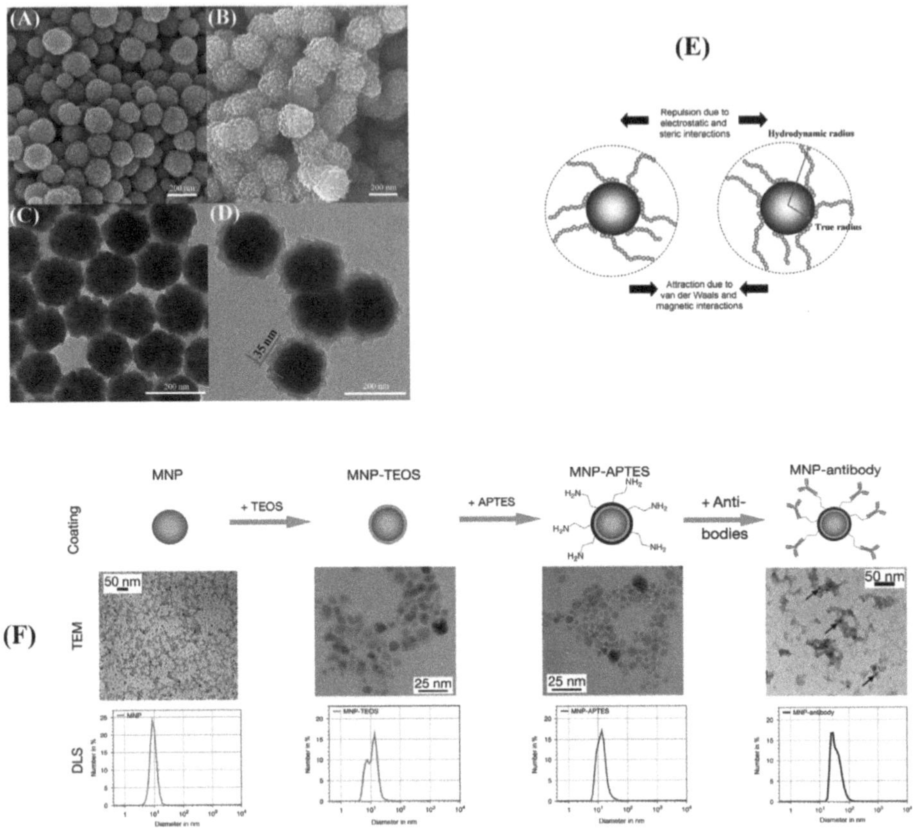

Figure 3.1 SEM images of (A) the bare Fe_3O_4 and (B) Fe_3O_4@TpPa-F_4; TEM images of (C) the bare Fe_3O_4 and (D) Fe_3O_4@TpPa-F_4 [6]; hydrodynamic radius versus true radius (E) [7]; and size determination with DLS measurements with a number distribution for the MNPs and according to TEM images (F). The core of the MNPs had a diameter of 5.8 ± 1.8 nm in TEM analysis and 9.7 ± 1.1 nm in DLS. MNP-TEOS were 11.8 ± 4.1 nm and MNP-APTES were 13.7 ± 3.6 nm in hydrodynamic diameter. The antibody-MNPs had a hydrodynamic diameter of 30.14 ± 13.08 nm. As an example, the TEM image of antiavidin MNPs is presented. Arrows indicate MNPs [8].

characterization of the formation of various superlattice nanocomposites both isostructural and several atomic crystal systems [15,16].

For instance, Zhang et al. applied the TEM technique for the characterization of the synthesized Fe_3O_4@TpPa-F_4 [6]. According to TEM images, the average size of Fe_3O_4 is estimated at approximately 200 nm (Fig. 3.1C and D). Also, the images confirm the formation of the core–shell structure of the Fe_3O_4@TpPa-F_4 with a uniform COF shell ($\sim$35 nm thickness) and a dark Fe_3O_4 core.

Over the past few years, a scientist has begun to see NPs as dynamic systems [17]. Therefore the study and characterization of their dynamic transformation are very important because it has a great impact on their efficiency in many applications.

3.2.1.3 High-resolution transition electron microscopy

High-resolution transition electron microscopy (HRTEM) is a phase-contrast imaging mode of TEM. In this mode, image is produced by combining both scattered and transmitted electrons [18]. HRTEM needs a larger objective aperture for collecting the scattered electrons in comparison with traditional TEM. Unlike conventional electron microscopes, which can provide the statistical assessment of NMs size and morphology, HRTEM provides sufficient resolution for imaging the crystal structure of a single particle. Therefore it provides important information about the structure of NMs. Also, arrays of atoms in crystalline structures can be detected by phase-contrast imaging capability of HRTEM that presents superior resolution. Therefore for the characterization of the internal structure of NMs HRTEM is the common and superior technique.

3.2.1.4 Dynamic light scattering

DLS is an extensively used technique for estimation of the size of NMs in colloidal suspensions [7,19]. In DLS, the scattered light from dispersed NMs in solution is measured as a function of time and used to estimate the hydrodynamic diameter of NMs by considering Stokes–Einstein assumption [1] (Fig. 3.1E). To avoid extensive scattering, a low concentration of NMs should be used in DLS techniques [20].

DLS presents advantages such as a fast, precise, and easy operation for suspensions. Besides, because DLS is an ensemble measurement technique, it provides a proper statistical representation of the sample. For homogeneous and monodisperse samples, this technique is very reproducible and sensitive. However, for DLS measurements, particles must be in suspension and undergoing Brownian motion.

However, being in suspension and having Brownian motion for particles are some of the limitations and conditions required for DLS measurements. Also, scattered light from smaller particles can be mostly obscured in the presence of large particles in suspension. Therefore DLS resolution is relatively low for heterogeneous and polydisperse suspension. Overall, the hydrodynamic radius can be accurately measured by DLS but the resolution of DLS is not enough for the detection of small aggregates. Therefore DLS measurements will be accurate and reliable when it has already been proven that there is no aggregation in the suspension. In this regard, differential centrifugal sedimentation (DCS) was applied for the detection of aggregation before DLS measurement [7,19]. As a result, measurements with DLS can only be accurate when DSC demonstrates that the samples are not aggregated.

To determine the thickness of the shell [21] and also to prove the successful surface functionalization of the surface of NMs, DLS is always applied as a common and useful technique. Therefore, by comparing the hydrodynamic size of the NPs before and after functionalization, the thickness of the functionalization layer can be evaluated [22].

As a good example, Rümenapp et al. applied DLS to investigate variation in the hydrodynamic size of magnetic nanoparticles (MNPs) after each functionalization step [8]. As expected, the hydrodynamic size of the originally MNPs increased from 9.7 to 11.8 nm after encapsulation with the tetraethoxysilane (TEOS) shell. Also, the hydrodynamic size of MNP-TEOS increased from 11.8 to 13.7 nm after functionalization with (3-aminopropyl)-triethoxysilane (APTES). After final functionalization with an antibody layer, the hydrodynamic size increased remarkably to 30.14 nm. Fig. 3.1F shows the changes in size distribution after the functionalization step obtained from their DLS measurements.

Other data that could be obtained from this technique is the polydispersity index (PDI). This parameter can be used to estimate the degree of nonuniformity in hydrodynamic diameter. A high PDI value shows high polydispersity and more significant heterogeneity in the hydrodynamic diameter of the particles. PDI = 0 means that all particles have the same size. When PDI is in the range of 0–0.1, particle system can still be considered as "monodispersed" because the polydispersity under this condition is negligible [23]. When PDI $<$ 0.1, for calculation and data analysis, it can be assumed to be uniform about its mean value. Besides, when PDI = 0.1–0.4, the effect of polydispersity is predominant, and it is very important to consider the size distribution during data analysis [7,19]. Finally, when PDI $>$ 0.4, other characterization techniques or further evaluations are required to confirm the reliability of the DLS results.

This is because high PDI may be resulted by other variables (such as interference of other species and multiple peaks). It should be noted that the PDI value is only defined for a single-mode (monomodal) particle system (particle distribution with a single peak). For a multimodal particle system, further analysis is required for the dispersity of the particles due to the loss of the importance of PDI value.

3.2.2 Surface properties

3.2.2.1 Fourier transform infrared spectroscopy

Fourier transform infrared spectroscopy (FTIR) is a technique based on the measurement of the absorption of electromagnetic radiation with wavelengths within the mid-infrared (IR) region (4000–400 cm^{-1}). In IR active compounds after absorption of IR radiation, their dipole moment is somehow changed. According to the position (related wavenumber) and intensity of observed bands in the IR spectrum, characterization of the functional groups and existing interactions in NMs can be performed. Since the FTIR technique is a very useful tool in identifying functional groups, this technique is widely applied to confirm the attachment of various compounds especially organic compounds with known functional groups at each step of MNP synthesis [1–5]. Due to advantages such as speed, low cost, and availability of the instruments, it is one of the first techniques selected for the characterization of NMs.

For instance, Mahpishanian and Sereshti applied FTIR for the characterization of synthesized β-cyclodextrin-functionalized magnetic-reduced graphene oxide [24]. The FTIR spectra of β-CD, ghraphene oxide (GO), Fe_3O_4, and β-CD/magnetic reduced graphene oxide (MRGO) are shown in Fig. 3.2A. The typical CD characteristic peaks at 1029, 1151, 1077, 1414, 2925, and 3429 cm^{-1} can be observed in

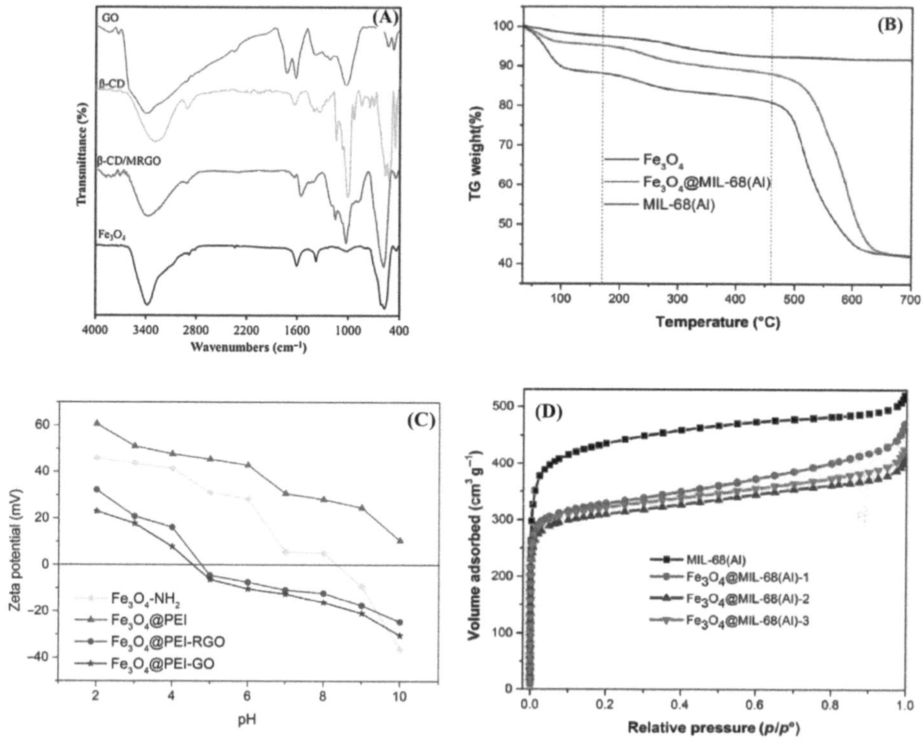

Figure 3.2 (A) FTIR spectra of GO, Fe_3O_4, β-CD, and β-CD/MRGO [24]; (B) TGA of Fe_3O_4, MIL-68(Al), Fe_3O_4@MIL-68(Al) [25]; (C) the zeta potential of Fe_3O_4-NH_2, Fe_3O_4@PEI, Fe_3O_4@PEI-GO, and Fe_3O_4@PEIRGO [26]; and (D) nitrogen adsorption isotherm for MIL-68(Al) and Fe_3O_4@MIL-68(Al)-*n* ($n = 1, 2, 3$) [25].

the FTIR spectrum of β-CD/MRGO. Accordingly attachment of β-CD molecules onto the graphene sheets can be confirmed. Also, observation of Fe—O band vibration as a strong peak at 583 cm^{-1} in the FTIR spectrum of β-CD/MRGO confirms entrapment of the existence of Fe_3O_4 NPs in the structure of the prepared nanocomposite. Also, the disappearance of the C=O vibration band (1740 cm^{-1}) and at the same time remaining O—H and the C—O stretching bands confirms the chemical reduction of GO nanosheets.

3.2.2.2 *Thermal gravimetric analysis*

While FTIR provides useful information about the attachment of the compounds to MNMs, it does not provide information about the extent of surface coating and/or mass-to-mass ratio of attached compounds to MNMs [1]. The amount of surface coating can strongly affect the magnetic saturation (Ms) of MNPs so that by increasing the mass fraction of the coating, the saturated magnetism can be reduced to such an extent that makes it separate magnetically [1]. TGA provides information concerning the mass and composition of the stabilizers. In this technique, with a gradual increase in temperature from ambient temperature up to ~800°C, a change in mass due to degradation, decomposition, and vaporization of the coatings is recorded. Accordingly, based on the initial mass of the sample, the amount and type of coatings or organic ligands can be determined [1–5].

Zhang et al. used TGA for the characterization of synthesized magnetic aluminum–based metal-organic framework (Fe_3O_4@MIL-68(Al)) [25]. Their obtained TGA curves for Fe_3O_4, MIL-68(Al), and Fe_3O_4@MIL-68(Al) are shown in Fig. 3.2B. According to these data, three regions of weight loss can be identified. Evaporation of water and residual dimethylformamide (DMF) leads to a weight loss of 4.62 wt.% in the first region (from 35°C to 180°C). A weight loss of 6.23 wt.% occurred in the second region from 180°C to 420°C. Crystal collapse and degradation of organic linker lead to a weight loss of 47 wt.% in the third region (from 420°C to 650°C). Accordingly, they found that Fe_3O_4@MIl-68(Al) has more thermal stability than MIL-68(Al) below 180°C.

3.2.2.3 Zeta potential (ζ-potential)

Stability of colloidal suspension of MNMs is severely affected by their ζ-potential. Stable colloidal suspensions with a low tendency to agglomeration have high negatively or positively charged particles that lead to high electrostatic repulsion between them. When the pH of the solution is much farther than the pH of the isoelectric point (pH value at which the zeta potential is zero), the particles will have a high positive and negative charge. In contrast, colloidal suspension with low the ζ-potential tends flocculation. In general, when colloids have ζ-potential in the range of ±20 to ±30 mV or higher, it can be very stable. This property is affected by surface modification, so to stabilize the suspensions, the surface modification must be done. Also, the ζ-potential is influenced by solvent composition, suspension concentration, and other additives. For complete characterization, this technique can be combined with DLS that provides evidence on aggregation tendency [27]. Besides, it is very important in analytical applications to know the surface charge of the synthesized NMs to interpret the extraction mechanism [28,29].

Also, this technique could be used to prove the successful synthesis of adsorbents. As a good example, Li et al. applied zeta potential analysis for the characterization of polyethyleneimine-modified magnetic-reduced graphene oxide (Fe_3O_4@PEI-RGO) [26]. The zeta potential of Fe_3O_4-NH_2, Fe_3O_4@PEI, Fe_3O_4@PEI-GO, and Fe_3O_4@PEI-RGO has been separately measured in solutions with different pHs. The results (Fig. 3.2C) showed that the pH of point of zero charge (PZC) is around 8.2 for the Fe_3O_4-NH_2; however, PEI attachments lead to increase PZC to above 10. As a result, due to such an increase, attachment of cationic PEI on the surface of Fe_3O_4–NH_2 can be confirmed. Further modification of Fe_3O_4@PEI with GO leads to change surface charge potential from positive to negative in the pH range from 2 to 10 due to ionization of the carboxylic acid groups of GO [30]. This observation also indicates the successful modification of Fe_3O_4@PEI with GO. Besides, functional groups can be reduced by reducing GO to RGO, which is confirmed by the results [31].

3.2.2.4 The Brunauer–Emmett–Teller

Brunauer–Emmett–Teller (BET) is used for the measurement of the surface area of materials. The physical absorption of gas on a solid surface is the principle of this method. It is extensively used to determine the surface area of NMs, as simple, fast, and a relatively accurate technique [32].

Zhang et al. used BET for characterization of the synthesized magnetic aluminum–based metal-organic framework (Fe_3O_4@MIL-68(Al)-*n*) [25]. The BET surface areas of prepared composites and MIL-68(Al) are shown in Fig. 3.2D. The pore volumes of Fe_3O_4@MIL-68(Al) and MIL-68(Al) are same, with distribution around 0.8 and 1.4 nm and after the incorporation of Fe_3O_4, the mesopore volumes increased in the Fe_3O_4@MIL-68(Al) composites (0.13, 0.29, 0.20, and 0.18 $cm^3\ g^{-1}$ for MIL-68 (Al), Fe_3O_4@MIL-68(Al)-1, Fe_3O_4@MIL-68(Al)-2, and Fe_3O_4@MIL-68(Al)-3, respectively). However, by increasing loading of Fe_3O_4 the surface areas and total pore volume of Fe_3O_4@MIL-68(Al)-*n* composites gradually decreased (1666.25, 1278.37, 1260.23, and 1209.72 $m^2\ g^{-1}$ for MIL-68(Al), Fe_3O_4@MIL-68(Al)-1, Fe_3O_4@MIL-68 (Al)-2, and Fe_3O_4@MIL-68(Al)-3, respectively), maybe due to the low surface area of Fe_3O_4 (9.8 $m^2\ g^{-1}$) [33]. However, they found that the mesoporous volume and the V_{meso}/V_{total} ratio of Fe_3O_4@MIL-68(Al)-*n* composites are higher than MIL-68(Al). Moreover, they estimated that based on the BET surface areas and pore volumes optimum Fe_3O_4 loading could be 5% that could be assigned to as Fe_3O_4@MIL-68(Al)-2.

3.2.2.5 Electrophoretic mobility

The surface charge of NMs can be measured by the electrophoretic mobility (EPM) technique. Low EPM values result in the formation of large aggregates, while very high EPM values result in very stable NPs suspensions for a prolonged time. The aggregation and disaggregation of iron oxide NPs in relation to pH, NP concentration, and natural organic matter were reported [34].

3.2.3 Chemical and structural properties

3.2.3.1 Energy-dispersive X-ray diffraction

Using this technique, information about structural details is improved and it is possible to perform on suspensions. Energy-dispersive X-ray diffraction (EDXD) can be applied to determine chemical composition as well as to determine the existing element in the synthesized MNMs. EDXD data can be used to estimate the structural characterization of various MNMs [1–5].

For instance, Fernandes et al. used EDXD spectra to characterize synthesized MNPs as a QuEChERS sorbent. They took the EDXD spectra at each step of the synthesis and observed that the composition of the NPs changes at each step, indicating successful synthesis at each step [35].

3.2.3.2 X-ray diffraction

XRD is one of the most widely used techniques to characterize MNMs. Information such as phase nature, crystal structure, lattice parameters, and crystal grain size can be obtained from XRD data [1–5]. The crystal grain size is estimated by the Scherrer equation using the width of the most extensive XRD peak for a particular sample.

One of the advantages of XRD techniques, which are usually done in powder form of the samples, is that it leads to statistically representative, volume-averaged values. By comparing the position and intensity of the obtained peaks for a sample with available reference patterns from the International Center for Diffraction Data (ICDD, previously known as Committee on Powder Diffraction Standards, JCPDS) database, the composition of MNM can be estimated. This technique is not suitable for particles smaller than 3 nm because of too broad peaks and also for amorphous materials [1]. XRD peak broadening is mainly caused by crystal/particle size and lattice strains other than instrument broadening.

The derivative size of XRD is usually larger than the so-called magnetic size because smaller domains are present in a particle in which all the moments are in the same direction, even in the case of single-domain particles. In contrast, for samples with very large particles, the derivative size of TEM is higher than the calculated size using XRD. When particles are larger than 50 nm in size, there is more than one crystal boundary on their surface. Such boundaries cannot be distinguished by XRD. Therefore the real size (TEM) can be greater than 50–55 nm, which is calculated by the Scherrer formula:

$$D = \frac{K\lambda}{\beta\cos\theta} \tag{3.1}$$

where D is the average particle diameter (in Å), K is a constant that is very close to unity and is related to the crystallite shape and to how β and D are defined, λ is the X-ray radiation wavelength in ångströms (Å), θ is the peak angle, and β is the width at half maximum (FHWM) of the respective XRD peak.

Yang et al. prepared a novel magnetic adsorbent by surface functionalization of Fe_3O_4@SiO_2 with 2-mercapto-4-methyl-5-thiazoleacetic acid [36]. They used XRD for the study of the crystal phases of products in every step. As shown in Fig. 3.3A, XRD pattern for Fe_3O_4 (a), Fe_3O_4@SiO_2 (b) matched well with that of cubic crystalline magnetite and there is no obvious difference between them. Also, obtained XRD patterns confirm that the coated layers of SiO_2 and 2-mercapto-4-methyl-5-thiazoleacetic acid (MMTA) are amorphous, and the crystal phase of Fe_3O_4 NPs does not change during the coating process. By considering Fig. 3.3A(c) and (d), it can be concluded that the functionalization of MMTA-Au on the surface of Fe_3O_4@SiO_2 successfully occurred because of the different spectra of MMTA-Au and Fe_3O_4 and also the presence of Fe_3O_4 and MMTA-Au peaks simultaneously in spectra of Fe_3O_4@SiO_2@ MMTA-Au (Fig. 3.3A(d)).

3.2.3.3 X-ray photoelectron spectroscopy

X-ray photoelectron spectroscopy (XPS) is a widely used characterization technique for surface chemical analysis of MNMs. The photoelectric effect is the main

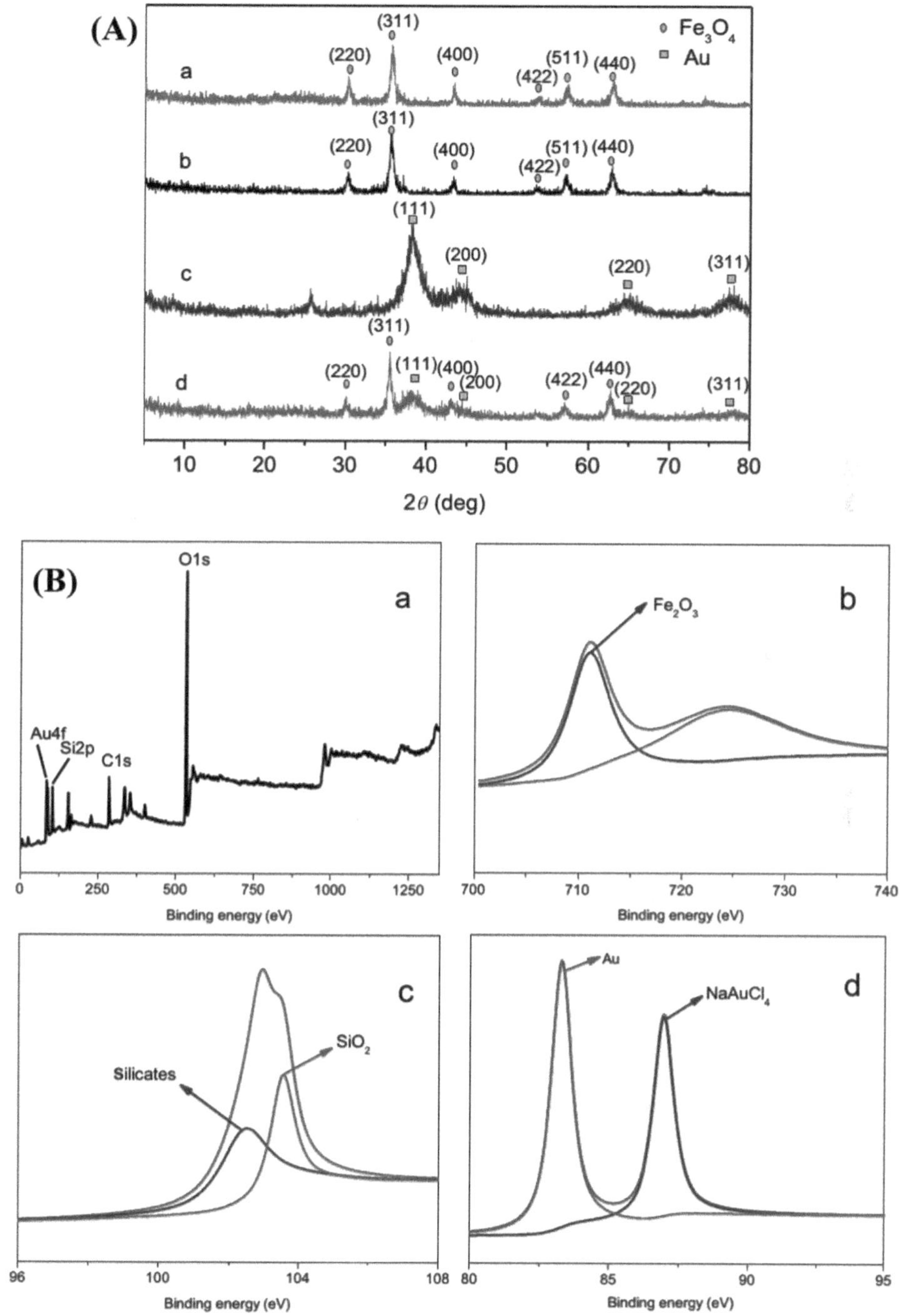

Figure 3.3 XRD patterns of all the materials in each step (A), (a) Fe_3O_4, (b) Fe_3O_4@SiO_2, (c) MMTA-Au, and (d) Fe_3O_4@SiO_2@ MMTA-Au [36] and XPS spectra of the final product (B) [36].

physical principle of XPS. XPS as a powerful quantitative technique provides useful information about the speciation of elements, oxidation states of elements, elemental composition, and electronic structure of MNMs [1–5].

It operates under ultrahigh vacuum conditions. In addition, the study of the surface coating such core/shell structures and ligand exchange processes can also be performed by this technique. In the case of core–shell NPs, Shard has published an article that reports a simple way to interpret XPS data for this type of particle. This includes a straightforward and accurate experimental method for converting XPS intensity to overlayer thickness, more suitable for spherical NPs [37].

As an added advantage of XPS, it provides depth information such as NPs size (up to 10 nm depth from the surface) and it is an almost nondestructive method. The main XPS drawbacks are [1]:

- The data interpretation and sample preparation (sample should be dried without any contamination) are complicated.
- Additional techniques are required for the characterization of the composition, structure, and other main properties of MNMs.

Yang et al. also used XPS analysis for characterization of the prepared magnetic adsorbent, Fe_3O_4@SiO_2@MMTA-Au [36]. The results are shown in Fig. 3.3B. According to obtained binding energies, the coating of Fe_3O_4 with SiO_2 and further functionalization of it with MMTA-Au can be demonstrated. They also applied a high-resolution scan for speciation of elements [36].

Karabelli et al. [38] used XPS to evaluate the mechanism of uptake Cu^{2+} by zerovalent iron. XPS results demonstrated that binding energy of 932.3 ± 0.1 eV is related to Cu $2p_{3/2}$ lines. Therefore the peaks can be mainly derived from Cu^{+} rather than Cu^{0}. Accordingly, the redox type mechanism for this uptake has been proven.

3.2.3.4 X-ray absorption spectroscopy

X-ray absorption near edge structure (XANES, also known as NEXAFS) and extended X-ray absorption fine structure (EXAFS) include X-ray absorption spectroscopy (XAS). In XAS techniques, the X-ray absorption coefficient of MNMs is measured as a function of energy. XAS techniques are selective because characteristic absorption edges related to different binding energies of electrons of each element are different and specific. EXAFS is a good way to detect the chemical speciation that may even occur at very low concentrations because it is a very sensitive technique. For the acquisition of XAS spectra, synchrotrons techniques are usually required; therefore such techniques are limited access and are not routine. XANES evaluates the density of states of partially filled/empty electronic states by investigating the excitation of an inner shell electron to states that are permitted by dipole selection rules [1].

Pearce et al. [39] applied XANES and EXAFS to investigate a series of $Fe_{3-x}Ti_xO_4$ NPs prepared. Their results showed that NP systems have a tunable built-in Fe^{2+}/Fe^{3+} ratio according to Ti^{4+} substitution for Fe^{3+} in the lattice, both on O_h and T_d sites.

3.2.4 Magnetic properties

3.2.4.1 Superconducting quantum interference device magnetometry

Superconducting quantum interference device magnetometry (SQUID) is a technique to measure the magnetic properties of materials in nanoscale. NMs due to sensitivity to local conditions and their small size present different properties compared to the bulk state. As the size of the material decreases, their particles change from multidomain, to singe domain, and ultimately to superparamagnetic state. Magnetization saturation (MS), blocking temperature, and magnetization remanence (MR) are properties that can be generally obtained from SQUID measurements [1,2].

SQUID can also be applied for the measurement of the magnetic response of individual molecules. Recently, a scanning magnetic microscope including a nano-SQUID has also been produced. Nano-SQUID is a superior probe for nanoscale spectroscopy and magnetic imaging. The main need for SQUID, which is designed to detect MNPs, is a very small SQUID area. Ideally, to get the best factor of coupling, the size of the loop must be comparable to those that couple with NPs directly [40]. Nano-SQUID presents the benefit of direct measurement of magnetization changes in systems with small spin due to the magneto-optic spin detection or magnetic resonance force microscopy [41].

3.2.4.2 Vibrating sample magnetometry

Vibrating sample magnetometry (VSM) is another technique used to evaluate the magnetization of the MNMs by recording M−H loops. Magnetization is usually investigated as a function of the external magnetic field applied (H) usually between −3 and 3 T at low and room temperature to obtain parameters such as MS and MR. The magnetic behavior of the MNMs can be characterized by VSM curves. For example, a hysteretic loop feature and zero magnetic remanences demonstrate that MNMs have superparamagnetic property. Also, saturation magnetization (M_s) can be obtained from the plateau part of the VSM curves [1−5]. In the case of the magnetic separation process, suitable magnetization and having superparamagnetism property are very important. For core/shell material, shell reduces MS of MNMs. In this context, Ma et al. [42] declared that saturation magnetization value ≥ 16.3 emu g^{-1} is required for magnetical separation of MNMs from solutions by an external magnet. Therefore the amount of coating should be adjusted so that the magnetic properties of NMs can be used in magnetic separation.

For instance, Mahpishanian and Sereshti [24] applied the VSM technique to estimate functionalization effect on M_S of Fe_3O_4 NPs in the synthesis of β-cyclodextrin-modified magnetic-reduced graphene oxide (β-CD/MRGO) (Fig. 3.4A). The value of saturation magnetization for Fe_3O_4, MRGO, and β-CD/MRGO has been reported as 69.1, 50.4, and 46.0 emu g^{-1}, respectively. Contribution of magnetically inactive β-CD molecules and GO nanosheets to the total magnetization leads to decreasing saturation magnetizations of MRGO and β-CD/MRGO.

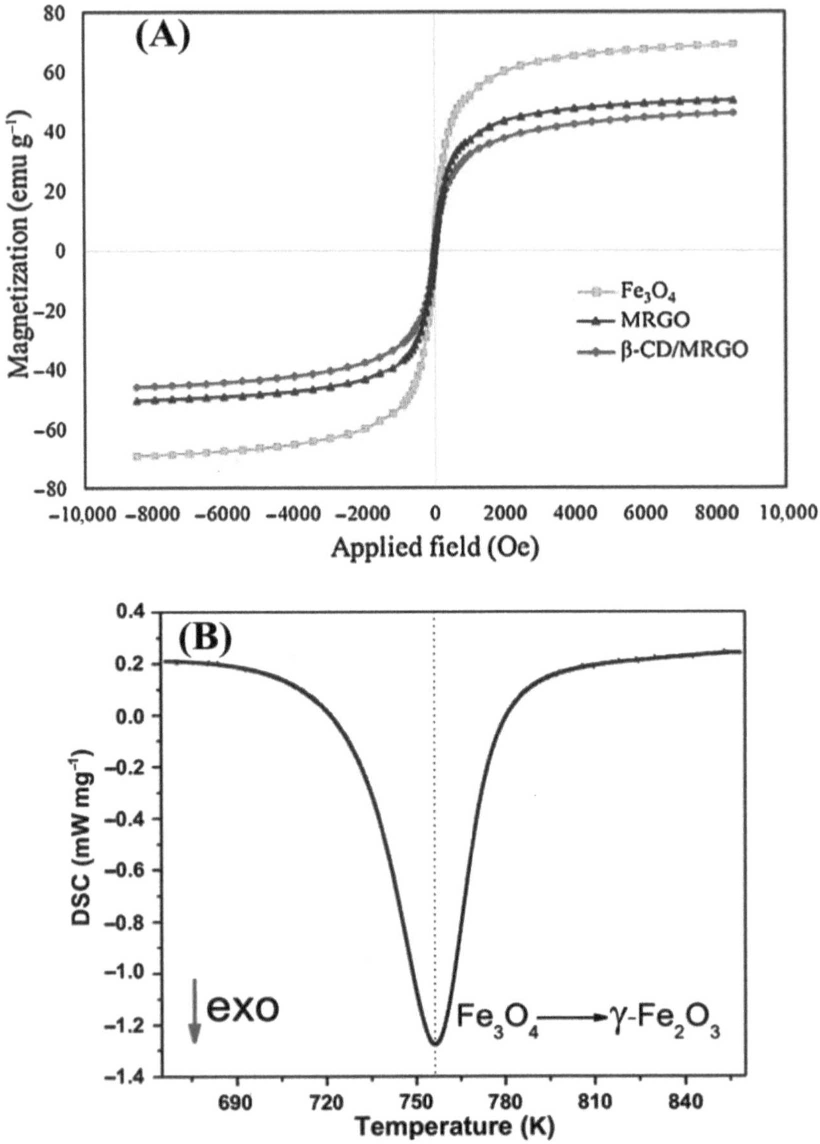

Figure 3.4 (A) Room-temperature hysteresis loop of Fe_3O_4, MRGO, and β-CD/MRGO [24] and (B) DSC analysis of synthesized particles [43].

3.2.4.3 Mössbauer spectroscopy

Mössbauer spectroscopy is a valuable technique that is based on the recoil-free resonance fluorescence of γ-photons in matter with Mössbauer-active elements, such as Fe [1]. This technique can be applied to assess the magnetic ordering of the Fe, symmetry, and spin state and oxidation state in MNMs, thereby determining the magnetic phase in a sample [1]. In addition, in the case of magnetically ordered materials, Mössbauer spectra recorded as a temperature function can be applied to quantify the thermal unblocking (superparamagnetism) and estimate the magnetic anisotropy energy [44].

3.2.5 Other common techniques

3.2.5.1 Differential scanning calorimetry

Differential scanning calorimetry (DSC) is a thermoanalytical method in which the difference in the heat amount required to increase the sample temperature and reference is recorded.

Mamani et al. [43] applied DSC for the thermal characterization of lauric acid–coated magnetic NPs (Fig. 3.4B). The oxidation of Fe_3O_4 to γ-Fe_2O_3 leads to an exothermic event at around 756K in DSC spectra.

3.2.5.2 UV–Vis spectroscopy

UV–Vis is another low-cost, fast, and facile characterization technique that is usually applied for NMs study. Some of the MNMs have optical features that are sensitive to shape, size, agglomeration state, concentration, and refractive index near the NM surface, making UV–Vis spectroscopy a very important technique to characterize such materials. Besides, it can be applied to evaluate the stability of NP colloidal solutions [45].

Fang et al. [46] applied the UV–Vis technique for the characterization of synthesized α-Fe_2O_3@SiO_2@Au NPs. They recorded UV–Vis spectra of α-Fe_2O_3@SiO_2@Au NPs, α-Fe_2O_3@SiO_2, and Au solutions. According to the maximum absorption wavelengths observed for α-Fe_2O_3 (370 nm), Au (519 nm), and also α-Fe_2O_3@SiO_2@Au NPs (370 and 524 nm), a strong interaction between Au and adjacent atoms is probable.

3.2.5.3 Raman spectroscopy

Raman spectroscopy can be applied to characterize the order of crystal structures. Mahpishanian and Sereshti [24] applied Raman spectroscopy to characterize synthesized β-cyclodextrin-functionalized magnetic-reduced graphene oxide (β-CD/MRGO). Fig. 3.5A shows the Raman spectra of GO and β-CD/MRGO. The spectrum of GO shows two main peaks at around 1340 and 1590 cm^{-1} that are assigned to the D band (caused by the disorder originates from structure defects) and G band (assigned to the first-order scattering of the E_{2g} photon of the sp^2 carbon domains), respectively. Estimation of disordering can be obtained from the ratio of the intensity of the D band and G band (I_D/I_G). The I_D/I_G for β-CD/MRGO is higher than GO, which indicates structural defects due to the incorporation of Fe_3O_4 and β-CD on the surface of the RGO.

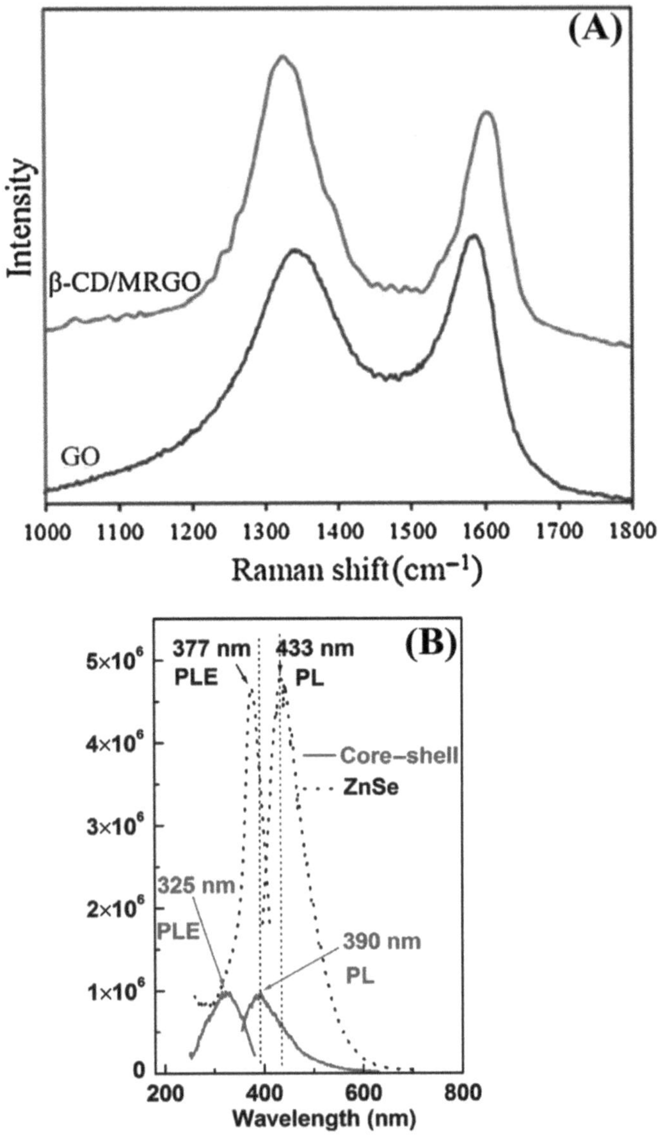

Figure 3.5 UV–Vis spectra of α-Fe_2O_3@SiO_2@Au nanocomposites, α-Fe_2O_3 nanocubes, and Au nanoparticles dissolved in distilled water. The inset shows a photograph of the samples [46]; (A) Raman spectra of GO and β-CD/MRGO [24]; (B) PLE and PL spectra measured at room temperature for the reference ZnSe and core/shell Fe_3O_4/ZnSe NP samples [47].

3.2.5.4 Photoluminescence spectroscopy

Photoluminescence (PL) records the light emitted from molecules or atoms or that have absorbed photons. PL is a characterization technique that is commonly useful for fluorescent MNMs, such as metal nanoclusters and quantum dots [1]. Recently, the inherent PL of metal NPs has received considerable attention. However, the quantum efficiency of the emission process is low; this inefficiency can be compensated by the large excitation phase in the plasmon resonance. Moreover, the PL of metal NPs has no photoblinking and photobleaching. Therefore, for optical labeling applications, PL has better efficiency than fluorescent molecules [48].

Vargas et al. synthesized Fe_3O_4/ZnSe NPs with good PL and magnetic responses [47]. Shift of PL emission/excitation spectra to shorter wavelengths and remarkably reduction of PL quantum yield in comparison with ZnSe NPs as reference sample (Fig. 3.5B) demonstrate incorporation of Fe_3O_4 and ZnSe NPs.

3.2.5.5 Inductively coupled plasma optical emission spectrometry

Inductively coupled plasma atomic emission spectrometry (ICP-AES) is a very sensitive instrument to determine inorganic materials (e.g., core of MNPs) in the structure of the MNMs [1]. ICP-AES offers advantages such as fast and simultaneous determination of multielements, the determination at trace-level concentrations, and tracing little concentration changes. In addition, ICP-AES has good reproducibility and wide dynamic linear range. In special biomedical applications, accurate iron concentration determination in iron oxide NPs is required. ICP-AES is one of the most extensively used techniques for the determination of iron concentration in worldwide method [1].

References

[1] S. Mourdikoudis, R.M. Pallares, N.T.K. Thanh, Characterization techniques for nanoparticles: comparison and complementarity upon studying nanoparticle properties, Nanoscale 10 (2018) 12871–12934.

[2] M. Faraji, Y. Yamini, M. Rezaee, Magnetic nanoparticles: synthesis, stabilization, functionalization, characterization, and applications, J. Iran. Chem. Soc. 7 (2010) 1–37.

[3] A.H. Lu, E.L. Salabas, F. Schüth, Magnetic nanoparticles: synthesis, protection, functionalization, and application, Angew. Chem. Int. Ed. 46 (2007) 1222–1244.

[4] B. Hu, M. He, B. Chen, Solid-Phase Extraction, first ed., Elsevier, 2020 (Chapter 9).

[5] M. Faraji, Recent analytical applications of magnetic nanoparticles, Nanochem. Res. 1 (2016) 264–290.

[6] M. Zhang, J. Li, C. Zhang, Z. Wu, Y. Yang, J. Li, et al., In-situ synthesis of fluorinated magnetic covalent organic frameworks for fluorinated magnetic solid-phase extraction of ultratrace perfluorinated compounds from milk, J. Chromatogr. A 1615 (2019) 460773.

[7] P.M. Carvalho, M.R. Felício, N.C. Santos, S. Gonçalves, M.M. Domingues, Application of light scattering techniques to nanoparticle characterization and development, Front. Chem. 6 (2018) 237.

[8] C. Rümenapp, B. Gleich, H.G. Mannherz, A. Haase, Detection of molecules and cells using nuclear magnetic resonance with magnetic nanoparticles, J. Magn. Magn. Mater. 380 (2015) 271–275.

[9] L. Reimer, H. Kohl, Transmission Electron Microscopy: Physics of Image Formation, first ed., Springer, 2008 (Chapter 1).

[10] L.E. Alexander, H.P. Klug, Determination of crystallite size with the X-Ray spectrometer, J. Appl. Phys. 21 (1950) 137–142.
[11] H. Borchert, E.V. Shevchenko, A. Robert, I. Mekis, A. Kornowski, G. Grübel, et al., Determination of nanocrystal sizes: a comparison of TEM, SAXS, and XRD studies of highly monodisperse $CoPt_3$ particles, Langmuir 21 (2005) 931–936.
[12] D. Mahl, J. Diendorf, W. Meyer-Zaika, M. Epple, Possibilities and limitations of different analytical methods for the size determination of a bimodal dispersion of metallic nanoparticles, Colloids Surf. A 377 (2011) 386–392.
[13] A. Albanese, W.C. Chan, Effect of gold nanoparticle aggregation on cell uptake and toxicity, ACS Nano 5 (2011) 5478–5489.
[14] R.M. Pallares, S.L. Kong, T.H. Ru, N.T.K. Thanh, Y. Lu, X. Su, A plasmonic nanosensor with inverse sensitivity for circulating cell-free DNA quantification, Chem. Commun. 51 (2015) 14524–14527.
[15] E.V. Shevchenko, D.V. Talapin, N.A. Kotov, S. O'Brien, C.B. Murray, Structural diversity in binary nanoparticle superlattices, Nature 439 (2006) 55–59.
[16] M.R. Jones, R.J. Macfarlane, B. Lee, J. Zhang, K.L. Young, A.J. Senesi, et al., DNA-nanoparticle superlattices formed from anisotropic building blocks, Nat. Mater. 9 (2010) 913–917.
[17] M.R. Wiesner, G.V. Lowry, E. Casman, P.M. Bertsch, C.W. Matson, R.T. Di Giulio, et al., Meditations on the ubiquity and mutability of nano-sized materials in the environment, ACS Nano 5 (2011) 8466–8470.
[18] D.B. Williams, C.B. Carter, Transmission Electron Microscopy, first ed., Springer, 2009 (Chapter 28).
[19] S.S. Leong, W.M. Ng, J.K. Lim, S.P. Yeap, Handbook of Materials Characterization, first ed., Springer, 2018 (Chapter 3).
[20] J. Lim, S.P. Yeap, H.X. Che, S.C. Low, Characterization of magnetic nanoparticle by dynamic light scattering, Nanoscale Res. Lett. 8 (2013) 381.
[21] J. Oberdisse, Adsorption and grafting on colloidal interfaces studied by scattering techniques, Curr. Opin. Colloid Interface Sci. 12 (2007) 3–8.
[22] G.P. Van der Beek, M.A. Cohen Stuart, The hydrodynamic thickness of adsorbed polymer layers measured by dynamic light scattering: effects of polymer concentration and segmental binding strength, J. Phys. France 49 (1988) 1449–1454.
[23] U. Nobbmann, A. Morfesis, Light scattering and nanoparticles, Mater. Today 12 (2009) 52–54.
[24] S. Mahpishanian, H. Sereshti, One-step green synthesis of β-cyclodextrin/iron oxide-reduced graphene oxide nanocomposite with high supramolecular recognition capability: application for vortex-assisted magnetic solid phase extraction of organochlorine pesticides residue from honey samples, J. Chromatogr. A 1485 (2017) 32–43.
[25] Y.Y. Zhang, Q. Liu, C. Yang, S.C. Wu, J.H. Cheng, Magnetic aluminum-based metal organic framework as a novel magnetic adsorbent for the effective removal of minocycline from aqueous solutions, Environ. Pollut. 255 (2019) 113226.
[26] N. Li, J. Chen, Y.P. Shi, Magnetic polyethyleneimine functionalized reduced graphene oxide as a novel magnetic solid-phase extraction adsorbent for the determination of polar acidic herbicides in rice, Anal. Chim. Acta 949 (2017) 23–34.
[27] F. Baldassarre, M. Cacciola, G. Ciccarella, A predictive model of iron oxide nanoparticles flocculation tuning Z-potential in aqueous environment for biological application, J. Nanopart. Res. 17 (2015) 377.
[28] M. Faraji, Y. Yamini, M. Rezaee, Extraction of trace amounts of mercury with sodium dodecyle sulphate-coated magnetite nanoparticles and its determination by flow

injection inductively coupled plasma-optical emission spectrometry, Talanta 81 (2010) 831–836.
[29] M. Faraji, Y. Yamini, A. Saleh, M. Rezaee, M. Ghambarian, R. Hassani, A nanoparticle-based solid-phase extraction procedure followed by flow injection inductively coupled plasma-optical emission spectrometry to determine some heavy metal ions in water samples, Anal. Chim. Acta 659 (2010) 172–177.
[30] T.H. Han, W.J. Lee, D.H. Lee, J.E. Kim, E.Y. Choi, S.O. Kim, Peptide/grapheme hybrid assembly into core/shell nanowires, Adv. Mater. 22 (2010) 2060–2064.
[31] K. Min, T.H. Han, J. Kim, J. Jung, C. Jung, S.M. Hong, et al., A facile route to fabricate stable reduced graphene oxide dispersions in various media and their transparent conductive thin films, J. Colloid Interface Sci. 383 (2012) 36–42.
[32] L.T. Lu, Water-Dispersible Magnetic NPs for Biomedical Applications: Synthesis and Characterization (Ph.D. thesis), University of Liverpool, 2011.
[33] G. Wu, J. Ma, S. Li, J. Guan, B. Jiang, L. Wang, et al., Magnetic copper-based metal organic framework as an effective and recyclable adsorbent for removal of two fluoroquinolone antibiotics from aqueous solutions, J. Colloid Interface Sci. 528 (2018) 360–371.
[34] M. Baalousha, Aggregation and disaggregation of iron oxide nanoparticles: Influence of particle concentration, pH and natural organic matter, Sci. Total Environ. 407 (2009) 2093–2101.
[35] V.C. Fernandes, M. Freitas, J.P.G. Pacheco, J.M. Oliveira, V.F. Domingues, C. Delerue-Matos, Magnetic dispersive micro solid-phase extraction and gas chromatography determination of organophosphorus pesticides in strawberries, J. Chromatogr. A 1566 (2018) 1–12.
[36] X. Yang, Y. Yin, Y. Zong, T. Wan, X. Liao, Magnetic nanocomposite as sorbent for magnetic solid phase extraction coupled with high performance liquid chromatography for determination of polycyclic aromatic hydrocarbons, Microchem. J. 145 (2019) 26–34.
[37] A. Shard, A straightforward method for interpreting XPS data from core–shell nanoparticles, J. Phys. Chem. C 116 (2012) 16806–16813.
[38] D. Karabelli, C. Züm, T. Shahwan, A.E. Erolu, T.B. Scott, K.R. Hallam, et al., Batch removal of aqueous Cu^{2+} ions using nanoparticles of zero-valent iron: a study of the capacity and mechanism of uptake, Ind. Eng. Chem. 47 (2008) 4758–4764.
[39] C.I. Pearce, O. Qafoku, J. Liu, E. Arenholz, S.M. Heald, R.K. Kukkadapu, et al., Synthesis and properties of titanomagnetite (Fe3xTixO4) nanoparticles: a tunable solid-state Fe(II/III) redox system, J. Colloid Interface Sci. 387 (2012) 24–38.
[40] R. Russo, E. Esposito, C. Granata, A. Vettoliere, M. Russo, C. Cannas, et al., Magnetic nanoparticle characterization using nano-SQUID based on niobium dayem bridges, Phys. Procedia 36 (2012) 293–299.
[41] C. Granata, R. Russo, E. Esposito, A. Vettoliere, M. Russo, A. Musinu, et al., Magnetic properties of iron oxide nanoparticles investigated by nanoSQUIDs, Eur. Phys. J. B 86 (2013) 272–276.
[42] Z. Ma, Y. Guan, H. Liu, Synthesis and characterization of micron-sized monodisperse superparamagnetic polymer particles with amino groups, J. Polym. Sci. Part A: Polym. Chem. 43 (2005) 3433–3439.
[43] J.B. Mamani, A.J. Costa-Filho, D.R. Cornejo, E.D. Vieira, L.F. Gamarra, Synthesis and characterization of magnetite nanoparticles coated with lauric acid, Mater. Charact. 81 (2013) 28–36.
[44] http://www.nanomag-project.eu/Mössbauer-spectroscopy.

[45] UV/Vis/IR Spectroscopy Analysis of NPs. NanoComposix (Nanocomposix.com), September 2012.

[46] C.L. Fang, K. Qian, J. Zhu, S. Wang, X. Lv, S.H. Yu, Monodisperse α-Fe_2O_3@SiO_2@Au core/shell nanocomposite spheres: synthesis, characterization and properties, Nanotechnology 19 (2008) 125601.

[47] J.M. Vargas, A.A. McBride, J.B. Plumley, Y. Fichou, T.A. Memon, V. Shah, et al., Synthesis and characterization of core/shell Fe_3O_4/ZnSe fluorescent magnetic nanoparticles, J. Appl. Phys. 109 (2011). 07B536-1.

[48] T. Zhang, G. Lu, H. Shen, K. Shi, Y. Jiang, D. Xu, et al., Photoluminescence of a single complex plasmonic nanoparticle, Sci. Rep. 4 (2014) 3867.

Toxicity, therapeutic applicability, and safe handling of magnetic nanomaterials

4

Yasser Al-Eryani[1], Maryam Dadashi[2], Sajjad Aftabi[3], Hedieh Sattarifard[3], Ghazal Ghavami[4], Zachary William Oldham[5], Arash Ghoorchian[6] and Saeid Ghavami[3,7,8,9]

[1]Department of Electrical and Computer Engineering, University of Manitoba, Winnipeg, MB, Canada, [2]Department of Oral Biology, Dr. Gerald Niznick College of Dentistry, Rady Faculty of Health Sciences, University of Manitoba, Winnipeg, MB, Canada, [3]Department of Human Anatomy and Cell Science, Max Rady College of Medicine, Rady Faculty of Health Sciences, University of Manitoba, Winnipeg, MB, Canada, [4]Faculty of Science, University of Waterloo, Waterloo, ON, Canada, [5]Faculty of Engineering, University of Waterloo, Waterloo, ON, Canada, [6]Faculty of Chemistry, Bu-Ali Sina University, Hamedan, Iran, [7]Research Institute in Oncology and Hematology, CancerCare Manitoba, University of Manitoba, Winnipeg, MN, Canada, [8]Autophagy Research Center, Shiraz University of Medical Sciences, Shiraz, Iran, [9]Faculty of Medicine, Katowice School of Technology, Katowice, Poland

4.1 Introduction

The application of engineered magnetic nanoparticles (MNPs) in a variety of research and industrial fields, including biomedical [1,2], environmental [3], and clinical disciplines [4], revealed several major health-related technical issues that are primarily related to the impact of these materials on overall human well-being. Therefore it is essential to know how MNPs interact with surrounding aquatic ecosystems and terrestrial environments. The fate of MNPs, their toxicity, and long-term behavior are highly dependent on their physical and chemical structures, surface properties, and the type of material they interact with.

Exposure to magnetic fields with varying strengths induces structural changes in biological macromolecules, as exemplified by the induction of the magnetothermodynamic phenomenon in conducting material. Such an effect can be detected by a change in the temperature of that material and may be controlled by engineering different properties of MNPs [5]. Physicists also refer to this magnetothermodynamic phenomenon as adiabatic demagnetization that plays a major role in the way MNPs interact with living biological matter and other materials in the surrounding aquatic ecosystems and terrestrial environments [6].

Magnetic Nanomaterials in Analytical Chemistry. DOI: https://doi.org/10.1016/B978-0-12-822131-0.00012-1

The biological and pharmacological properties of MNPs are directly affected by their physical characteristics, such as composition, overall particle size, and surface chemistry [7]. These properties can be effectively utilized in many clinical applications, for example, by using the superparamagnetic iron oxide nanoparticles (SPIONs) in the design of magnetic resonance imaging (MRI) devices. Besides, regarding their superparamagnetic, biocompatible, and biodegradable features, MNPs can be considered for cancer treatment [e.g., photodynamic therapy [8], chemotherapy [9], photothermal therapy [10], and magnetic hyperthermia (MHT) [11]] and targeted delivery of drugs or genes [12]. As sensory materials, MNPs can also be used in the detection of metastatic lymph nodes, inflammatory and/or neural degenerative diseases, and liver metastases [13–15]. Regardless of these potential benefits, uncontrolled exposure to MNPs may cause serious undesirable damage to biological molecules and tissue. For instance, various constructions of SPIONs may cause oxidative stress and DNA damage that results in cellular perturbations, including alterations of gene expression profiles, actin cytoskeleton modulation, disruption of iron homeostasis, impairment of cell-cycle regulation, and activation of (detrimental) signaling pathways [16–18].

When different types of nanoparticles (NPs), including MNPs, are exposed to tissue and/or fluid, some of the macromolecules are absorbed into their surface. Specific surface characteristics, such as surface energy and chemistry, will determine the main features of this absorption process; these features may be modeled to conduct an intentional functionalization and/or modification of the MNP surface [19]. Using this principle, the surface of MNPs can be engineered and adjusted to match with a certain protein and/or other macromolecules, which allows it to target a particular assembly of the biomolecule, either within a membrane or inside a specific cell surface or organelle. A surface adaptation process may also be triggered by an outer unintentional hook that leads to a molecular-level interaction between MNPs and the exposed material. Such an outer trigger defines the way MNPs ultimately behave, before reaching a stabilized state.

Even though different substances within an environment have distinct bulk chemical compositions, surface properties, sizes, and shapes, the adverse effects of MNPs on these materials will have basic common features related to how the exposed material responds at the molecular level. Shared mechanistic origins stem from the fact that MNPs act through changing the ionization properties of atoms. The extent and predictability of such an effect depend on the atomic structure of the biological and nonbiological substances present in the subjected area, which would warrant a high-level classification scheme of all existing materials [18,20,21]. The physical, chemical, and structural characteristics of MNPs and their potential undesirable adverse effects have opened the door for the design of efficient recycling and/or disposal techniques that comply with health and environmental safety regulations and standards. In this chapter, we discuss the main features of the interaction between MNPs and the surrounding environment. Besides, we will focus on general MNP toxicity and desired (controlled) toxic effects of MNPs in cancer treatment. Lastly, the long-term behavior of MNPs and emerging state-of-the-art techniques to recycle and dispose of them will be addressed.

4.2 MNPs and the ecosystem

Recently, the interaction between MNPs and the environment at the nanolevel has drawn significant interest from both public and governmental authorities. This is due to the recent broad exploitation of engineered MNPs in a variety of practical applications [22].

Generally, MNP interactions with the environment, especially terrestrial and aquatic ecosystems (Fig. 4.1), encompass all physicochemical reactions triggered by contact with abiotic components, such as dissolved organic matter and surfactants. The physicochemical characteristics of MNPs are found to change dramatically at the moment they are discharged or released into the surrounding environment.

As an example, the aggregation/agglomeration status of MNPs can be significantly affected by several parameters, such as ionic strength, solution pH, surfactants, organic matter included within aqueous columns, soil matrices' clay, and temperature. Such an environmental dependence of MNPs plays a major role in changing the predestination and long-term behavior of MNPs. Furthermore, the nature of the interaction between MNPs and colloidal substances is associated with collision efficiency and possibly steric or electrostatic dynamics. In addition to aggregation and agglomeration, the surface chemistry of MNPs can also alter once they are discharged or released into the environment since the high specific surface area will cause them to collide under strong cohesion. For example, when dissolving different organic matter and minerals into water, surface properties change, depending on the size of the particles [23].

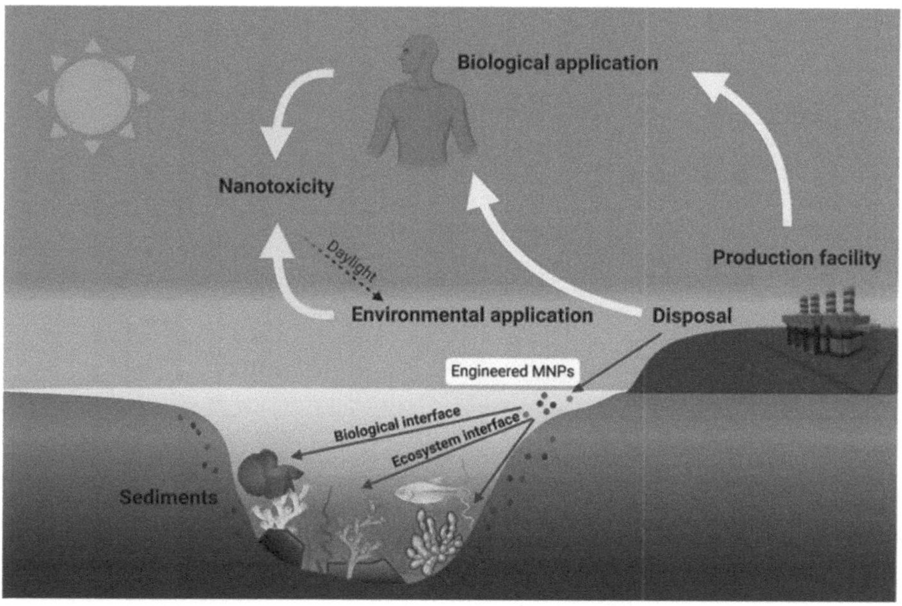

Figure 4.1 Interaction between MNPs and terrestrial and aquatic ecosystems. Different forms of MNPs-bio-eco interactions take place in three stages of their handling: (1) manufacturing of MNPs, (2) practical use of different forms of MNPs, and (3) recycling/disposal of consumed MNPs.

4.3 MNPs and biological life

As MNPs are the only clinically approved metal oxide NPs and are increasingly being applied, accurate characterization and understanding of engineered SPIONs are essential [24]. When a living cell is exposed to iron oxide nanoparticles (IONPs), toxicity is generated by a series of biochemical reactions, including (1) the production of reactive oxygen species (ROS) through diffusion, (2) endocytosis with protein coordination due to the release of metal ions, and (3) depolarization of the membrane, which results in a severe loss in cell integrity [24]. These events are summarized in Fig. 4.2.

Iron oxide MNPs can be used as a local heat source implanted within defected cancer tissue in the body. Such targeted localization is first achieved by diffusing the IONPs particles throughout Brownian motion and/or relaxation mechanisms. Once properly positioned, a magnetic field is applied to the desired area so that IONPs heat up through the MHT phenomenon and terminate the targeted cancer tissue [23]. Endocytosis, on the other hand, represents a basic biological process in which cells take up macromolecules and IONPs via transfer vesicles formed from the cell membrane. The kinetics of endocytosis are highly dependent on the pathways involved,

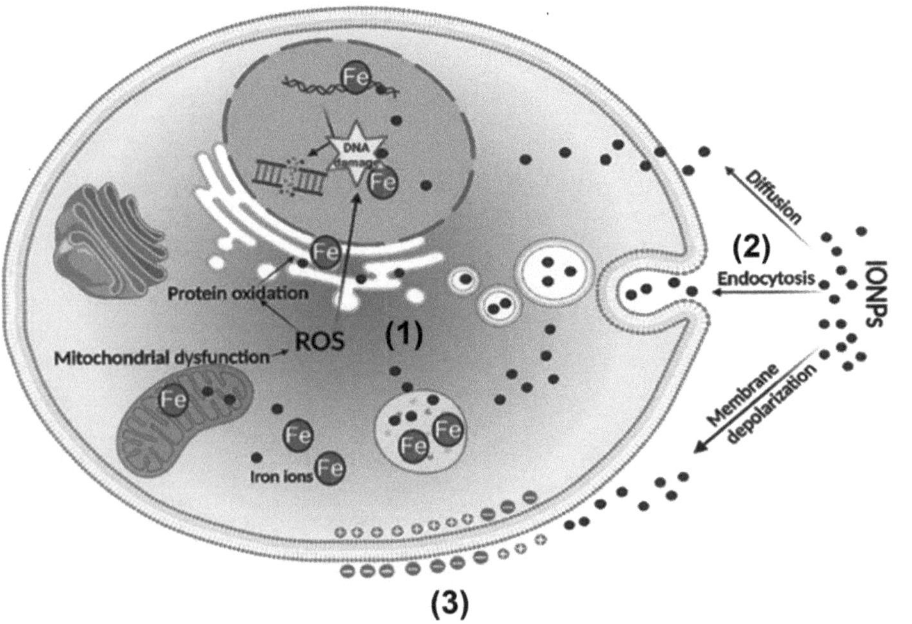

Figure 4.2 Schematic representations of the toxic responses of biological cells to IONPs exposure. (1) Intracellular ROS production: this toxic response can be measured by dichlorofluorescein diacetate, an oxidation-sensitive fluoroprobe that is nonfluorescent and is freely taken up by cells and hydrolyzed by esterases that remove the dichlorofluorescein diacetate group [25], (2) endocytosis reaction, and (3) membrane depolarization.

which in turn rely on the charge, size, and coating of IONPs, and the type of the cell being labeled [26].

Another important effect of IONPs on biological membranes is the ability of IONPs to transfer an electron from the inner to the outer side of the membrane, resulting in less negative charge inside the cell. When a negative charge leaves the cell membrane, a positive balancing charge gets inside the membrane simultaneously. Initially balancing ions and electron transport are out of phase, a sudden change in the membrane potential is induced by the addition of a reduceable substrate. Such a change results in hyperpolarization when electron transport is slightly delayed as compared to ion balancing. Conversely, it may also cause depolarization if the primary process is represented by electron transport [27,28].

4.4 Toxic effects of MNPs

The toxicity of MNPs is associated with their small particle size, shape, and, by extension, the larger surface area for a given volume of NPs. The small particle size allows the MNPs to interact with and enter cells, while the shape and surface area of a given NP can allow for greater specific interaction with surface charges for both the cell's components and the NP [29]. One of the main mechanisms underlying the cytotoxicity of MNPs in the production of ROS then interacts with biomolecules, such as lipids, nucleic acids, and proteins. These interactions can negatively affect essential functions of the cell, including the electron transport chain or mitochondrial structure, which can eventually lead to cell death [29]. Fig. 4.3 illustrates the relationship between MNPs, ROS generation, and cytotoxicity [30].

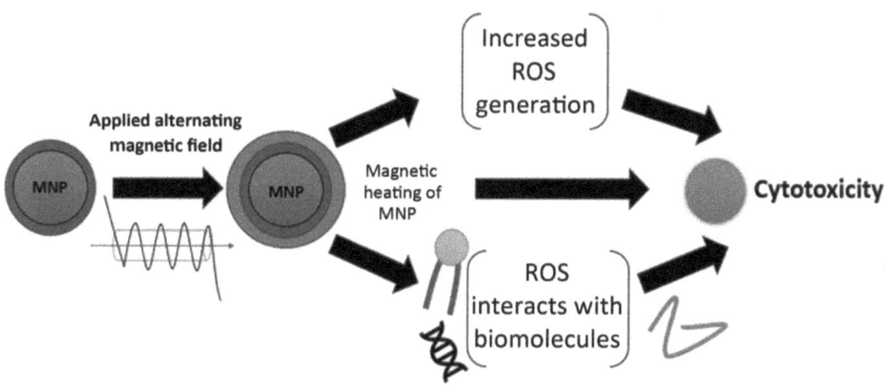

Figure 4.3 MNP-induced ROS production and cytotoxicity in mammalian cells. An applied alternating magnetic field on an MNP leads to a magnetic heating response generating more ROS. The increase in ROS generation leads to more interactions with various biomolecules, creating a cytotoxic response.

Nanotoxicity in the body may also result in cell cycle arrest, which can lead to apoptosis or necrosis [29]. MNPs often activate oxidative stress, resulting in cell inflammation and damaging the DNA of the biological system. Their toxicity is initiated by the increase in ROS in the host cell. Generally, when considering the effects of MNPs on an organ system, the type of MNP and route of exposure will affect where the MNPs accumulate in the body [29]. The build-up of MNPs can lead to toxic effects depending on the cellular environment and total accumulation volume. Site-specific aggregation relies on the size of the MNPs, as larger particles can be more easily detected and cleared by the body's natural defense systems. For example, MNPs above a specific size threshold would not traverse from the liver or the spleen and were stopped by macrophage-mediated phagocytosis [31].

With regards to specific organs and biological systems, the toxic effects of MNPs are generally related to the induction of oxidative stress. For example, in the cardiovascular system, cardiac arrhythmia could be seen as the main resultant effect of MNP exposure; when mice were subjected to IONPs with an LD_{50} of 0.1636 grams of NP per kilogram body weight, necrosis and denaturation of the myocardium were observed [31]. Although primarily observed in the myocardium, necrosis and denaturation could also be detected in other major organs to some extent. This is not to say that all the effects that MNPs produce will have a net negative or harmful impact. Different types of MNPs and their respective coatings can provide protection, hindering nonspecific toxicity that would otherwise harm "normal/healthy" cells and DNA [32]. The toxic effects of MNPs can also be used/directed toward cancerous cells, implicating a therapeutic application, as will be discussed later in this chapter.

MNPs can be widely applied in biomedicine, largely due to their unique magnetic properties and small size. The latter allows biomolecules to cover the NP surface, whereas control by an external magnetic field regulates the transport and immobilization of MNPs in human tissues. Notably, MNPs can transfer heat to a specific location in response to the cyclic variance of the external magnetic field [33]. IONPs are the most commonly used MNPs for therapeutic applications [31]. They are made of magnetite or maghemite [33] and have several desirable properties, such as superparamagnetism and high surface to volume ratio, and can be easily separated [31].

MNPs are widely used in the detection of diseases and can be used to treat tumors with heat (hyperthermia) [34]. One of the clinical applications of MNPs is magnetic bioseparation, a technique in which bioentities attached to nanomaterial surfaces are removed from their bioenvironment by the use of a magnetic force. Magnetic separation can be used in clinical diagnoses, as it allows for cells and bacteria to be easily separated from blood samples, bone marrow, tissue grinds, or culture media. This technique inflicts minimal stress and damage to the separated cells and bacteria, which favors the use of MNPs in separation assays [33]. SPIONs have been extensively used in MRI contrast enhancement [15].

These applications are possible in part due to the tendency of MNPs to agglomerate because of their small size and large surface to volume ratio [35]. If the target location in the body is hydrophobic, SPIONs may agglomerate due to opsonin

Table 4.1 Summary of the desirable and undesirable toxic effects of magnetic nanoparticles [15,28–31].

Properties of MNPs	Reasons for MNP toxicity	Use of the effects in biomedicine	Toxic effects of MNPs in the human body
Small size	Ability to interact with and enter cells	Tumor treatment (hyperthermia)	Inflammation or apoptosis/ necrosis of cells
Large surface area	Greater interaction potential depending on the shape	Magnetic bioseparation	DNA damage, oxidative stress
Magnetic field response	Heating and associated damaging of cells by MNPs	MRI contrast enhancement, magnetic hyperthermia	Cardiac arrhythmia

binding [36]. The administration of IONPs in rats did not affect the growth process and no significant body weight changes were observed. Food and water intake decreased early in the study, with no significant changes later on [37]. In terms of specific body systems, the administration of 200 mg kg^{-1} of zinc oxide NPs decreased in heart tissue without changing its weight [31]. Table 4.1 summarizes several toxic effects of MNPs and their potential usage/harm.

4.5 Toxicity of MNPs in cancer treatment

Cancer can be defined as the uncontrollable growth of abnormal cells in the human body [38]. It is one of the leading threats to public health, contributing to approximately 14 million deaths globally per year. Chemotherapy, immunotherapy, and surgical removal of cancerous tumors are the general techniques used to treat individuals afflicted by cancer [36,39]. As cancer research continues to advance, the use of NPs has become an important part of theranostics [40], an emerging field of medicine in which drugs and/or techniques are combined for the simultaneous or sequential diagnosis and treatment of human diseases. The main objectives of theranostic nanomedicine are to improve efficiency in cancer detection and treatment and decrease the toxicity of treatment [41]. Chemotherapy is used to stop the growth of cancer cells by using chemicals that inhibit cellular function; however, more traditional chemotherapy alone may not be (sufficiently) effective [15].

The use of MNPs in cancer theranostics has many advantages. With regards to diagnosis, MNPs can be used to image specific areas of the human body to locate cancerous cells. Also, MNP-enhanced MRI provides extremely high image contrast that allows for the detection of smaller cancers that would otherwise be overlooked and missed [42]. For the application in MRI, IONPs are modified with ligands that bind to cancer biomarkers or the surface of cancer cells [38]. The enhancing effect of MNPs on MRI allows for accurate drug delivery monitoring, treatment response evaluation, and external magnetic field–controlled delivery and release of treatments [35]. In comparison to other chemotherapy agents, a highly controlled MNP-based system likely has fewer side effects and is capable of targeted delivery down to the scale of ligand binding. This binding can occur in the form of bioconjugation or encapsulation in the MNP polymer coating [43].

As eluded to previously, MNPs can also be used in cancer treatment by destroying cancer tissue through hyperthermia. This technology uses a magnetic field oscillating iron oxide MNPs to generate heat, which in turn decreases the size of the tumor [42]. The MNPs are placed inside or around the tumor, increasing its temperature. If a temperature of 45°C is reached, tumor cells may be killed by necrosis, while chemotherapy efficiency is already enhanced at 42°C. By using MHT, the cancer cells can be killed without damaging surrounding healthy cells [33]. Also, MHT provides the advantage of unlimited penetration through tissue and the potential to deliver multiple cycles of hyperthermia [23].

As MNPs transition from bench to bedside, several factors must be considered to assess feasibility. The human body can have a multitude of responses to any foreign particle that is not intended for the specific treatment the individual is undergoing. As such, it is important to understand the pathways, reactivity, and specific mechanisms involved in MNP use [15]. Looking ahead, while the expectations for MNP applicability are high, the undesirable toxic effects, including oxidative stress, DNA alterations, and nonspecific cellular uptake, must be addressed and understood in greater depth to ensure the safety and effectiveness of MNP use [32]. Demonstrating the general safety of MNPs and expanding our knowledge regarding their specific health effects will also play an important role in the public's perception of these applications, especially for individuals with options regarding disease treatment [44].

4.6 Long-term behavior of MNPs

The agglomeration of MNPs, which may lead to deactivation, is a serious challenge in the application of these materials. It is a time-dependent process in which individually dispersed particles aggregate into a large cluster. As time passes, the number of dispersed particles decreases and more agglomerates are formed [45]. The occurrence of MNP agglomeration is considered an undesirable process, as it decreases the specific surface area and, as such, dramatically reduces reactivity [46]. The precipitation of formed agglomerates from the MNP suspension is another disadvantage that limits the applicability of MNPs [47].

In general, the high surface free energy of NPs results in overall high surface energy. Hence, considering thermodynamic aspects, NPs tend to aggregate to reduce the surface energy [48]. Phenrat et al. investigated the time-dependent agglomeration of magnetic zero-valent iron oxide using an in situ micrograph approach [49] and demonstrated that the agglomeration of primary NPs occurred as time elapsed. Two different kinetic domains have been proposed for the agglomeration of MNPs, as illustrated in Fig. 4.4. The first domain, the collision step, corresponds to the Brownian motion of dispersed MNPs and the formation of μ-sized agglomerates and is governed by the balance between attractive and repulsive energies of NPs. The second domain, the adhesion step, indicates further chain reaction–like agglomeration of these μ-sized agglomerates [50] and is critically influenced by particle–particle interactions.

Moreover, the use of MNPs in various applications depends on their chemical stability in different mediums [51]. The chemical stability of Fe_3O_4, the most studied MNP, in different environmental solutions was comprehensively presented by Kalska-Szostko and colleagues [52]. X-ray diffraction, transmission electron microscopy, Mössbauer spectroscopy, Fourier transform infrared, and atomic absorption spectroscopy approaches were employed to investigate the influences of several media on unmodified MNPs; water, ethanol, and an aqueous NaCl solution had the least effect on Fe_3O_4 stability. The structure of NPs significantly changes in the presence of these solutions. In contrast, exposure to acidic media, such as acetic and citric acid, leads to the dissolution of the iron oxide core. Interestingly, commercially available white wine increases the stability of MNPs. The formation of a stabilizing shell protects MNPs from dissolution. It should be pointed out that the presence of cations and anions can affect the chemical stability of MNPs as well [53–55]. In 2010 the agglomeration kinetics of bare and modified MNPs were

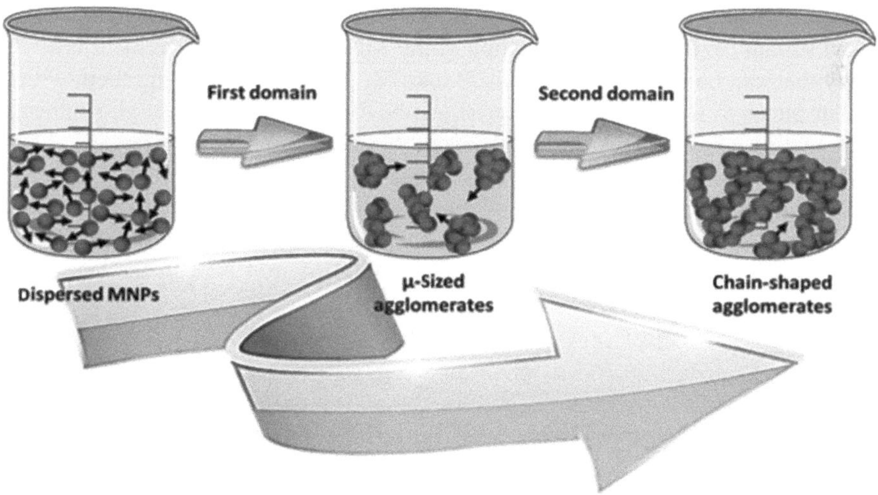

Figure 4.4 Schematic illustration of the agglomeration process of MNPs. Two different kinetic domains may be identified for the agglomeration of MNPs. The first domain, collision step, corresponds to the Brownian motion of dispersed MNPs and the formation of μ-sized agglomerates. The second domain, adhesion step, indicates the further agglomeration of these μ-sized agglomerates, like a chain reaction.

studied in the presence of Na^+, Mg^{2+}, and Ca^{2+} ions, which constitute the most abundant cations in the natural aquatic environments [54]. The chemical stability of MNPs was found to decrease in the following order: $Ca^{2+} > Mg^{2+} > Na^+$. In a similar study, the chemical stability of MNPs was shown to decrease in the presence of divalent cations under both acidic and basic conditions [55]. To date, little effort has been made on identifying the effects of anions on the chemical stability of MNPs. However, the aggregation behavior of citric acid–coated Fe_3O_4 NPs in the presence of some anions (phosphate, sulfate, and chloride) has been investigated, and anions were found to enhance the chemical stability of MNPs because of the formation of a more negative layer [55].

Thus it appears that bare MNPs are not the preferred choice for use in chemical, biochemical, and biomedical applications. To boost chemical stability, two main strategies, the modification of surface or viscosity increment, can be implemented. Surface modification has been widely applied to improve the dispersibility of MNPs. In this strategy, the MNPs surface is coated with an inorganic or organic layer. The MNPs can be modified with a passive layer, such as 3-aminopropyltrimethoxysilane [56], silica [57], gold [58], carbon [59], metal oxides [60–62], thiols [63], amines [64], and carboxylic acids [64,65]. Such a passive layer protects MNPs from the environment; for example, the silica shell was used to enhance electrostatic repulsion between individually dispersed MNPs [66]. Moreover, polymers and surfactants are good choices for surface modification of MNPs due to the existence of electrostatic and steric repulsion that stabilizes the bare NPs. Many researchers have utilized different synthetic or natural polymers, such as humic acid, chitosan, starch, carboxymethyl cellulose, alginate, poly (acrylic acid), polyaspartate, sodium polystyrene sulfonate, and polyacrylamide [67,68], and various types of surfactants (e.g., sodium dodecyl sulfate [69]) for the stabilization of the MNPs. Viscosity increment is another strategy that can be employed to stabilize bare MNPs by encapsulating them within a nonpolar phase to form an oil-in-water emulsion. Sakulchaicharoen et al. synthesized and characterized FePd NPs in the presence of guar gum as a stabilizer [70]. Due to the gelation that occurred during the synthesis process, Brownian motion was reduced, and, consequently, the agglomeration of MNPs was mitigated. Results indicated that the FePd NP suspension with guar gum stabilized within 48 hours. Similar observations were made for the preparation of iron NPs in the presence of xanthan gum [71].

4.7 Disposal, recycling, and handling techniques for MNPs

NPs have revolutionized our daily life in various application areas, such as electronics, cosmetics, energy, and medicine [72–75]. Their market grows significantly every year and it is expected to reach over $55 billion by 2022 [76]. Besides all benefits that they have brought into our lives, knowing their potential hazard and consequently establishing safe disposal methods have vital importance to avoid any potential threats to our environment and public health.

Tracking NPs is a challenge due to their size (typically 1–100 nm) [77,78]. They can release hazardous compounds during construction and disposal phases as listed in Table 4.2 [72,79]. Therefore ecosystems and public health might be negatively affected by the degradation of nanoparticle wastes (NPWs) and their potential toxicity [80]. Thus it is of paramount importance to establish protocols that are focused on safe handling and disposal of NPWs [81–84].

MNPs, as one of the most widely used types of NPs [75,85,86], play an essential role in several technological applications, such as biomedical tools, target drug delivery, medical diagnostics and treatments, cancer therapy, waste management, genetic engineering, immunoassays, and MRI, as has been described in previous sections [73,75,87–94]. MNPs are synthesized based on shape, size, and composition, and are classified into different types, including Fe NPs, cubic Ni NPs, hcp Co NPs, cubic Fe_3O_4 NPs, core–shell FePt/PbS NPs, and dumbbell Fe_3O_4–Au–Fe_3O_4 NPs [95–97].

Our knowledge and existing protocols are currently limited and immature regarding nanowastes and safe disposal of MNPs. Several guidelines for safely handling and using nanomaterials have been published by various organizations. The International Organization for Standardization (ISO) and OECD [79] were the pioneers in introducing the standards required for the accomplishment of safety in nanotechnology applications (ISO/TS 80004). Other organizations such as the European Commission (EC), the United Nations Environmental Programme (UNEP), United Nations Environmental Programme (UNEP), the International Union for Conservation of Nature (IUCN), and US Environmental Protection Agency (EPA) have later joined to further establish programs and guidelines for safe disposal of nanomaterials [98,99]. Table 4.3 summarizes the most commonly used guidelines for handling NP products in different countries.

Table 4.2 Possible leakage routes of waste containing nanomaterials from different waste disposal procedures [79].

Disposal method	Possible leakages
Recycling	Imbedded in secondary materials produced after recycling
Incineration	Flue gas emissions to the environment
	Fly and bottom ash to landfills
	Fly and bottom ash to storage facilities
	Bottom ash to industrial applications (e.g., roads)
Landfilling	Landfill gas emissions to the environment
	Landfill surface emissions to the environment
	Leachate to leachate treatment facilities
	Leachate to wastewater treatment facilities
Wastewater treatment	Emissions to surface water
	Wastewater sludge to incinerators
	Wastewater sludge to landfills
	Wastewater sludge used for agricultural applications

Table 4.3 Most commonly used guidelines for handling nanoparticle products in several countries [100].

Country	United States	European Union (EU)	Korea	Australia	Japan	Thailand
Regulation guidelines	TSCA FIFRA FFDCA	REACH CLP RoHS CR	Comprehensive plans of nano-safety (2011.10)	NICNAS	MOE-METI including information	NANOTEC

Although each type of MNP has unique chemical and physical characteristics, most of them have similar properties and functions in an analogous way [101]. Thus disposal methods could be universal for most of the MNPs. Here, we briefly mention currently used methods for the disposal of MNPs as suggested by the governing bodies.

4.7.1 Incineration

Incineration is one of the most practical waste handling techniques with high capability to reduce NPWs to harmless compounds [102]. MNP materials are burnt in an oxidative environment with extremely high temperatures (around 850°C–1200°C). Incineration could be used for disposal of several types of nanomaterial wastes: (1) consumer products in municipal solid waste, (2) wastes from nanotechnology research and development, (3) medical and infectious wastes, and (4) hazardous wastes.

The incineration procedure is as follows: MNP materials are burnt in an oxidative environment with extremely high temperatures (around 850°C–1200°C) and are mixed and placed in incineration plants with high temperatures. Flammable parts will burn, and other remains will deposit in the combustion chamber as slag or dust in the flue gas. Nowadays, several flue gas filters are introduced to reduce the volume of toxic material to a minimum [79]. Incinerators collect MNPs through such filters and accumulate them as fly/bottom ash, which is then sent to landfills for final disposal as shown in Fig. 4.5.

4.7.2 Landfilling

Landfilling waste disposal (dumping) is one of the most commonly used and cheapest procedures for waste management. The main drawback of this method is the production of several greenhouse gases (i.e., methane and carbon dioxide) that can affect

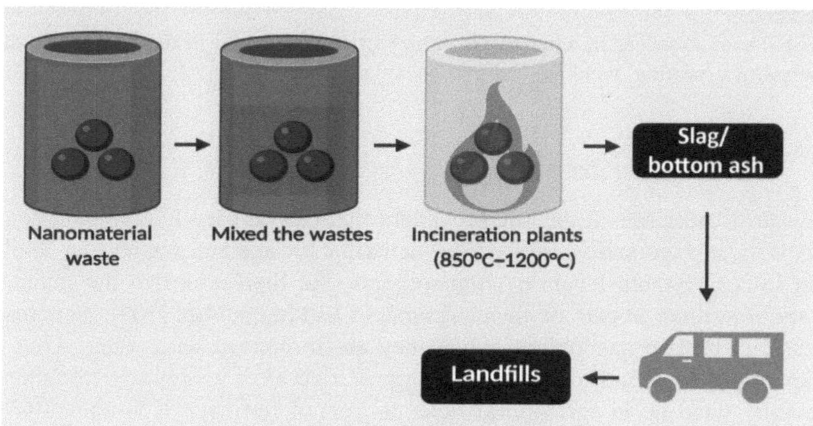

Figure 4.5 Schematic representation of the incineration process.

global warming [103,104]. The nanomaterial wastes that go through the landfill process for disposal have the potential to contribute to leachate. However, establishing highly specialized and controlled engineered landfills has improved/modernized waste handling at these sites and reduced the movement of both municipal solid waste and NPs to leachate. Several industrial countries use geosynthetic barriers to line the bottom of a landfill and incorporate collection systems for leachate and landfill gases. As such, geosynthetic barriers effectively prevent leaching the gases and leachate into the atmosphere and surface grounds, respectively [105,106].

4.7.3 Recycling

Recycling is a common method to convert waste into reusable materials, and markedly reduces the amount of waste sent to landfills. As mentioned, landfills are capable of producing several greenhouse gases. Therefore recycling can play an essential role in reducing the production of such gases [107]. Unfortunately, little information is available on the final fate of magnetic nanomaterials in current recycling procedures, as it is still challenging to measure the long-term effects of nanomaterial particles obtained after recycling. In general, NP recycling is performed in three stages [79].

1. Artificial resources and other metals are shredded with shredding machines to homogenize the waste particles. After shredding, reusable waste particles are segregated from unwanted materials and sorted based on size.
2. Particles are released and dust containing the NPs is created.
3. All the processes should be performed under highly controlled conditions to prevent the leakage of dust clouds to the environment.

These are the main concerns regarding exposure to nanomaterials during the NPWs recycling procedure [108]:

1. Possibility of exposure to the emission of NP dust during transport, sorting, grinding, and shredding.
2. Possibility of exposure to nanoobjects, which may be released in the flue gas or thermal progressions (heating, welding, and pyrolysis).

4.7.4 Wastewater treatment

Wastewater treatment is a specific procedure that can be used to eliminate contaminations from sewage and wastewater by detoxing sewage and wastewater until they contain low/acceptable levels of contaminants and then return to the main water cycle and minimum impact on the environment and human life [109]. Nanomaterial products can release nanoobjects when they are in contact with water. Thus these particles might be found in recycled sewage sludge after wastewater treatment and accidentally wind up in agricultural fields as part of fertilizer. Consequently, ultramodern wastewater treatments have been developed that can effectively eliminate the number of NPs ending up in agriculture areas [110,111].

4.7.5 Separation of magnetic parts

Recovery and extraction of the magnetic parts of MNPs offer an effective method for their recycling [112]. Although magnetic separation is currently the most effective available approach, the state-of-the-art NP technology is required to design more advanced separators for this method [84].

4.8 MNP stability: measurement and importance

Based on the properties and applications of MNPs, the core, shape, size, surface chemistry, composition, and aggregation should be precisely measured and controlled; such control processes should guarantee the so-called "MNPs stability" [113]. The term stability in MNPs refers to preserving the physical and chemical properties of the particles over time and in particular during their use in a certain application. In this section, we address some of the main characteristics that affect the stability of MNPs, such as aggregation, morphology and size, metal/metal oxidation, and surface distribution. First, thermodynamics of surface atoms (e.g., surface excitation and oxidation [114,115]) are found to cause a series of variations in MNP phases (solid, liquid, and gas). This may benefit the use and handling of MNPs in some situations (such as recycling dissolved MNPs into dried pellets through either lyophilization or evaporation) but may complicate their use as well (stability of MNPs coating shell in cancer treatment) [113]. Next, morphological analysis of complex nanostructures cannot be satisfactorily achieved by using traditional 2D electron microscopy imaging [116]. At present, widely established and standard imaging procedures simply do not provide enough information about the morphology of the MNPs and require advancements [113]. Third, the dimensions of MNPs are to be better understood by boxplots or histograms showing the size distributions or by whisker plots, rather than merely reporting the average sizes and standard deviations [113]. Nanostructures tend to ripen and grow to larger particles that disturb the culmination and evaluation of their sizes. The detailed information on the size of MNPs is not sophisticated if it does not include the distribution of small and large particles [117]. Last, to boost the initial retention of NPs and lower the chance of aggregation, novel strategies, such as using new capping agents that quench the high surface energy, stabilize the surface chemistry [118], retain morphology, and elevate the entropic interplay between NPs [119,120], can be applied. Altogether, proper stability measurements, reproducible quantifying analyses, and reports need to be specified, considering all stability-dependent aspects, including electrical, magnetic, mechanical, optical, catalytic, and thermal properties. In a study by Chen et al., at least one measure of the stability of Cr(VI) on polyethylenimine MNPs was introduced as a reference for effective removal of Cr(VI) from wastewater [121]. Taking the stability specifications into account leads to more efficient waste removal and proper disposal of MNPs from terrestrial and aquatic environments. Ideally, specific MNP characteristics, such as the core, shape, size, surface chemistry, composition, and aggregation, should be specified and guaranteed as "nanoparticle stability" [113].

4.9 Conclusions

MNPs have recently emerged in many industrial applications related to biomedical, environmental, and clinical disciplines. Such a variety of inapplicability of MNPs was accompanied by new technical challenges related to their fate, toxicity, long-term behavior, disposal and/or recycling, and stability. We described how different structures of MNPs interact with biological life at all levels based on their physical and chemical structures. Toxicity of MNPs can have severe detrimental effects that may even lead to serious damage and collapse of (human) organic systems, but when applied strategically it can be used for several advanced medical procedures, such as MNP-enhanced MRI and cancer treatment through MNP-induced local hyperthermia. Over time, dispersed MNPs tend to aggregate into a large cluster that causes a severe shortage of individual particles and a loss of stable MNP characteristics. This may affect their efficiency in clinical applications that require a large amount of precision, such as drug delivery in cancer treatment procedures. Upon fully utilizing MNPs, many techniques have been developed or customized to either dispose of or recycle them. MNPs will play a major role in modern diagnostic and therapeutic medicine (e.g., theranostics). Careful and well-controlled use of MNPs will certainly improve the overall quality of life and decrease mortality in several diseases, including different types of cancers, and cardiovascular and developmental disorders. By facilitating smart targeting and controlled release of medication, MNPs will be able to decrease the side effects of many medications and improve their therapeutic efficiency. However, further development of safe and efficient technologies for disposal and recycling of these compounds is crucial to keep our ecosystems and the environment free of any damage caused by their accumulation.

Acknowledgments

The authors would like to thank Science Impact (Winnipeg, Canada) for (post-) editing the manuscript.

References

[1] C. Xu, S. Shi, L. Feng, F. Chen, S.A. Graves, E.B. Ehlerding, et al., Long circulating reduced graphene oxide-iron oxide nanoparticles for efficient tumor targeting and multimodality imaging, Nanoscale 8 (2016) 12683–12692.

[2] C.C. Berry, A.S.G. Curtis, Functionalisation of magnetic nanoparticles for applications in biomedicine, J. Phys. D: Appl. Phys. 36 (2003) R198–R206.

[3] A. Pratt, Chapter 7 - Environmental applications of magnetic nanoparticles, in: C. Binns (Ed.), Frontiers of Nanoscience, Elsevier, 2014, pp. 259–307.

[4] A.K. Gupta, M. Gupta, Cytotoxicity suppression and cellular uptake enhancement of surface modified magnetic nanoparticles, Biomaterials 26 (2005) 1565–1573.

[5] V. Franco, J.S. Blázquez, J.J. Ipus, J.Y. Law, L.M. Moreno-Ramírez, A. Conde, Magnetocaloric effect: from materials research to refrigeration devices, Prog. Mater. Sci. 93 (2018) 112–232.
[6] Fundamentals of magnetic field effects, in: M. Yamaguchi, Y. Tanimoto (Eds.), Magneto-Science: Magnetic Field Effects on Materials: Fundamentals and Applications, Springer Berlin Heidelberg, Berlin, Heidelberg, 2006.
[7] C. Chouly, D. Pouliquen, I. Lucet, J.J. Jeune, P. Jallet, Development of superparamagnetic nanoparticles for MRI: effect of particle size, charge and surface nature on biodistribution, J. Microencapsul. 13 (1996) 245–255.
[8] L. Li, M. Nurunnabi, M. Nafiujjaman, Y.Y. Jeong, Y.K. Lee, K.M. Huh, A photosensitizer-conjugated magnetic iron oxide/gold hybrid nanoparticle as an activatable platform for photodynamic cancer therapy, J. Mater. Chem. B 2 (2014) 2929–2937.
[9] M. Torres-Lugo, C. Rinaldi, Thermal potentiation of chemotherapy by magnetic nanoparticles, Nanomed. (Lond.) 8 (2013) 1689–1707.
[10] S. Shen, S. Wang, R. Zheng, X. Zhu, X. Jiang, D. Fu, et al., Magnetic nanoparticle clusters for photothermal therapy with near-infrared irradiation, Biomaterials 39 (2015) 67–74.
[11] P.T. Yin, B.P. Shah, K.B. Lee, Combined magnetic nanoparticle-based microRNA and hyperthermia therapy to enhance apoptosis in brain cancer cells, Small 10 (2014) 4106–4112.
[12] J.W. Bulte, T. Douglas, B. Witwer, S.C. Zhang, E. Strable, B.K. Lewis, et al., Magnetodendrimers allow endosomal magnetic labeling and in vivo tracking of stem cells, Nat. Biotechnol. 19 (2001) 1141–1147.
[13] M. Wu, S. Huang, Magnetic nanoparticles in cancer diagnosis, drug delivery and treatment, Mol. Clin. Oncol. 7 (2017) 738–746.
[14] L. Bacigalupo, S. Aufort, M.C. Eberlé, E. Assenat, M. Ychou, B. Gallix, Assessment of liver metastases from colorectal adenocarcinoma following chemotherapy: SPIO-MRI versus FDG-PET/CT, Radiol. Med. 115 (2010) 1087–1100.
[15] R.A. Revia, M. Zhang, Magnetite nanoparticles for cancer diagnosis, treatment, and treatment monitoring: recent advances, Mater. Today (Kidlington) 19 (2016) 157–168.
[16] W.G. Kuschner, H. Wong, A. D'Alessandro, P. Quinlan, P.D. Blanc, Human pulmonary responses to experimental inhalation of high concentration fine and ultrafine magnesium oxide particles, Environ. Health Perspect. 105 (1997) 1234–1237.
[17] G. Oberdörster, A. Maynard, K. Donaldson, V. Castranova, J. Fitzpatrick, K. Ausman, et al., Principles for characterizing the potential human health effects from exposure to nanomaterials: elements of a screening strategy, Part. Fibre Toxicol. 2 (2005) 8.
[18] N. Singh, G.J. Jenkins, R. Asadi, S.H. Doak, Potential toxicity of superparamagnetic iron oxide nanoparticles (SPION), Nano Rev. 1 (2010).
[19] E.A. Schellenberger, F. Reynolds, R. Weissleder, L. Josephson, Surface-functionalized nanoparticle library yields probes for apoptotic cells, Chembiochem 5 (2004) 275–279.
[20] A. Espinosa, A. Muñoz-Noval, M. García-Hernández, A. Serrano, J. Jiménez de la Morena, A. Figuerola, et al., Magnetic properties of iron oxide nanoparticles prepared by seeded-growth route, J. Nanopart. Res. 15 (2013) 1514.
[21] K. Woo, J. Hong, S. Choi, H.-W. Lee, J.-P. Ahn, C.S. Kim, et al., Easy synthesis and magnetic properties of iron oxide nanoparticles, Chem. Mater. 16 (2004) 2814–2818.
[22] OECD, Current Developments/Activities on the Safety of Manufactured Nanomaterials, 2007.

[23] A. Espinosa, R. Di Corato, J. Kolosnjaj-Tabi, P. Flaud, T. Pellegrino, C. Wilhelm, Duality of iron oxide nanoparticles in cancer therapy: amplification of heating efficiency by magnetic hyperthermia and photothermal bimodal treatment, ACS Nano 10 (2016) 2436–2446.

[24] L.S. Arias, J.P. Pessan, A.P.M. Vieira, T.M.T. Lima, A.C.B. Delbem, Iron oxide nanoparticles for biomedical applications: a perspective on synthesis, drugs, antimicrobial activity, and toxicity, Antibiotics (Basel) 7 (2018).

[25] H.L. Karlsson, P. Cronholm, J. Gustafsson, L. Möller, Copper oxide nanoparticles are highly toxic: a comparison between metal oxide nanoparticles and carbon nanotubes, Chem. Res. Toxicol. 21 (2008) 1726–1732.

[26] O. Osman, L.F. Zanini, M. Frénéa-Robin, F. Dumas-Bouchiat, N.M. Dempsey, G. Reyne, et al., Monitoring the endocytosis of magnetic nanoparticles by cells using permanent micro-flux sources, Biomed. Microdevices 14 (2012) 947–954.

[27] P.C. Sijmons, F.C. Lanfermeijer, A.H. de Boer, H.B. Prins, H.F. Bienfait, Depolarization of cell membrane potential during trans-plasma membrane electron transfer to extracellular electron acceptors in iron-deficient roots of *Phaseolus vulgaris* L, Plant Physiol. 76 (1984) 943–946.

[28] R.H. Fang, A.V. Kroll, W. Gao, L. Zhang, Cell membrane coating nanotechnology, Adv. Mater. 30 (2018) e1706759.

[29] Y.W. Huang, M. Cambre, H.J. Lee, The toxicity of nanoparticles depends on multiple molecular and physicochemical mechanisms, Int. J. Mol. Sci. 18 (2017).

[30] R.J. Wydra, P.G. Rychahou, B.M. Evers, K.W. Anderson, T.D. Dziubla, J.Z. Hilt, The role of ROS generation from magnetic nanoparticles in an alternating magnetic field on cytotoxicity, Acta Biomater. 25 (2015) 284–290.

[31] Z. Jiang, K. Shan, J. Song, J. Liu, S. Rajendran, A. Pugazhendhi, et al., Toxic effects of magnetic nanoparticles on normal cells and organs, Life Sci. 220 (2019) 156–161.

[32] Y.F. Yang, Y.J. Lin, C.M. Liao, Toxicity-based toxicokinetic/toxicodynamic assessment of bioaccumulation and nanotoxicity of zerovalent iron nanoparticles in *Caenorhabditis elegans*, Int. J. Nanomed. 12 (2017) 4607–4621.

[33] A. Wu, P. Ou, L. Zeng, Biomedical applications of magnetic nanoparticles, Nano 05 (2010) 245–270.

[34] X. Mou, Z. Ali, S. Li, N. He, Applications of magnetic nanoparticles in targeted drug delivery system, J. Nanosci. Nanotechnol. 15 (2015) 54–62.

[35] I.O. Wulandari, V.T. Mardila, D.J.D.H. Santjojo, A. Sabarudin, Preparation and characterization of chitosan-coated Fe_3O_4 nanoparticles using ex-situ co-precipitation method and tripolyphosphate/sulphate as dual crosslinkers, IOP Conf. Series: Mater. Sci. Eng. 299 (2018) 012064.

[36] Y. Bao, T. Wen, A.C. Samia, A. Khandhar, K.M. Krishnan, Magnetic nanoparticles: material engineering and emerging applications in lithography and biomedicine, J. Mater. Sci. 51 (2016) 513–553.

[37] U.S. Gaharwar, R. Meena, Biodistribution, clearance and morphological alterations of intravenously administered iron oxide nanoparticles in male Wistar rats, Int. J. Nanomed. 14 (2019) 9677–9692.

[38] L. Zhu, Z. Zhou, H. Mao, L. Yang, Magnetic nanoparticles for precision oncology: theranostic magnetic iron oxide nanoparticles for image-guided and targeted cancer therapy, Nanomed. (Lond.) 12 (2017) 73–87.

[39] G.-T. Yu, L. Rao, H. Wu, L.-L. Yang, L.-L. Bu, W.-W. Deng, et al., Myeloid-derived suppressor cell membrane-coated magnetic nanoparticles for cancer theranostics by

inducing macrophage polarization and synergizing immunogenic cell death, Adv. Funct. Mater. 28 (2018) 1801389.
[40] V.S. Madamsetty, A. Mukherjee, S. Mukherjee, Recent trends of the bio-inspired nanoparticles in cancer theranostics, Front. Pharmacol. 10 (2019) 1264.
[41] O.L. Gobbo, K. Sjaastad, M.W. Radomski, Y. Volkov, A. Prina-Mello, Magnetic nanoparticles in cancer theranostics, Theranostics 5 (2015) 1249–1263.
[42] S. Fathi Karkan, M. Mohammadhosseini, Y. Panahi, M. Milani, N. Zarghami, A. Akbarzadeh, et al., Magnetic nanoparticles in cancer diagnosis and treatment: a review, Artif. Cells Nanomed. Biotechnol. 45 (2017) 1–5.
[43] L. Hajba, A. Guttman, The use of magnetic nanoparticles in cancer theranostics: toward handheld diagnostic devices, Biotechnol. Adv. 34 (2016) 354–361.
[44] J.A. Fogarty, J.R. Swartz, The exciting potential of modular nanoparticles for rapid development of highly effective vaccines, Curr. Opin. Chem. Eng. 19 (2018) 1–8.
[45] S.C. Tang, I.M. Lo, Magnetic nanoparticles: essential factors for sustainable environmental applications, Water Res. 47 (2013) 2613–2632.
[46] D. Dickson, G. Liu, C. Li, G. Tachiev, Y. Cai, Dispersion and stability of bare hematite nanoparticles: effect of dispersion tools, nanoparticle concentration, humic acid and ionic strength, Sci. Total Environ. 419 (2012) 170–177.
[47] W. Yu, H. Xie, A review on nanofluids: preparation, stability mechanisms, and applications, J. Nanomater. 2012 (2012) 435873.
[48] A.R. Petosa, D.P. Jaisi, I.R. Quevedo, M. Elimelech, N. Tufenkji, Aggregation and deposition of engineered nanomaterials in aquatic environments: role of physicochemical interactions, Environ. Sci. Technol. 44 (2010) 6532–6549.
[49] T. Phenrat, N. Saleh, K. Sirk, R.D. Tilton, G.V. Lowry, Aggregation and sedimentation of aqueous nanoscale zerovalent iron dispersions, Environ. Sci. Technol. 41 (2007) 284–290.
[50] C. Henry, J.P. Minier, J. Pozorski, G. Lefèvre, A new stochastic approach for the simulation of agglomeration between colloidal particles, Langmuir 29 (2013) 13694–13707.
[51] J.L. Lyon, D.A. Fleming, M.B. Stone, P. Schiffer, M.E. Williams, Synthesis of Fe oxide core/Au shell nanoparticles by iterative hydroxylamine seeding, Nano Lett. 4 (2004) 719–723.
[52] B. Kalska-Szostko, U. Wykowska, K. Piekut, D. Satuła, Stability of Fe_3O_4 nanoparticles in various model solutions, Colloids Surf. A: Physicochem. Eng. Asp. 450 (2014) 15–24.
[53] S.E. Mylon, K.L. Chen, M. Elimelech, Influence of natural organic matter and ionic composition on the kinetics and structure of hematite colloid aggregation: implications to iron depletion in estuaries, Langmuir 20 (2004) 9000–9006.
[54] X. Liu, M. Wazne, Y. Han, C. Christodoulatos, K.L. Jasinkiewicz, Effects of natural organic matter on aggregation kinetics of boron nanoparticles in monovalent and divalent electrolytes, J. Colloid Interface Sci. 348 (2010) 101–107.
[55] H. Wang, X. Zhao, X. Han, Z. Tang, S. Liu, W. Guo, et al., Effects of monovalent and divalent metal cations on the aggregation and suspension of Fe3O4 magnetic nanoparticles in aqueous solution, Sci. Total. Environ. 586 (2017) 817–826.
[56] B. Feng, R.Y. Hong, L.S. Wang, L. Guo, H.Z. Li, J. Ding, et al., Synthesis of Fe_3O_4/APTES/PEG diacid functionalized magnetic nanoparticles for MR imaging, Colloids Surf. A: Physicochem. Eng. Asp. 328 (2008) 52–59.
[57] S. Santra, R. Tapec, N. Theodoropoulou, J. Dobson, A. Hebard, W. Tan, Synthesis and characterization of silica-coated iron oxide nanoparticles in microemulsion: the effect of nonionic surfactants, Langmuir 17 (2001) 2900–2906.

[58] W. Wu, Q. He, H. Chen, J. Tang, L. Nie, Sonochemical synthesis, structure and magnetic properties of air-stable Fe_3O_4/Au nanoparticles, Nanotechnology 18 (2007) 145609.
[59] Z. Wang, H. Guo, Y. Yu, N. He, Synthesis and characterization of a novel magnetic carrier with its composition of Fe3O4/carbon using hydrothermal reaction, J. Magn. Magn. Mater. 302 (2006) 397–404.
[60] N.O. Núñez, P. Tartaj, M.P. Morales, P. Bonville, C.J. Serna, Yttria-coated FeCo magnetic nanoneedles, Chem. Mater. 16 (2004) 3119–3124.
[61] C.L. Carnes, K.J. Klabunde, Unique chemical reactivities of nanocrystalline metal oxides toward hydrogen sulfide, Chem. Mater. 14 (2002) 1806–1811.
[62] R.H. Castro, P. Hidalgo, J.A. Coaquira, J. Bettini, D. Zanchet, D. Gouvêa, Surface segregation in SnO_2–Fe_2O_3 nanopowders and effects in Mössbauer spectroscopy, Eur. J. Inorg. Chem. 2005 (2005) 2134–2138.
[63] C. Dai, A.G. Stack, A. Koishi, A. Fernandez-Martinez, S.S. Lee, Y. Hu, Heterogeneous nucleation and growth of barium sulfate at organic–water interfaces: interplay between surface hydrophobicity and Ba^{2+} adsorption, Langmuir 32 (2016) 5277–5284.
[64] S. Djuric, J. Venit, P. Magnus, Silicon in synthesis: stabase adducts - a new primary amine protecting group: alkylation of ethyl glycinate, Tetrahedron Lett. 22 (1981) 1787–1790.
[65] K.A. Huynh, K.L. Chen, Aggregation kinetics of citrate and polyvinylpyrrolidone coated silver nanoparticles in monovalent and divalent electrolyte solutions, Environ. Sci. Technol. 45 (2011) 5564–5571.
[66] H.L. Ding, Y.X. Zhang, S. Wang, J.M. Xu, S.C. Xu, G.H. Li, Fe_3O_4@SiO_2 core/shell nanoparticles: the silica coating regulations with a single core for different core sizes and shell thicknesses, Chem. Mater. 24 (2012) 4572–4580.
[67] T. Phenrat, N. Saleh, K. Sirk, H.-J. Kim, R.D. Tilton, G.V. Lowry, Stabilization of aqueous nanoscale zerovalent iron dispersions by anionic polyelectrolytes: adsorbed anionic polyelectrolyte layer properties and their effect on aggregation and sedimentation, J. Nanopart. Res. 10 (2008) 795–814.
[68] C.M. Cirtiu, T. Raychoudhury, S. Ghoshal, A. Moores, Systematic comparison of the size, surface characteristics and colloidal stability of zero valent iron nanoparticles pre- and post-grafted with common polymers, Colloids Surf. A: Physicochem. Eng. Asp. 390 (2011) 95–104.
[69] Y.W. Choi, H. Lee, Y. Song, D. Sohn, Colloidal stability of iron oxide nanoparticles with multivalent polymer surfactants, J. Colloid Interface Sci. 443 (2015) 8–12.
[70] N. Sakulchaicharoen, D.M. O'Carroll, J.E. Herrera, Enhanced stability and dechlorination activity of pre-synthesis stabilized nanoscale FePd particles, J. Contam. Hydrol. 118 (2010) 117–127.
[71] S. Comba, R. Sethi, Stabilization of highly concentrated suspensions of iron nanoparticles using shear-thinning gels of xanthan gum, Water Res. 43 (2009) 3717–3726.
[72] H. Bahadar, F. Maqbool, K. Niaz, M. Abdollahi, Toxicity of nanoparticles and an overview of current experimental models, Iran. Biomed. J. 20 (2016) 1.
[73] Y. Zhang, L. Zhang, X. Song, X. Gu, H. Sun, C. Fu, et al., Synthesis of superparamagnetic iron oxide nanoparticles modified with MPEG-PEI via photochemistry as new MRI contrast agent, J. Nanomater. 2015 (2015).
[74] S. Marchesan, M. Prato, Nanomaterials for (Nano) Medicine, ACS Publications, 2013.
[75] Y. Yin, A.P. Alivisatos, Colloidal nanocrystal synthesis and the organic–inorganic interface, Nature 437 (2005) 664–670.

[76] E. Inshakova, O. Inshakov, World market for nanomaterials: Structure and trends, MATEC web of conferences, EDP Sci. (2017) 02013.
[77] M. Vert, Y. Doi, K.-H. Hellwich, M. Hess, P. Hodge, P. Kubisa, et al., Terminology for biorelated polymers and applications (IUPAC Recommendations 2012), Pure Appl. Chem. 84 (2012) 377–410.
[78] S. Laurent, D. Forge, M. Port, A. Roch, C. Robic, L. Vander Elst, et al., Magnetic iron oxide nanoparticles: synthesis, stabilization, vectorization, physicochemical characterizations, and biological applications, Chem. Rev. 108 (2008) 2064–2110.
[79] OECD, Nanomaterials in Waste Streams, 2016.
[80] J. Corbett, P. McKeown, G. Peggs, R. Whatmore, Nanotechnology: international developments and emerging products, CIRP Ann. 49 (2000) 523–545.
[81] S. Ibbotson, S. Kara, Environmental Footprint of Magnetite Nanoparticle Application: Iron Oxide Case, Leveraging Technology for a Sustainable World, Springer, 2012, pp. 113–118.
[82] X. Zhao, M. Yu, D. Xu, A. Liu, X. Hou, F. Hao, et al., Distribution, bioaccumulation, trophic transfer, and influences of CeO_2 nanoparticles in a constructed aquatic food web, Environ. Sci. Technol. 51 (2017) 5205–5214.
[83] M. Gao, Y. Yang, M. Lv, W. Song, Z. Song, Oxidative stress and DNA damage in zebrafish liver due to hydroxyapatite nanoparticles-loaded cadmium, Chemosphere 202 (2018) 498–505.
[84] L.M. Rossi, N.J. Costa, F.P. Silva, R. Wojcieszak, Magnetic nanomaterials in catalysis: advanced catalysts for magnetic separation and beyond, Green Chem. 16 (2014) 2906–2933.
[85] B.L. Cushing, V.L. Kolesnichenko, C.J. O'connor, Recent advances in the liquid-phase syntheses of inorganic nanoparticles, Chem. Rev. 104 (2004) 3893–3946.
[86] I. Khan, K. Saeed, I. Khan, Nanoparticles: properties, applications and toxicities, Arab. J. Chem. 12 (2019) 908–931.
[87] Y. Wu, F. Zhang, J. Ma, X. Zhang, L. Wu, B. Qu, et al., Association of large intergenic noncoding RNA expression with disease activity and organ damage in systemic lupus erythematosus, Arthritis Res. Ther. 17 (2015) 131.
[88] X. Gu, Y. Zhang, H. Sun, X. Song, C. Fu, P. Dong, Mussel-inspired polydopamine coated iron oxide nanoparticles for biomedical application, J. Nanomater. 2015 (2015).
[89] I. Sharifi, H. Shokrollahi, S. Amiri, Ferrite-based magnetic nanofluids used in hyperthermia applications, J. Magn. Magn. Mater. 324 (2012) 903–915.
[90] K. Wu, D. Su, J. Liu, R. Saha, J.-P. Wang, Magnetic nanoparticles in nanomedicine: a review of recent advances, Nanotechnology 30 (2019) 502003.
[91] S. Arsalani, E.J. Guidelli, M.A. Silveira, C.E. Salmon, J.F. Araujo, A.C. Bruno, et al., Magnetic Fe_3O_4 nanoparticles coated by natural rubber latex as MRI contrast agent, J. Magn. Magn. Mater. 475 (2019) 458–464.
[92] M. Neamtu, C. Nadejde, V.-D. Hodoroaba, R.J. Schneider, L. Verestiuc, U. Panne, Functionalized magnetic nanoparticles: Synthesis, characterization, catalytic application and assessment of toxicity, Sci. Rep. 8 (2018) 1–11.
[93] H. Zhang, X.L. Liu, Y.F. Zhang, F. Gao, G.L. Li, Y. He, et al., Magnetic nanoparticles based cancer therapy: current status and applications, Sci. China Life Sci. 61 (2018) 400–414.
[94] Z.M. Avval, L. Malekpour, F. Raeisi, A. Babapoor, S.M. Mousavi, S.A. Hashemi, et al., Introduction of magnetic and supermagnetic nanoparticles in new approach of targeting drug delivery and cancer therapy application, Drug Metab. Rev. 52 (2020) 157–184.

[95] J. Kudr, Y. Haddad, L. Richtera, Z. Heger, M. Cernak, V. Adam, et al., Magnetic nanoparticles: from design and synthesis to real world applications, Nanomaterials 7 (2017) 243.
[96] B. Duong, H. Khurshid, P. Gangopadhyay, J. Devkota, K. Stojak, H. Srikanth, et al., Enhanced magnetism in highly ordered magnetite nanoparticle-filled nanohole arrays, Small 10 (2014) 2840–2848.
[97] M.R. Zamani Kouhpanji, B.J. Stadler, A guideline for effectively synthesizing and characterizing magnetic nanoparticles for advancing nanobiotechnology: a review, Sensors 20 (2020) 2554.
[98] M. Duvall, A. Wyatt, Regulation of Nanotechnology and Nanomaterials at EPA and Around the World: Recent Developments and Context, Beveridge & Diamond, PC, Washington, DC, 2011.
[99] M.M. Lasat, K.F. Chung, J. Lead, S. McGrath, R.J. Owen, S. Rocks, et al., Advancing the understanding of environmental transformations, bioavailability and effects of nanomaterials, an International US Environmental Protection Agency—UK Environmental Nanoscience Initiative Joint Program, J. Environ. Prot. 9 (2018) 385.
[100] H.-N. Umh, B.-C. Lee, Y.-H. Kim, New paradigm for nanowastes treatment, Clean. Technol. 18 (2012) 250–258.
[101] M. Tadic, S. Kralj, M. Jagodic, D. Hanzel, D. Makovec, Magnetic properties of novel superparamagnetic iron oxide nanoclusters and their peculiarity under annealing treatment, Appl. Surf. Sci. 322 (2014) 255–264.
[102] A. Wagner, A.P. White, M.C. Tang, S. Agarwal, T.A. Stueckle, Y. Rojanasakul, et al., Incineration of nanoclay composites leads to byproducts with reduced cellular reactivity, Sci. Rep. 8 (2018) 1–15.
[103] C. Zhang, T. Xu, H. Feng, S. Chen, Greenhouse gas emissions from landfills: a review and bibliometric analysis, Sustainability 11 (2019) 2282.
[104] Y. Wang, J.W. Levis, M.A. Barlaz, An assessment of the dynamic global warming impact associated with long-term emissions from landfills, Environ. Sci. Technol. 54 (2019) 1304–1313.
[105] K.P. von Maubeuge, Geosynthetic barriers in regulations and recommendations in line with the ISO design guide? Innov. Infrastruct. Solut. 3 (2018) 75.
[106] D.-J. Kong, H.-N. Wu, J.-C. Chai, A. Arulrajah, State-of-the-art review of geosynthetic clay liners, Sustainability 9 (2017) 2110.
[107] V. Ram, K.C. Kishore, S.N. Kalidindi, Environmental benefits of construction and demolition debris recycling: Evidence from an Indian case study using life cycle assessment, J. Clean. Prod. 255 (2020) 120258.
[108] J. Struwe, E. Schindler, Bedeutung von Nanomaterialien beim Recycling von Abfällen, Arbeitspapier (2012).
[109] J.M. Paraskeva, Curriculum and the Generation of Utopia: Interrogating the Current State of Critical Curriculum Theory, Routledge, 2020.
[110] OECD Publishing, Nanomaterials in Waste Streams-Current Knowledge on Risks and Impacts, OECD Publishing, 2016.
[111] G. Yang, G. Zhang, H. Wang, Current state of sludge production, management, treatment and disposal in China, Water Res. 78 (2015) 60–73.
[112] C. Sappino, L. Primitivo, M. De Angelis, M.O. Domenici, A. Mastrodonato, I.B. Romdan, et al., Functionalized magnetic nanoparticles as catalysts for enantioselective Henry reaction, ACS Omega 4 (2019) 21809–21817.
[113] H.T. Phan, A.J. Haes, What does nanoparticle stability mean? J. Phys. Chem. C 123 (2019) 16495–16507.

[114] C. Levard, E.M. Hotze, G.V. Lowry, G.E. Brown Jr, Environmental transformations of silver nanoparticles: impact on stability and toxicity, Environ. Sci. Technol. 46 (2012) 6900–6914.
[115] F. Noor, H. Zhang, T. Korakianitis, D. Wen, Oxidation and ignition of aluminum nanomaterials, Phys. Chem. Chem. Phys. 15 (2013) 20176–20188.
[116] H. de Puig, J.O. Tam, C.-W. Yen, L. Gehrke, K. Hamad-Schifferli, Extinction coefficient of gold nanostars, J. Phys. Chem. C 119 (2015) 17408–17415.
[117] R. Bienert, F. Emmerling, A.F. Thünemann, The size distribution of 'gold standard' nanoparticles, Anal. Bioanal. Chem. 395 (2009) 1651.
[118] G. Cao, Nanostructures and Nanomaterials: Synthesis, Properties and Applications, World Scientific, 2004.
[119] H.T. Phan, A.J. Haes, Impacts of pH and intermolecular interactions on surface-enhanced Raman scattering chemical enhancements, J. Phys. Chem. C 122 (2018) 14846–14856.
[120] M.C.S. Pierre, P.M. Mackie, M. Roca, A.J. Haes, Correlating molecular surface coverage and solution-phase nanoparticle concentration to surface-enhanced Raman scattering intensities, J. Phys. Chem. C 115 (2011) 18511–18517.
[121] B. Chen, X. Zhao, Y. Liu, B. Xu, X. Pan, Highly stable and covalently functionalized magnetic nanoparticles by polyethyleneimine for Cr (VI) adsorption in aqueous solution, RSC Adv. 5 (2015) 1398–1405.

Section II

Application of magnetic nanomaterials as sorbents

Application of magnetic nanomaterials in food analysis

5

Mohammad Faraji[1] *and Yadollah Yamini*[2]
[1]Research Group of Food, Halal and Agricultural Products, Research Department of Food Technology and Agricultural Products, Standard Research Institute (SRI), Karaj, Iran,
[2]Department of Chemistry, Faculty of Sciences, Tarbiat Modares University, Tehran, Iran

5.1 Introduction

Analyzing food samples is one of the most attractive research areas because of the importance of ensuring food quality and safety in the community. Usually, direct analysis of foodstuff is very difficult without any sample preparation method due to the trace level of the analytes and complexity of the matrices. Therefore sample preparation for extraction, cleanup, or preconcentration of analytes from the food samples is needed before instrumental analysis. Solid-phase extraction (SPE) is one of the oldest and most widely used sample preparation techniques. It includes the use of sorbents that separate the target compounds from other interference in the food samples. In this technique, the unwanted impurities or target compounds can be retained in the stationary phase. Then, the stationary phase is eluted by a proper eluent and the eluent is discarded (in the case of cleanup purposes) or collected (in the case of preconcentration purposes).

In recent years, many efforts have been made to miniaturize this technique as well as increasing the efficiency and selectivity of the adsorbents used for a broad range of analytes with different polarities. Therefore the use of nanomaterials and especially magnetic nanomaterials (MNMs) as excellent alternatives for conventional adsorbents has been strongly considered.

Nowadays, MNMs are considered to be advancing sorbents in sample preparation techniques of food samples for the extraction and enrichment of food components and pollutants at trace levels [1−8]. MNMs, as adsorbents, incorporate the unique properties of nanomaterials and advantages of magnetic carrier technology, compared to traditional micrometer-size adsorbents. The former can afford it when applied in complex sample matrices. As a result, the use of such sorbents in sample preparation methods results in a good and rapid dispersion as well as efficient recovery and reuse when used for food samples.

5.2 Magnetic solid-phase extraction

In 1999 Šafaříková and Šafařik [9] introduced the first application of magnetic solid-phase extraction (MSPE) in analytical chemistry. They used silanized

Magnetic Nanomaterials in Analytical Chemistry. DOI: https://doi.org/10.1016/B978-0-12-822131-0.00003-0

magnetite for MSPE of a copper-phthalocyanine dye. In MSPE techniques, the first magnetic sorbent is added to the sample solution. After the desired compound is absorbed on the magnetic sorbent, the sorbent is separated from the sample solution using an external magnetic field. Then, the extracted analytes are eluted using a proper solvent from the sorbent. Finally, the eluent is separated from the sorbent by the magnet and introduced to analytical instruments. MSPE procedure is schematically shown in Fig. 5.1 [10].

Since the introduction of the MSPE, over time, the primary micrometer magnetic sorbents have been replaced with nanosized and very diverse magnetic adsorbents. Nowadays, nanosized magnetic sorbents consist of advanced materials and hybrids of two or more of them. MSPE with MNMs as the sorbent has several advantages over traditional SPE based on nonmagnetic micrometer sorbents:

1. It avoids time consumption and tedious on-column SPE procedures.
2. By eliminating the filtration and centrifugation steps, the simplicity and speed of the method have been increased especially for large sample volumes.
3. The range of sorbents with the appropriate variety and selectivity has been increased.
4. The use of organic solvents has been drastically reduced.
5. The possibility of achieving higher cleanup efficiency by applying magnetic separation due to the majority of sample impurities is diamagnetic.

These advantages could be achieved when magnetic sorbent ideally possess the following features [2]:

- High specific surface area with proper porosity, and with easy surface modification; to obtain high extraction efficiency or adsorption capacity with a small amount of the sorbent.

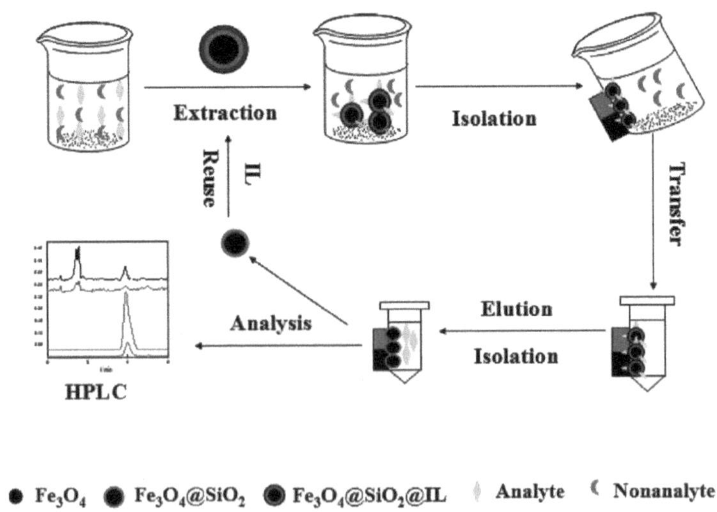

Figure 5.1 Schematic of the MSPE method. Procedure for the MSPE of RhB using Fe_3O_4@SiO_2@IL MNPs [10].

- Good dispersibility; to obtain fast and efficient adsorption/desorption.
- Proper selectivity; to obtain accurate results from complex matrices with interference.
- Suitable magnetic property; to perform fast and easy separation of the sorbent.
- Suitable durability; stability of the sorbent in acidic and basic conditions as well as under stirring and ultrasound process.
- Reusability.
- Easy and low-cost preparation of the sorbent with accessible raw materials as well as suitable batch-to-batch reproducibility.

5.3 Novel sorbents based on advanced magnetic nanomaterials

Up to date, a variety of MNMs have been used as sorbents in MSPE techniques [1–8]. In this sense, by coupling magnetic nanoparticles with advanced materials, such as metal-organic frameworks (MOFs), graphene/graphene oxides, carbon nanotubes (CNTs), molecular/ion-imprinted polymers (MIPs/IIPs), layered double hydroxides (LDHs), mesoporous materials, porous organic polymers (POPs), restricted access materials, and aptamers, magnetic nanosorbents with magnetic properties and unique properties of advanced materials are produced. In some cases, hybrid magnetic nanocomposites are synthesized simultaneously with the integration of two or more advanced materials to benefit from the features of them. In some recent studies, magnetic nanocomposites have also been prepared by combining two or more advanced materials to allow researchers to take their unique advantages simultaneously. In the following sections, some of these advanced materials and their important features will be reviewed.

5.3.1 Metal-organic frameworks

MOFs are advanced materials that are formed as crystal networks through inorganic metals and multidentate organic ligands through coordinated bonds [11,12]. The combination of inorganic and organic units in their construction results in favorable features, such as large specific surface area and high porosity and for use as selective sorbents. The porosity of some MOFs may reach 90% of their volume with specific surface areas more than 7000 $m^2\,g^{-1}$ [13]. Another feature of MOFs is the designable structure. First, various inorganic metal and organic ligands can be chosen for producing MOFs for specific applications [14]. The pore size can be controlled by changing the ligands and metals, as well as by controlling the reaction conditions [15]. However, poor dispersion of MOFs in aqueous media limits their use in the efficient enrichment of target compounds [16,17]. To solve such limitations, MOFs can be further functionalized to meet the requirements of specific applications. By postsynthetic modification, functional groups with different properties such as graphene oxide (GO) [16] or polyethylenimine (PEI) can be introduced [17]. These characteristics allow MOFs to be utilized as an efficient sorbent in MSPE.

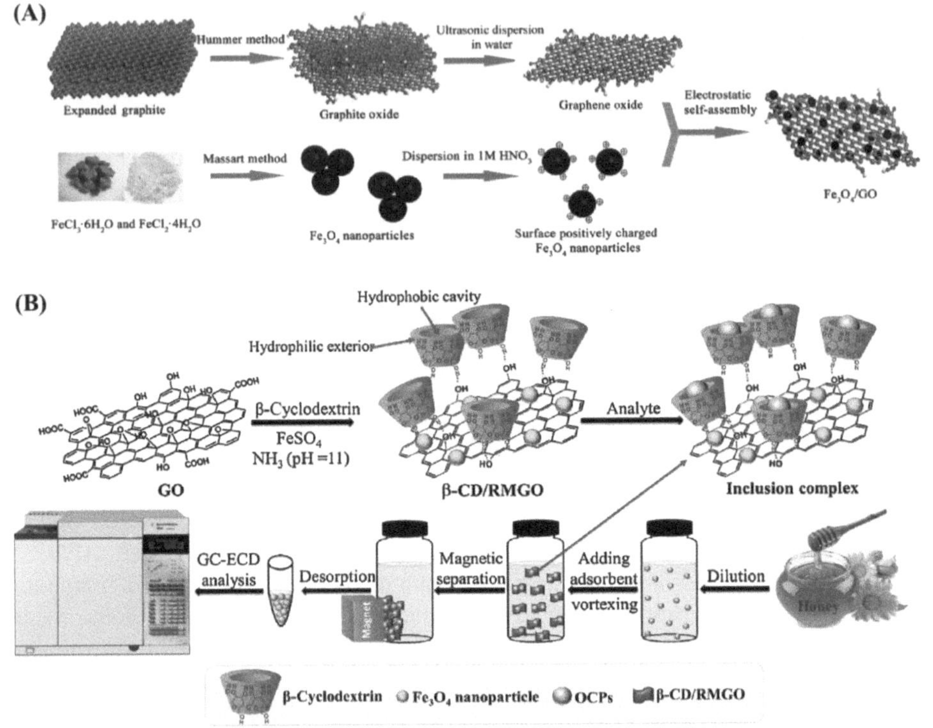

Figure 5.2 (A) Schematic illustration of the fabrication of the Fe_3O_4/GO nanocomposite [19]. (B) Schematic illustration of the procedure for the synthesis of β-CD/MRGO and MSPE steps for OCPs analysis in honey [20].

Li et al. prepared a metal-organic framework/graphene oxide (Fe_3O_4/MIL-100 (Fe)/GO) as a new sorbent for MSPE of enrofloxacin (ENR) in milk samples followed by spectrofluorimetry [16]. The sorbent (Fe_3O_4/MIL-100(Fe)/GO) exhibited extraction efficiency higher than that of Fe_3O_4/GO. They attributed this better performance to the formation of coordination bonds between the carboxyl group in ENR and free Fe(III) sites. Also, using GO in the prepared composite leads to improve the dispersivity of Fe_3O_4/MIL-100(Fe) because of the excellent hydrophilic character of GO.

5.3.2 Graphene/graphene oxides

Graphene (G) is a carbon-based material that has a single-layer two-dimensional honeycomb network structure. As a result of G unique structure, it has many superior chemical and physical features, such as large surface area, easy chemical functionalization, high thermal and electronic conductivity, and great mechanical strength [2,3,18,19]. GO is the oxidized form of G. Preparation of the Fe_3O_4/GO from graphite is schematically shown in Fig. 5.2A [20]. GO compared to G or reduced GO has more oxygen-containing groups, such as hydroxyl, carboxyl, epoxy groups, and carbonyl on its surface. Besides, GO presents a strong affinity for compounds containing aromatic ring via $\pi-\pi$ stacking interaction, Van der Waals, and other noncovalent interactions [18]. Preparation of the sorbents based on magnetic GO nanocomposites integrates the superior properties of GO by easily and quickly separating the MNMs. The introduction of magnetic properties into GO could combine both advantages of the GO with the convenience of the magnetic separation, resulting in an efficient magnetic adsorbent [20,21].

Mahpishanian and Sereshti [21] synthesized β-cyclodextrin/iron oxide–reduced graphene oxide (β-CD/MRGO) as a novel magnetic nanocomposite for vortex-assisted MSPE of 16 organochlorine pesticides (OCPs) from honey. The introduction of β-CD leads to excellent water dispersibility and molecular selectivity of the nanocomposite. Also, the saturation magnetism of the nanocomposites has been excellent (46.0 emu g^{-1}), leading to its rapid magnetic separation. In addition, the magnetic responsivity of the nanocomposite is excellent (46.0 emu g^{-1}). Synthesis routes of β-CD/MRGO and steps of MSPE for OCPs extraction are shown in Fig. 5.2B.

5.3.3 Carbon nanotubes

CNTs are usually curled by a single (SWCNTs) or multiple layers (MWCNTs) of graphene to form hollow tubular structures [2,3]. Magnetic carbon nanotubes (MCNTs) are usually prepared by depositing MNPs on CNTs through one-step or two-step reactions. The resulting nanocomposites can not only be evenly dispersed in the sample solution but can also be easily separated and collected with the assistance of magnets [2,3]. MCNTs due to their high adsorption capacity and large specific surface area are investigated as excellent sorbents in MSPE for different organic compounds in low concentrations in food samples [2,3]. However, further functionalization is usually required with hydrophilic functional groups because of the stability and poor dispersibility of MCNTs in aqueous samples [22–24].

Zhao et al. developed a new MSPE sorbent based on functionalization of MCNTs with polyethylene glycol (PEG) (PEG-MWCNTs-MNP) for extraction of various classes of mycotoxins [aflatoxins (AF), ochratoxin A (OTA), and zearalenone (ZEN)] in milk samples before ultrahigh-performance liquid chromatography Q-Exactive high-resolution mass spectrometry [22]. They reported that MCNTs, when modified with PEG, dispersibility, stability, and homogeneity of the prepared nanocomposite, can be significantly increased in aqueous samples without aggregation.

5.3.4 Molecular/ion-imprinted polymer

MIPs/IIPs are emerging and novel sorbents with excellent selectivity for template ions or molecules synthesized by imprinting techniques. Under certain synthesis conditions, in the presence of template ions or molecules, monomers, cross-linking reagents, and initiators, highly cross-linked 3D network polymers are formed. Then, template molecules or ions are removed from 3D network polymers leading to create a cavity in the structure of the polymer that corresponds to the spatial structure, size, and binding sites of the template. Such IIPs/MIPs have excellent selectivity to template ions or molecules. They are stable and can be relatively easily prepared for diverse applications in complicated food matrices [2,3]. However, separation of IIPs/MIPs sorbents from sample solutions often requires multiple centrifugations and filtration steps making their fast preparation and separation not convenient. In this regard, the introduction of magnetism to MIPs/IIPs can solve this problem [2]. Therefore surface imprinting on MNPs with high specific surface area and saturation magnetization presents the advantages of both MIPs and magnetic separation such as selective enrichment of target analyte, efficient cleanup, without the need to centrifugation or filtration steps [2,25].

Zhou et al. prepared a novel sorbent for MSPE of tetracyclines from untreated milk and egg samples before HPLC/UV determination based on restricted access media-magnetic molecularly imprinted polymers (RAMe-MMIPs) (Fig. 5.3). The prepared RAMe-MMIPs provide an excellent selectivity for the target molecules in the samples as well as excellent efficiency in the elimination of protein macromolecules at the same time. They reported a protein exclusion rate of 99.4%. Besides, RAMe-MMIPs have fast binding kinetics (35 minutes), a uniform imprinted and hydrophilic layer (600 nm), greater absorption capacity, and high selectivity. A combination of RAMe and MMIPs for preparation of such sorbents provides an upgrade strategy for the selective extraction and detection of analytes directly from complicated food samples containing high proteins with excellent selectivity in extraction and cleanup [26].

5.3.5 Layered double hydroxides

In recent years, LDHs, also known as anionic clays and/or hydrotalcite-like compounds, have received much attention as extraction sorbents [2–4,27]. LDHs are materials with 2D nanostructured layers, which are composed of positively charged layers of metal hydroxides separated by charge-balancing anions in addition to water molecules [28]. LDHs have many advantageous attributes as an adsorbent, such as large interlayer spaces (high porosity), large specific surface area, good anion-exchange capacity (huge number of exchangeable anions), water-resistant structure, and easy synthesis/preparation [2–4,27,29]. However, LDHs cannot be directly applied for enriching pollutants from sample solutions due to their poor dispersive characteristics and difficult separation from aqueous solution. Thus some modifications with hydrophilic compounds such as polydopamine [28] or tannic acid [30] (Fig. 5.4A) are required to improve the efficiency of LDHs sorbents.

Du et al. prepared a new three-layer nanocomposite based on magnetic LDHs for the extraction of organophosphorus pesticides in juice samples [28]. The magnetic nanocomposite has been fabricated by coating polydopamine and Mg/Al LDHs onto $NiFe_2O_4$ nanoparticles to overcome the shortcomings of low surface area and easy agglomeration associated with conventional Fe-based crystal nanoparticles. Additionally, the higher dispersibility of polydopamine and larger specific

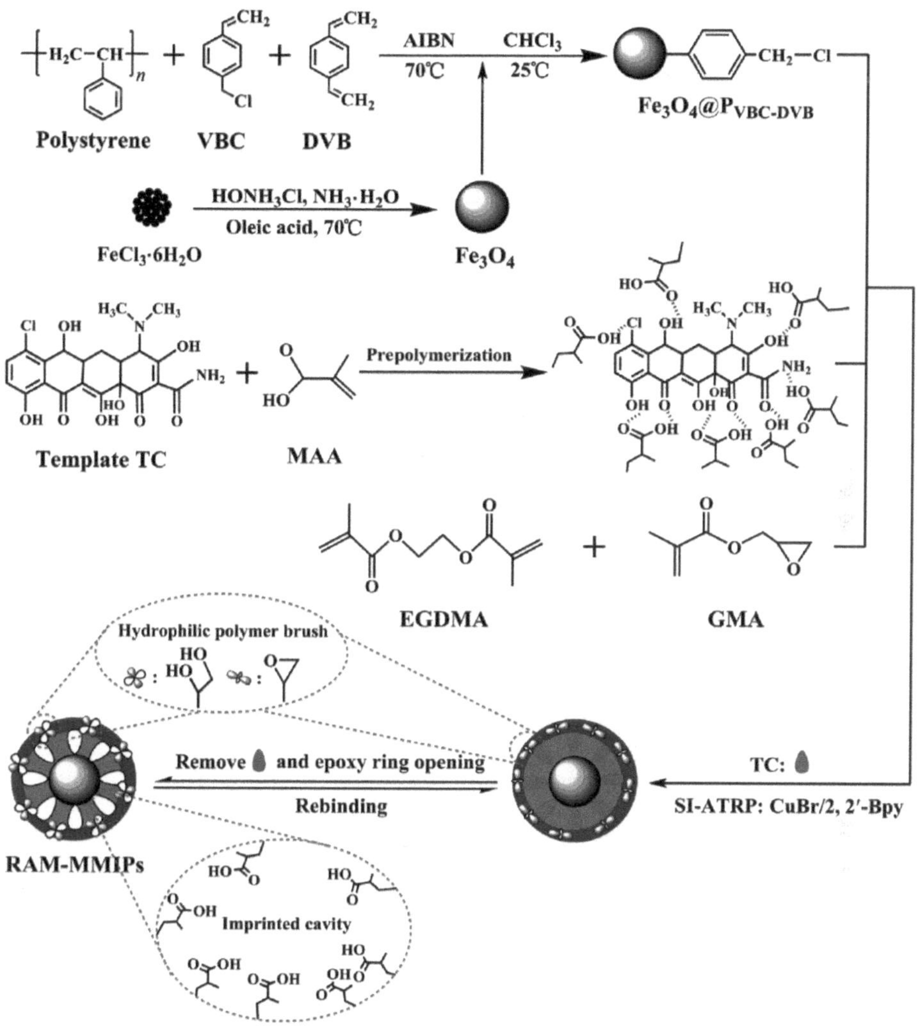

Figure 5.3 Procedure for preparing restricted access media-magnetic molecularly imprinted polymers [26].

surface area of Mg/Al LDHs are conducive to rapid dispersion and efficient adsorption of organophosphorus pesticides by the synthesized nanocomposites, thereby greatly improving the efficiency of MSPE.

5.3.6 Mesoporous materials

Mesoporous materials are porous materials with pore diameters in the range of 2–50 nm. These materials have unique features such as large specific surface areas, large pore volume, easy surface modification, and adjustable mesoporous

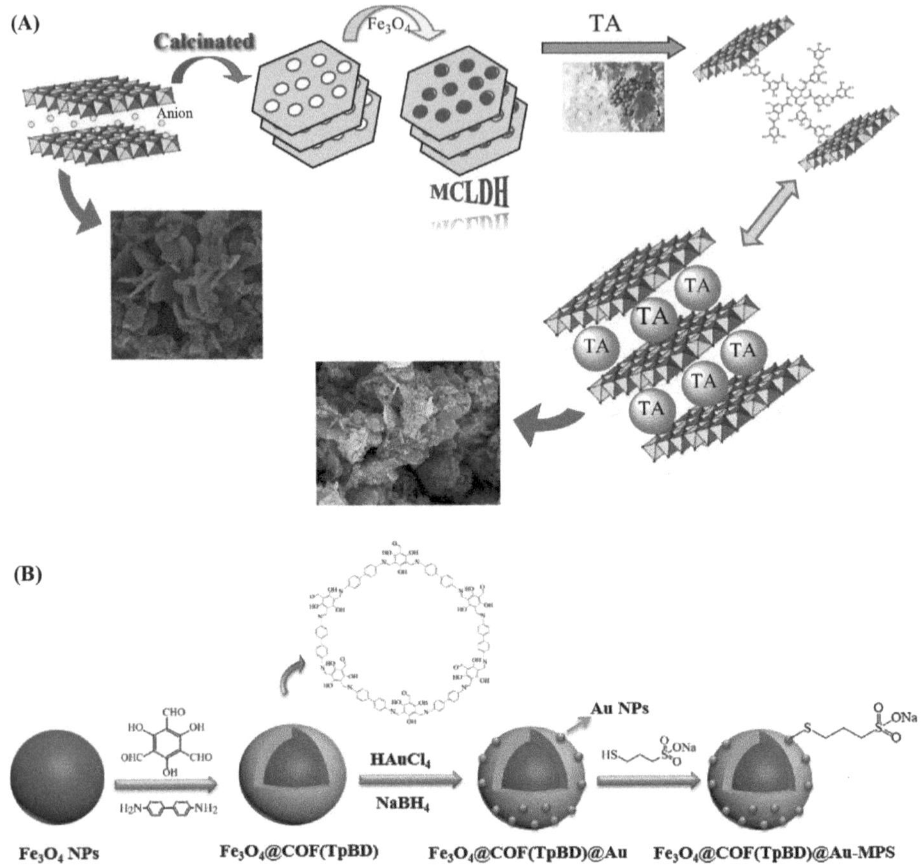

Figure 5.4 (A) Schematic of the synthesized CLDH@Fe_3O_4@Tanic acid [30]. (B) Schematic fabrication process of Fe_3O_4@COF(TpBD)@Au-MPS [31].

channels [2]. Because of the low cost and simple preparation of mesoporous silica gel, it is a widely used mesoporous material. The general route for the preparation of magnetic mesoporous materials consists of the sol–gel reaction on the surface of MNPs in the presence of a template such as surfactants and then removal of the template [32]. Mesoporous silica MCM-41 and MCM-48 are the most commonly used silica nanoporous materials. MCM-48 has a cubic array with a 3D pore network structure that leads to fast and favorable mass transfer kinetics compared to MCM-41 with the hexagonal array of unidirectional pore structure. Therefore sorbent MCM-48 usually provides more efficiency as absorbent support than MCM-41 [33].

Molaei et al. prepared a new magnetic mesoporous silica (m-MCM-41) as a sorbent for MSPE of four biogenic amines (BAs) including histamine, putrescine, tyramine, and cadaverine in Kashk samples [32]. Also, Liu et al. synthesized a magnetic cubic mesoporous silica (Fe_3O_4@14-3-14-MCM-48) nanocomposite using a template of cationic Gemini surfactant for extraction of nine endocrine-disrupting compounds (EDCs) from the migrants of food contact materials [33].

5.3.7 Porous organic polymers

POPs are materials based on porous network polymer synthesized by covalent bonding of organic monomers containing elements such as C, H, O, N, and B. POPs can be divided into different groups according to their crystalline characteristics and structures [2]. Covalent organic frameworks (COFs) are interesting and emerging group of POPs with large specific surface area, porous and periodic structures, low density, superior functional properties, and thermal stability and which are formed by covalent attachment of different types of building blocks, typically such as TpBD [34,35]. POPs have large specific surface areas, low skeleton density, good stability, high porosity, and more adsorption sites per unit mass without metal ions compared to inorganic porous materials. COFs present different interactions such as $\pi-\pi$ stacking interaction, hydrogen bonding, and hydrophobic interaction at the same time that exhibited excellent potential as a sorbent in extraction techniques [34]. However, the use of POPs in extraction techniques is associated with some limitations such as poor dispersion due to their high hydrophobicity as well as the difficulty of the separation of the sorbent because of their lightweight [31]. Introduction of MNPs to POP materials leads to solving POPs limitations by improving their dispersion and separation properties making them attractive sorbents with abundant adsorption sites and pores and favorable properties [2].

Wen et al. [31] synthesized sulfonate-functionalized magnetic COFs (Fe_3O_4@COF(TpBD)@Au-MPS nanocomposites) as a novel sorbent for MSPE of fluoroquinolones (FQs) (Fig. 5.4B). The results show that Fe_3O_4@COF(TpBD) @Au-MPS exhibit a higher affinity to FQs compared with Fe_3O_4@COF(TpBD). Also, sulfonate functionalization not only allows strong electrostatic interactions of the sorbent with the analytes but also greatly improves the dispersibility of the sorbent.

5.3.8 Restricted access materials (RAMs)

Determining target compounds in complicated real samples containing macromolecules is commonly associated with major challenges. In this regard, in recent years, RAMs have been proposed to solve this problem. Macromolecules removal by RAM is performed though a size-exclusion mechanism, in which a hydrophilic barrier excludes macromolecules but allows small molecules to permeate into hydrophobic pores of the sorbent. In other words, RAMs can exclude proteins and retain small target molecules at the same time, which facilitates direct HPLC injection of proteinous samples [36]. As a result, the application of RAM sorbents leads to avoid protein precipitation step in sample preparation of biological, which may cause analytes losing. Therefore, after simple dilution, they can be directly used for sample preparation, leads to simplification and saving time in sample preparation methods [2,36]. RAM is usually prepared by attachment of a porous hydrophilic layer on the surface of some sorbents such as CNTs, silica, and MNPs. In recent years, magnetic RAMs are used in the analysis of complicated biological samples.

Florez et al. prepared a new RAM for MSPE of penicillin G, trimethoprim, oxytetracycline, and doxycycline in milk samples by HPLC. The RAM is obtained by attachment of hydrophilic monomers and casein onto magnetic mesoporous polyaniline (RA-MMPAni-HM-CAS) [37]. The results showed that the method and the technique proved to be economical, fast, and easy to operate, in addition, to be a viable procedure and to use minimum volumes of organic solvents. RA-MMPAni-HMCAS demonstrated excellent extraction performance (approximately 100%) for all tested antibiotics, and also was able to effectively eliminate protein, leading to its rapid, easy, and efficient operation.

5.3.9 Aptamers

Aptamers are single-stranded DNA or RNA that may be used to recognize different target compounds by providing specific and/or high-affinity binding sites via electrostatic and hydrophobic interactions and hydrogen bonding to their targets [4,38]. These sorbents are superior because of advantages such as binding affinity and high specificity, easy and controllable synthesis and modification, good stability, and low cost [39]. In addition, aptamers have a better affinity toward target compounds, specificity and stability with targets, high affinity constant, and low-cost and easy preparation and modification methods compared to antibody-based sorbents. These features make aptamer-modified sorbents desirable for a variety of applications for recognizing particular target compounds, ranging from small molecules to proteins, cells, and even tissues [39].

Tu et al. synthesized an aptamer-functionalized Fe_3O_4/GO as a selective sorbent for MSPE of chloramphenicol in honey and milk samples before HPLC [40]. The aptamer was covalently bonded to the Fe_3O_4/graphene oxide complex by 1-ethyl-3-(3-dimethylaminopropyl)carbodiimide hydrochloride and *N*-hydroxysuccinimide.

5.4 Magnetic nanomaterials as sorbents for preconcentration purposes

MNMs are successfully applied in the sample preparation of various food and beverage samples. Some examples are briefly described in the following paragraphs.

5.4.1 Food safety applications

Food safety analysis is a major concern in today's society because it is directly linked to everyone's safety and health. The development of a sample preparation method that enables selective and sensitive determination of complicated matrices and checks hazardous pollutants is of particular importance for food safety. To this end, the use of MNMs in MSPE techniques has emerged as an effective tool to achieve this goal [7,8].

5.4.1.1 Analysis of pesticides

Pesticides have been dramatically and undeniably used in the production of agriculture against pests that can damage crops to prevent or reduce losses and thus increase the product yield of crops, fruits, and vegetables [41].

Food contamination with pesticides is considered one of the major global concerns in the field of food safety as a result of improper or excessive use of them [7,8]. Some of the important effects of pesticides on human health include chronic neurological deficits, acute neurotoxicity, and endocrine systems or cancer [42]. Given the complexity of the food samples and trace level of pesticide residues, it is highly desirable to develop an accurate and reliable sample preparation method with high selectivity and sensitivity to isolate and determine pesticide residues from real foodstuffs before instrumental analysis. Thus, in recent years, application of MNMs as novel sorbents in MSPE techniques has been extensively investigated in the field of pesticide residues analysis. In this sense, the most recent and significant examples are shown in Table 5.1. Table 5.1 shows that different types of pesticide residues from various highly complex foods, such as oils, honey, milk, tea, different vegetables, fruits, and juices, can be analyzed via MSPE followed with various analytical instruments [21,28,30,43–52].

5.4.1.2 Analysis of veterinary drug residues

In recent years, the residue of veterinary drugs in animal-originated foods (eggs, milk, honey, meat, etc.) has become a growing problem since they cause allergic reactions and toxic effect to the consumers and induce the drug-resistant pathogens [16,23,31,92]. Given the complexity of the food samples and trace level of the drugs, it is highly desirable to develop an accurate and reliable sample preparation method with high selectivity and sensitivity to isolate and determine drugs from real foodstuffs. Thus, in recent years, application of MNMs as novel sorbents in MSPE techniques has been investigated for the analysis of drug residue in food samples. In this sense, the most recent and significant examples are shown in Table 5.1. Table 5.1 exhibited that MNMs have been efficiently used for MSPE of various types of pesticides in different complicated food samples such as different meat samples, honey, egg, and milk [16,23,26,31,37,53–64].

5.4.1.3 Analysis of heavy metal ions

Food contamination with heavy metal ions by the accumulation of these ions in food samples has become a major concern worldwide. Air, soil, and water pollution with heavy metals lead to heavy metal accumulation in agriculture and animal-originated food products. Because most of them are toxic to human health and even at trace concentration levels can cause mutagenicity and carcinogenicity [93]. In this sense, the most recent and significant examples are shown in Table 5.1.

Table 5.1 The most recent and significant examples of magnetic nanomaterials as sorbents in food safety applications.

Analyte	Sorbent	Sample	Detection system	LOD (μg kg^{-1})	Eluent	V_{sample} (mL)	Ref.
Pesticides residue							
Diazinon and metalaxyl	Calcined-layered double hydroxide sorbent modified with the tannic acid	Fruit juice	GC-FID	0.2 and 0.8	100 μL of ethanol	10	[30]
Imidacloprid and 2,4-dichlorophenoxyacetic acid	Triazine-based polymeric network–modified magnetic nanoparticles/graphene oxide (TPN/Fe_3O_4 NPs/GO) nanocomposite	Cucumber, tomato, and water	HPLC-UV	0.17 and 1.7	220 μL acetone	50	[43]
Organophosphorus pesticides	$NiFe_2O_4$@polydopamine@Mg/Al–layered double hydroxides nanocomposites	Fruit juices	HPLC-DAD	0.06–0.13	1.6 mL ACN	6.0	[28]
Fenazaquin	Oleic acid–functionalized magnetic nanoparticles	Almond	GC-MS	0.06	100 μL ethanol	30	[44]
Organophosphates pesticides	Restricted access carbon nanotubes	Bovine milk	HPLC-UV	0.36–0.95	1 mL of ACN	40	[45]
Polar acidic herbicides	Polyethylenimine-modified reduced graphene oxide (Fe_3O_4@PEI-RGO)	Rice	HPLC-DAD	0.67–2.0	200 μL of alkaline methanol	10	[46]
Organochlorine pesticides residue	β-Cyclodextrin/iron oxide–reduced graphene oxide nanocomposite	Honey	GC-ECD	0.00052–0.00321	0.5 mL of ACN and dichloromethane (4:1, v/v)	50	[21]
Benzoylurea insecticides	Magnetic metal-organic framework composite (Fe_3O_4@MOF-808)	Tea beverages and juice	HPLC-UV	0.04–0.15	2 mL of acetonitrile	70	[47]
Acaricides	C16mimBr-coated attapulgite/polyaniline-polypyrrole/Fe_3O_4 (ATP/PANI-PPY/Fe_3O_4)	Fruit juice	HPLC-DAD	0.16–0.57	1 mL of ACN	8.0	[48]
Ziram	Magnetic-carboxylated nanodiamonds	Food	FAAS	5.3	1 mL of 0.25 M HNO_3 in acetone	10	[49]

Pyrethroids	Ionic liquid–coated and cyclodextrin-functionalized magnetic core dendrimer nanocomposites	Juices	HPLC-DAD	0.36–1.3	1 mL of acetone	8	[50]
Pesticide residues	Magnetic multiwalled carbon nanotubes	Apple, pear, carrot, and cucumber	HPLC-DAD	20–50	3 mL of ACN	10	[51]
Six chiral pesticides	Fe_3O_4-modified carbon nanospheres	Functional foods	LC-MS/MS	1–4	4 mL ACN	10	[52]
Drug residues							
Five penicillins	DABCO-based ionic liquid–modified Fe_3O_4 nanoparticles ([DABCO–C_3OH]Cl@MNPs)	Milk	UPLC-MS/MS	0.03–0.20	500 μL mobile phase	50	[53]
Antibiotics (doxycycline, oxytetracycline, penicillin G, and trimethoprim)	Restricted access material based on mesoporous polyaniline coated with hydrophilic monomers and casein	Milk	HPLC-DAD	8.0	800 μL acetonitrile: acetic acid (7:3, v/v)	1.0	[37]
Aminoglycosides	Magnetic nanoparticles decorated with D-galactose and galactitol (Fe_3O_4@SiN-galactose and Fe_3O_4@SiN-galactitol)	Honey	LC-MS/MS	1–28	150 μL ACN	4.0	[54]
Fluoroquinolones	Zr-Fe-C MNPs	Meat-based baby food	HPLC-FLD	1.5–3.0	0.4 mL of 0.02 mol L^{-1} KOH in methanol	2.7	[55]
Aminoglycosides	Poly(*N*-acryloyl-glucosamine-*co*-methylenebisacrylamide) (poly(AGA-*co*-MBA))-modified Fe_3O_4 nanoparticles	Meat	HILIC-MS/MS	0.6–23.6	5.0 mL of FA/ACN/H_2O (0.1/10/90, v/v/v)	10	[56]
Aminoglycoside antibiotics	Polyvinyl alcohol–coated core–shell magnetic nanoparticle	Honey	HILIC-MS/MS	0.993–1.23	1 mL eluent (water/ACN/FA = 80:19:1)	10	[57]
Sulfonamide antibiotics	Molybdenum disulfide (MoS_2)-based core–shell magnetic nanocomposite	Milk, pork, and fish meat	HPLC-MS/MS	0.40–1.5	3 mL methanol containing 1% NH_3	25	[58]

(*Continued*)

Table 5.1 (Continued)

Analyte	Sorbent	Sample	Detection system	LOD ($\mu g\ kg^{-1}$)	Eluent	V_{sample} (mL)	Ref.
Sulfonamides	Magnetic, mesoporous core/shell structured Fe_3O_4@JUC-48 nanocomposite	Meat	HPLC-UV	1.73–5.23	0.8 mL methanol containing 5% AA	8.0	[59]
Fluoroquinolones	Sulfonate-functionalized covalent organic framework (Fe_3O_4@COF(TpBD)@Au-MPS)	Meat	HPLC-MS/MS	0.1–1.0	1 mL elution solvent (FA: methanol = 4:6)	10	[31]
Tetracycline veterinary drugs	Restricted access media-magnetic molecularly imprinted polymers	Milk and egg	HPLC/UV	1.03–1.31	5 mL of AA–methanol (3:7, v/v)	5.0	[26]
Cephalexin	Magnetic molecular-imprinted polymers (MMIPs)	Pork and milk	HPLC-UV	5.00	3.0 mL of methanol/AA (9:1, v/v)	10	[60]
17β-Estradiol	Zipper-like magnetic molecularly imprinted microspheres	Milk and pork	HPLC-UV	2.52	Methanol/acetic acid solutions (1 mL, v/v, 9/1) three times	50	[61]
Sulfonamides	Amino-terminated carbon nanotubes (CNT-NH_2) by using isocyanates	Milk	LC-HRMS	0.002–0.010	0.5 mL 80% ACN	100	[23]
Sulfonamide antibiotics	Thiol-functionalized magnetic carbon nanotubes	Milk and commercial chicken meat	HPLC-DAD	0.02–1.5	200 μL ACN and 1% NH_3	30	[62]
Three tetracyclines	β-Cyclodextrin-functionalized silica-coated magnetic graphene oxide	Bovine milk	HPLC-UV	1.8–2.9	1 mL of MeOH	40	[63]
Chloramphenicol	MMIPs	Pork and honey	HPLC-UV	10	3 mL of a methanol–AA mixture (8/2, v/v)	10	[64]

Enrofloxacin	Magnetic metal-organic framework/graphene oxide (Fe_3O_4/MIL-100(Fe)/GO)	Milk	Spectrofluorimetry	0.65	4.00 mL of methanol containing 15% formic acid	100	[16]
Heavy metals							
Mercury	Graphene/$ZnFe_2O_4$ nanocomposite	Water and fish	CV-AAS	0.010	5 mL of 2 mol L^{-1} HCl	150	[65]
Cobalt and nickel	Sodium dodecyl sulfate–coated Fe_3O_4 nanoparticles	Spinach, tea, and water	FAAS	0.09 and 0.40	1.0 mL of the ethanolic solution of 10% HNO_3	300	[66]
Cadmium	SDS-PAN immobilized on the Fe_3O_4@Al_2O_3 core–shell MNPs	Edible oils	GFAAS	0.006	270 μL 2.0 mol L^{-1} HNO_3	NR	[67]
Zinc ions	Zn(II)-imprinted polymer grafted on graphene oxide/magnetic chitosan nanocomposite	Well water, drinking water, black tea, rice, and milk	FAAS	0.09	0.5 mL of EDTA solution (0.005 mol L^{-1}, pH = 5.5)	100	[68]
Arsenic	Fe_3O_4@AuNPs magnetic particles	Natural water	HG-AFS	0.0037	5 mL of 0.5 mol L^{-1} HCl	200	[69]
Speciation of mercury	Fe_3O_4@SiO_2@γ-mercaptopropyl trimethoxysilane magnetic nanoparticles	Water and fish	HPLC-ICP-MS	0.49–0.74	0.5 mL 0.1 mol L^{-1} HNO_3 and 4% (m/v) thiourea	100	[70]
Ziram	Magnetic-carboxylated nanodiamonds	Rice and cracked wheat	FAAS	5.3	1 mL of 0.25 M HNO_3 in acetone	10	[49]
Heavy metal ions (Cd, Co, Cr, Cu, Mn, Ni, and Pb)	Fe_3O_4@catechol nanoparticles	Water and fruit	ICP-OES	0.2–0.9	250 μL of 0.3 M EDTA	60	[71]
Copper	*N,N'*-bis (acetylacetone) ethylenediimine (Fe_3O_4@SiO_2-EDN)	Cereal and water	FAAS	1.5	1.8 mL of 2.5 mol L^{-1} HNO_3	50	[72]

(Continued)

Table 5.1 (Continued)

Analyte	Sorbent	Sample	Detection system	LOD ($\mu g\ kg^{-1}$)	Eluent	V_{sample} (mL)	Ref.
Lead	Stearic acid–coated magnetic nanoparticle	Red pepper	SQT-FAAS	10	200 μL of concentrated HNO_3 (65%)	30	[73]
Lead	Cobalt magnetic particles	Black tea	SQT-FAAS	7.77	4.0 mol L^{-1} of HCl	30	[74]
Total selenium	Polystyrene-*g*-polyoleic acid-*g*-polyethylene glycol graft copolymer	Milk and water	ETAAS	0.00606	2 mL of 0.5 mol L^{-1} HCl	100	[75]
Cr(VI) and Cr(III)	Dextran-modified magnetic nanoparticles	Tea, coffee, and mineral water	Linear sweep voltammetry	0.01 μM	250 μL of 0.1 mol L^{-1} NaOH	10	[76]
Hg(II)	Magnetic graphene oxide modified by 2-pyridine carboxaldehyde thiosemicarbazone groups (Fe_3O_4@GO/2-PTSC)	Fish (Tilapia), rice, tea, and milk	ICP-OES	0.0079	0.5 mL of HCl, 0.1 mol L^{-1}	100	[77]
Polycyclic aromatic hydrocarbons (PAHs)							
PAHs	2-Mercapto-4-methyl-5-thiazoleacetic acid bonding Au (MMTA-Au) deposited on Fe_3O_4@SiO_2 (Fe_3O_4@SiO_2@MMTA-Au)	Water	HPLC-FLD	0.25×10^{-3} to 37.5×10^{-3}	1.0 mL ACN	100	[78]
16 PAHs	Three-dimensional ionic liquid–functionalized magnetic graphene oxide nanocomposite	Vegetable oils	GC-MS	0.05–0.30	0.5 mL toluene	20	[18]
16 PAHs	[60]Fullerene-functionalized magnetic nanoparticles (Fe_3O_4@SiO_2@C_{60})	Tea	GC-MS	0.0008–0.0143	0.5 mL *n*-hexane and 0.5 mL acetone	100	[79]
PAHs	Magnetic poly(β-cyclodextrin-ionic liquid) nanocomposites	Rice	GC-FID	0.01–0.18	200 μL of ACN	25	[80]

Endocrine-disrupting chemicals (EDCs)							
Steroidal and phenolic EDCs	Magnetic covalent organic frameworks (Fe_3O_4@COF(TpBD))	Raw meat	HPLC-UV	1.4–8.7	1.0 mL of ACN	10	[81]
EDCs	Cationic Gemini surfactant–templated magnetic cubic mesoporous silica	Food contact materials	HPLC-DAD	1.0–2.5	1.0 mL methanol/ ACN	10	[33]
Bisphenols (BPs)	Ni-based N-doped graphene tubes (Ni@N-GrTs)	Milky tea and infant milk powder	HPLC-FLD	0.1–0.2	2.0 mL acetone	5.0	[82]
BP endocrine disruptors	Magnetic nitrogen–doped reduced graphene oxide (Fe_3O_4@N-RGO)	Carbonated beverages	HPLC-UV	0.1 to 0.2	200 μL of methanol/ acetone (1:1)	10	[83]
EDCs	Magnetic amino-functionalized microporous organic network composites	Beverage bottle and juice	HPLC-UV	0.015–0.030	0.4 mL of methanol (0.2 mL × 2)	20	[84]
BPs	Magnetic N-doped carbon nanotubes	Fruit juices	LC-MS	0.43×10^{-3} to 2.47×10^{-3}	2 mL methanol	10	[24]
Alkylphenols	$CoFe_2O_4$/oleic acid nanoparticles	Baby foods	GC-MS	0.4–1.7	3 mL ethanol	25	[85]
Mycotoxins							
Multimycotoxins	PEGylated multiwalled carbon nanotubes magnetic nanoparticles	Milk	UHPLC-Q-HRMS	0.005–0.050	5 mL of ethyl acetate with 1% formic acid (FA)	40	[22]
Ochratoxin A	Aptamer-targeted magnetic nanospheres	Wheat flour	HPLC-FLD	0.3	2.5 mL of 95% ACN (containing 1% AA)	10	[86]
Patulin	MMIP	Fruit juice	LC-MS/MS	0.1	1 mL acetonitrile	1.0	[87]
Aflatoxin B1	Immunomagnetic sorbent	Vegetable oils	Flurometer	0.0048	Methanol (2 mL)	20	[88]
Zearalenone and its derivatives	Multiwalled carbon nanotube-magnetic nanoparticles	Maize	UHPLC-MS/MS	0.03–0.04	10 mL of acetone (0.5% FA)	5.0	[89]

(Continued)

Table 5.1 (Continued)

Analyte	Sorbent	Sample	Detection system	LOD (μg kg^{-1})	Eluent	V_{sample} (mL)	Ref.
Other contaminates							
Perfluorinated compounds	Fluorinated magnetic covalent organic frameworks (Fe_3O_4@TpPa-F_4)	Milk	HPLC-MS/MS	0.005×10^{-3} to 0.05×10^{-3}	1.5 mL methanol	20	[90]
Acrylamide	Dummy molecularly imprinted polymers	Biscuit	HPLC-UV	1.3	2 mL MeOH/ NH_4OH (90:10 v/v)	15	[91]
Biogenic amines	Magnetic mesoporous silica (m-MCM-41)	Kashk	HPLC-UV	8.0–16	0.5 mL ACN	10	[32]
Malachite green and crystal violet	Magnetic mesoporous metal-organic framework-5 (Fe_3O_4@PEI-MOF-5)	Fish	UHPLC-MS/MS	0.3 and 0.08	1.0 mL methanol containing 1% FA	30	[17]

NR, Not reported.

Table 5.1 shows that different types of heavy metals from various highly complex foods, such as fish, edible oils, cereals, milk, tea, coffee, and vegetable samples, can be efficiently extracted at trace levels via MSPE [23,49,65–77].

5.4.1.4 Analysis of polycyclic aromatic hydrocarbons

Polycyclic aromatic hydrocarbons (PAHs) are well-known mutagenic and carcinogenic compounds that contain two or more fused aromatic rings [94]. They are generally caused by high-temperature pyrolysis or incomplete combustion of biomass and fossil fuels, oil spills, and other industrial processes [95]. Food contamination by PAHs could be occurred through the environment (the main source) and/or some food processing such as smoking, roasting, barbecue, and grilling. The trace amounts of PAHs have been identified in different food samples such as coffee, vegetable oils, roasted meat, seafood, milk, tea, and fruits [96]. Food ingestion is the main source of exposure to PAHs, especially in edible vegetable oils [97].

The US Environmental Protection Agency (USEPA) has classified 16 PAHs as "priority pollutants" [98]. However, all the PAHs have not been included in food legislation particularly those compounds with low molecular weights. Under Commission Regulation (EU) No 835/2011 as regards maximum levels (MLs) for PAHs in foodstuffs, four heavy molecular PAHs, namely PAH4 (benzo[*a*]pyrene, benz[*a*]anthracene, benzo[*b*]fluoranthene, and chrysene) have been recommended as useful indicators in analysis of the food samples with MLs in the range of 1–30 $\mu g\,kg^{-1}$ [99]. In this sense, the most recent and significant examples of PAHs extraction in different food samples are shown in Table 5.1 [18,78,79,100].

5.4.1.5 Analysis of endocrine-disrupting chemicals

EDCs are synthetic chemicals that can disrupt the endocrine system and damage reproductive health, even at very low concentrations [80,101]. As a result, the monitoring of EDCs has received great concerns recently. Bisphenols, alkylphenols, and ultraviolet filters are important EDCs that are widely used in food contact materials. Packaged foods due to the migration of antioxidants, plasticizer, flame retardants, and other polymer additives in food packaging materials are the main source of exposure to EDCs in humans. Four most important phenolic EDCs are bisphenol A (BPA), 4-*tert*-octyl phenol (4-*t*-OP), 4-nonylphenol (4-NP), and 4-alpha-cumyl phenol (4-α-CP). BPA is one of the highest yield chemicals in the world (about 5–6 billion pounds per year) [102]. BPA can be found in canned drinks and foods, bottled drinking, microwaveable food products, and released from polycarbonate baby bottles [103–105]. Also, ultraviolet filters are extensively used in plastic food packaging materials due to their excellent antioxidant ability [106]. Moreover, alkylphenols such as 4-*tert*-butyl phenol and 4-octyl phenol may also be used in the plastics production chain. Given the complexity of the food samples and trace level of the EDCs, it is necessary to establish a reliable and accurate sample preparation method for EDCs determination at trace concentration levels. In this sense, the most recent and significant examples are shown in Table 5.1. Table 5.1 demonstrates that various kinds of EDCs can be efficiently analyzed via MSPE from different complicated food samples such as raw meat, baby foods, fruit juice, milk, milky tea, and infant milk powder samples [24,33,81–85].

5.4.1.6 Analysis of mycotoxins

Mycotoxins are naturally occurring secondary metabolites of fungal species (*Fusarium*, *Aspergillus*, and *Penicillium* genera) that can grow in a wide variety of foods and agricultural commodities worldwide [107]. AF, ZEN, OTA, patulin, fumonisins, deoxynivalenol, and trichothecenes are the most important classes of mycotoxins. Carcinogenic, teratogenic, immunosuppressive, and mutagenic effects in humans and animals have been demonstrated in the presence of mycotoxins in food [108]. Thus the European Union has been established mycotoxins MLs in some food samples such as edible oils, nuts, cereals, and milk [109].

Given the complexity of the food samples and trace level of the mycotoxins, it is necessary to establish a reliable and accurate sample preparation method for the determination of them at trace levels. In this sense, the most recent and significant examples are shown in Table 5.1. Table 5.1 shows that different types of mycotoxins can be efficiently analyzed via MSPE in various complicated food samples such as vegetable oils, milk, wheat flour, maize, and fruit juice [22,86–89].

5.4.1.7 Analysis of other contaminants

Apart from their previously mentioned food safety applications (pesticides, drug residue, heavy metals, PAHs, endocrine-disrupting chemicals, and mycotoxins), MNMs have also been used for the determination of other different families of food contaminates such as perfluorinated compounds [90], acrylamide [91], BAs [32], and crystal violet and malachite green [17] in food samples (Table 5.1).

5.4.2 Food quality and authenticity applications

Today, ensuring food quality and authenticity is one of the important aspects alongside food safety. Most applications of MNMs as sorbents in food analysis are in the field of food safety. Also MNMs have been used for the analysis of natural ingredients and food additives to evaluate the quality and authenticity of food samples. The most recent and significant examples are shown in Table 5.2 [10,25,110–123].

In edible oils, free fatty acids (FFAs) analysis can provide useful information for oil authentication and quality control. The analysis can provide important information. Zhu et al. applied MSPE for extraction of 64 FFAs in edible oils followed by ($GC \times GC$-MS). They first applied the method for analysis of seven targeted FFAs and then performed comprehensive profiling of untargeted FFAs in oil by coupling the extraction method with $GC \times GC$-MS. They detected 64 FFAs positively or putatively [111].

Wu et al. prepared a magnetic graphene oxide as a sorbent for MSPE of lignans in sesame oil [119]. Their obtained results showed that in comparison with

Table 5.2 The most recent and significant examples of magnetic nanomaterials as sorbents in food quality and authenticity applications.

Analyte	Sorbent	Sample	Detection system	LOD (μg kg^{-1})	Eluent	V_{sample} (mL)	Ref.
Ethyl vanillin, trans-cinnamic acid, methyl cinnamate, ethyl cinnamate, and benzyl cinnamate	Fe_3O_4@GO nanocomposite	Orange juice, chocolate, and fruit sugar	HPLC-UV	20–40	2 mL ACN containing 10% HAc	2.0	[110]
64 Free fatty acids	Fe_3O_4@SiO_2	Edible oils	GC × GC-MS	5.6–25.8	0.4 mL of acetone	0.5	[111]
Phenolic compounds	Magnetic-carboxylated multiwalled carbon nanotubes	Sesame oil	LC-MS/MS	0.03–43.00	3.0 mL ethanol	4.0	[112]
19 Phenolic compounds	Magnetic graphene (Fe_3O_4-r-GO)	Oilseeds	LC-MS/MS	0.02–90.00	6 mL of 0.1% FA–methanol	5.0	[113]
Sudan dyes	Magnetic nanocomposite of naphthalene-based polyimide polymer-modified Fe_3O_4 (Fe_3O_4@NPI)	Chili sauce	HPLC-DAD	0.02–0.05	2.0 mL of ethyl acetate	5.0	[114]
Melamine	Magnetic molecularly imprinted polymers	Milk	HLPC-UV	67.0	2 mL of methanol/NH_3 (95:5 v/v)	5	[25]
Tartrazine	Polyethylenimine magnetic nanoparticles (Co-Fe_2O_4@PEI)	Saffron spray and cotton candy	UV-Vis	5.09	150 μL of (9:1 v/v%, ethanol/NaOH)	10	[115]
Caffeine	Three-dimensional graphene-Fe_3O_4 nanoparticles	Teas, cocoa powder, coffees, and chocolate	GC-FID	100	1 mL of chloroform	50	[116]
4-Hydroxybenzoic acid and 3,4-dihydroxybenzoic acid	Graphene-magnetite composite (G-Fe_3O_4)	Stingless bee honey	HPLC-UV/Vis	80–140	200 μL methanol	30	[117]
Sudan dyes and Para Red	γ-Fe_2O_3 magnetic nanoparticle functionalized with carboxylated multiwalled carbon nanotube	Chili tomato sauces and ketchup	HPLC-UV-Vis	0.16–84	0.5 mL of acetone (twice)	50	[118]

(*Continued*)

Table 5.2 (Continued)

Analyte	Sorbent	Sample	Detection system	LOD ($\mu g\ kg^{-1}$)	Eluent	V_{sample} (mL)	Ref.
Rhodamine B	Ionic liquid–coated core–shell magnetic nanoparticles	Chili powder and Chinese prickly ash	HPLC-Vis	0.08	4.0 mL ethanol	100	[10]
Lignans	Magnetic graphene oxide	Sesame oil	HPLC-DAD	20–50	1.7 mL methanol	4	[119]
Synthetic colorants	Polyethylenimine-functionalized magnetic nanoparticles (PEI@PDA@Fe_3O_4 NPs)	Candies and beverages	HPLC-DAD	0.15–0.20	1 mL ACN–$NH_3 \cdot H_2O$ (70:30, v/v; $n = 3$)	100	[120]
Trans-resveratrol	Alendronate sodium–grafted mesoporous magnetic nanoparticle	Peanut oils	HPLC-UV	0.3	1 mL of *n*-hexane	6	[121]
Six Sudan dyes	Magnetic trimeric chromium octahedral metal-organic framework (Fe_3O_4-NH_2@MIL-101)	Tomato sauce	HPLC-DAD	0.5–2.5	1 mL × 2 ethyl acetate	1.0	[122]
Quercetin	MMIPs with core–shell structure	Apple	HPLC-UV	200	200 μL methanol-CH_2Cl_2 (4:1, v/v) three times	50	[123]

traditional methods such as alumina column chromatography and saponification method, their developed method has benefits such as good cleanup, simple operation, and low cost and rapidity, which has led to introducing of a practical and important method for sesame oil analysis and quality assessment of sesame oil.

Zhao et al. prepared alendronate sodium−grafted mesoporous MNPs (Fe_3O_4@ANDS) as a novel MPSE sorbent for facile and effective extraction of trans-resveratrol in vegetable oils followed by HPLC [121]. They declared that the proposed method can be used to investigate adulteration and quality control of peanut oils. Besides, other polar target compounds have also been extracted from edible oil by Fe_3O_4@ANDS.

5.5 Magnetic nanomaterials as sorbent for cleanup purposes

Although most of the applications of MNMs as sorbents in MSPE are related to extraction and preconcentration of different species from food matrices, the applications of such sorbents for only cleanup purposes have also been investigated. Due to the complexity of the food sample matrices, a cleanup step is usually required before instrumental analysis of the analytes. One of the efficient cleanup procedures in food sample analysis is QuEChERS (quick, easy, cheap, effective, rugged, and safe) methodology due to their simplicity and excellent interferences removal [41]. However, the use of a cleanup step using different sorbents (e.g., C18, PSA, GCB, and zirconium materials) is essential to ensure that interferences are effectively removed. In some researches coupling of QuEChERS and MSPE is being developed by using MNMs as sorbents for cleanup step [89,124−129]. The benefits of these approaches can be their speed, easy separation steps, the variety of available MNMs adsorbents, and high efficiency.

Fernandes et al. [124] evaluated the application of MNMs (Fe_3O_4@TEOS, Fe_3O_4@TEOS@MPS, and Fe_3O_4@TEOS@MPS @PMAA) in cleanup step of QuEChERS for trace determination of brominated flame retardants in red fruit samples (strawberries, raspberries, and blueberries) using GC-MS. The fewer amount of MNPs as sorbents gets excellent cleanup efficiency for pigments and other co-extracts of the matrix.

Zheng et al. developed a "one-step" QuEChERS method for the determination of 11 pesticides in vegetables and freshly squeezed fruits juice by GC-MS analysis [125]. By integrating the "extraction/partitioning" and "cleanup" steps in one step, the purification of analytes is performed without any centrifugation. The magnetic sorbents facilities the phase separation and can simultaneously adsorb impurities. Compared to common QuEChERS methods, extraction/partitioning and cleanup are integrated into one step, resulting in time-saving (within 3.5 minutes) with keeping good cleanup performance (Fig. 5.5).

Rashidi Nodeh et al. proposed a graphene (G) modified with Fe_3O_4 and sol−gel hybrid tetraethoxysilane-methyltrimethoxysilane (TEOS-MTMOS) as a cleanup

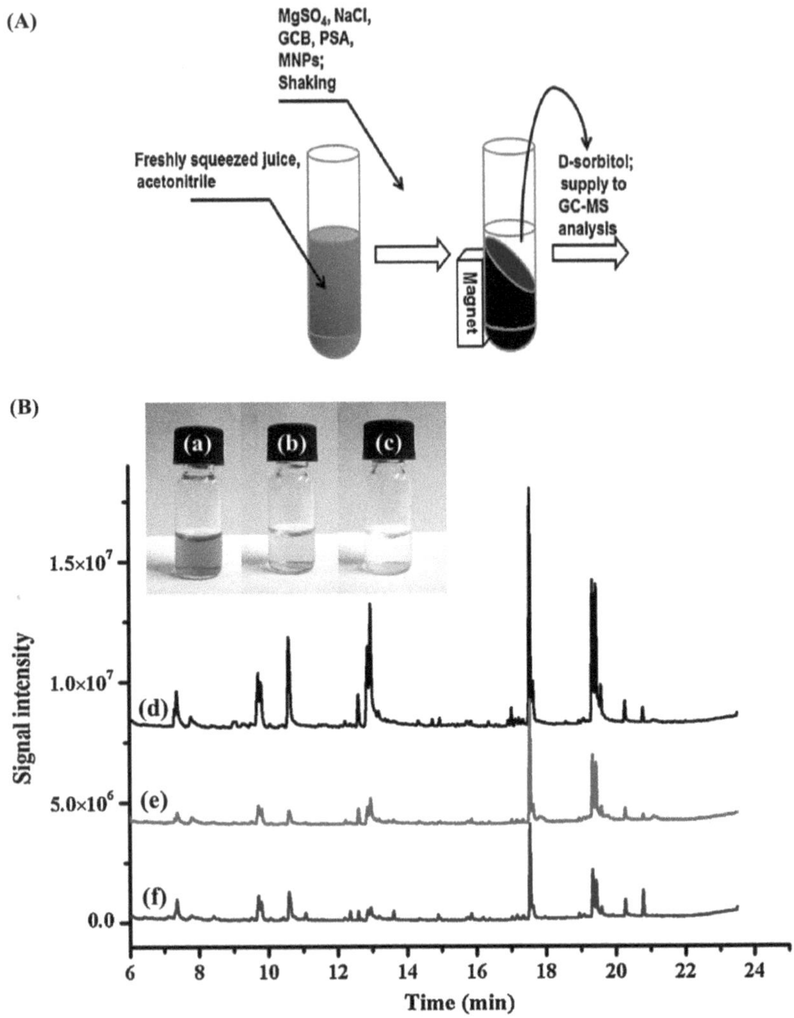

Figure 5.5 (A) Schematic of magnetic "one-step" QuEChERS. (B) Photography and chromatograms of cleanup performance by different QuEChERS: (a) extract for cucumber-squeezed juice without cleanup procedure; (b) extract for cucumber-squeezed juice with cleanup procedure by QuEChERS based on C18 and PSA; (c) extract for cucumber-squeezed juice with cleanup procedure by magnetic "one-step" QuEChERS; the relevant full scan chromatograms for (a)–(c) samples are (d)–(f), respectively [125].

sorbent for MSPE of acrylamide in various food samples (potato, fried eggplant, fried banana, potato chips, cheese snack, and toasted beards) before GC-MS [126]. Their results showed that after extraction of the samples with ACN, this phase can be effectively cleanedup (remove color and interference). In addition, as a result of the possibility of magnetic separation, the Fe_3O_4@G-TEOS-MTMOS has been reused up to seven times (recovery > 85%). However, the common adsorbents used in the QuEChERS methods are disposable.

5.6 Conclusions

Nowadays, MNMs are widely used in food analysis as novel, efficient, and low-cost sorbents in preconcentration or cleanup modes to assess food safety or food quality. Although much of the early researches on food analysis by MSPE has been related to the extraction of food contaminants, in recent years, the use of these substances in food quality control and authenticity has also been widely considered. MNMs offer several advantages such as high surface-area-to-volume ratio, fast extraction kinetic, reusability, and easy collection of sorbent that lead to improvement of analysis throughput, analysis cost, and simplicity of MSPE techniques compared to the traditional SPE column operation. Various advanced nanomaterials such as MOFs, G, CNTs, COFs, and LDHs are used to prepare magnetic nanocomposites in food sample pretreatment. However, the synthesis of new hybrid sorbents is continuously growing to find more efficient sorbents and introduce different extraction mechanisms at the same time. One of the most important trends in this field is the preparation of novel and emerging sorbent materials that are selective or ideally specific toward target compounds and have high sorption capacity. In this regard, the development of novel nanocomposites based on MIPs and RAMs is an attractive approach in this field that leads to efficient and selective extraction of target analytes even in complex matrices without the need of any sample pretreatment. Also, miniaturization and automation of MSPE techniques are other attractive trends.

References

[1] M. Faraji, Y. Yamini, M. Rezaee, Magnetic nanoparticles: synthesis, stabilization, functionalization, characterization, and applications, J. Iran. Chem. Soc. 7 (2010) 1–37.
[2] B. Hu, M. He, B. Chen, Solid-Phase Extraction, first ed., Elsevier, 2020 (Chapter 9).
[3] M. Faraji, Recent analytical applications of magnetic nanoparticles, Nanochem. Res. 1 (2016) 264–290.
[4] M. Faraji, Y. Yamini, M. Gholami, Recent advances and trends in applications of solid-phase extraction techniques in food and environmental analysis, Chromatographia 82 (2019) 1207–1249.
[5] J. Płotka-Wasylka, N. Szczepańska, M. de la Guardia, J. Namieśnik, Modern trends in solid phase extraction: new sorbent media, Trends Anal. Chem. 77 (2016) 23–43.
[6] J. González-Sálamo, B. Socas-Rodríguez, J. Hernández-Borges, M.Á. Rodríguez-Delgado, Nanomaterials as sorbents for food sample analysis, Trends Anal. Chem. 85 (2016) 203–220.
[7] H.L. Jiang, N. Li, L. Cui, X. Wang, R.S. Zhao, Recent application of magnetic solid phase extraction for food safety analysis, Trends Anal. Chem. 120 (2019) 115632.

[8] A.A. Hernández-Hernández, G.A. Álvarez-Romero, E. Contreras-López, K. Aguilar-Arteaga, A. Castañeda-Ovando, Food analysis by microextraction methods based on the use of magnetic nanoparticles as supports: recent advances, Food Anal. Methods 10 (2017) 2974–2993.
[9] M. Šafaříková, I. Šafařík, Magnetic solid-phase extraction, J. Magn. Magn. Mater. 194 (1999) 108–112.
[10] J. Chen, X. Zhu, Magnetic solid phase extraction using ionic liquid-coated core–shell magnetic nanoparticles followed by high-performance liquid chromatography for determination of Rhodamine B in food samples, Food Chem. 200 (2016) 10–15.
[11] G. Li, L. Xia, J. Dong, Y. Chen, Y. Li, Solid-Phase Extraction, first ed., Elsevier, 2020 (Chapter 10).
[12] J.R. Li, J. Sculley, H.C. Zhou, Metal-organic frameworks for separations, Chem. Rev. 112 (2012) 869–932.
[13] Y. Wang, M. Rui, G. Lu, Recent applications of metal-organic frameworks in sample pretreatment, J. Sep. Sci. 41 (2018) 180–194.
[14] O.M. Yaghi, M. O'Keeffe, N.W. Ockwig, H.K. Chae, M. Eddaoudi, J. Kim, Reticular synthesis and the design of new materials, Nature 423 (2003) 705–714.
[15] S. Kitagawa, R. Matsuda, Chemistry of coordination space of porous coordination polymers, Coord. Chem. Rev. 251 (2007) 2490–2509.
[16] Z. Li, M. Qi, C. Tu, W. Wang, J. Chen, A. Wang, Magnetic metal-organic framework/graphene oxide-based solid-phase extraction combined with spectrofluorimetry for the determination of enrofloxacin in milk sample, Food Anal. Methods 10 (2017) 4094–4103.
[17] Z. Zhou, Y. Fu, Q. Qin, X. Lu, X. Shi, C. Zhao, et al., Synthesis of magnetic mesoporous metal-organic framework-5 for the effective enrichment of malachite green and crystal violet in fish samples, J. Chromatogr. A 1560 (2018) 19–25.
[18] Y. Zhang, H. Zhou, Z. Zhang, X. Wu, W. Chen, Y. Zhu, et al., Three-dimensional ionic liquid functionalized magnetic graphene oxide nanocomposite for the magnetic dispersive solid phase extraction of 16 polycyclic aromatic hydrocarbons in vegetable oils, J. Chromatogr. A 1489 (2017) 29–38.
[19] N. Li, H. Jiang, X. Wang, X. Wang, G. Xu, B. Zhang, et al., Recent advances in graphene-based magnetic composites for magnetic solid-phase extraction, Trends Anal. Chem. 102 (2018) 60–74.
[20] Q. Han, Z. Wang, J. Xia, S. Chen, X. Zhang, M. Ding, Facile and tunable fabrication of Fe_3O_4/graphene oxide nanocomposites and their application in the magnetic solid-phase extraction of polycyclic aromatic hydrocarbons from environmental water samples, Talanta 101 (2012) 388–395.
[21] S. Mahpishanian, H. Sereshti, One-step green synthesis of β-cyclodextrin/iron oxide-reduced graphene oxide nanocomposite with high supramolecular recognition capability: application for vortex-assisted magnetic solid phase extraction of organochlorine pesticides residue from honey samples, J. Chromatogr. A 1485 (2017) 32–43.
[22] Y. Zhao, Y. Yuan, X. Bai, Y. Liu, G. Wu, F. Yangd, et al., Multi-mycotoxins analysis in liquid milk by UHPLC-Q-Exactive HRMS after magnetic solid-phase extraction based on PEGylated multi-walled carbon nanotubes, Food Chem. 305 (2020) 125429.
[23] L. Fu, H. Zhou, E. Miao, S. Lu, S. Jing, Y. Hu, et al., Functionalization of amino terminated carbon nanotubes with isocyanates for magnetic solid phase extraction of sulfonamides from milk and their subsequent determination by liquid chromatography-high resolution mass spectrometry, Food Chem. 289 (2019) 701–707.

[24] H. Jiang, Y. Lin, N. Li, Z. Wang, M. Liu, R. Zhao, et al., Application of magnetic N-doped carbon nanotubes in solid-phase extraction of trace bisphenols from fruit juices, Food Chem. 269 (2018) 413–418.
[25] H. Gholami, M. Arabi, M. Ghaedi, A. Ostovan, A.R. Bagheri, Column packing elimination in matrix solid phase dispersion by using water compatible magnetic molecularly imprinted polymer for recognition of melamine from milk samples, J. Chromatogr. A 1594 (2019) 13–22.
[26] Y. Zhou, H. Liu, J. Li, Z. Sun, T. Cai, X. Wang, et al., Restricted access magnetic imprinted microspheres for directly selective extraction of tetracycline veterinary drugs from complex, samples, J. Chromatogr. A 1613 (2020) 460684.
[27] M. Sajid, C. Basheer, Layered double hydroxides: emerging sorbent materials for analytical extractions, Trends Anal. Chem. 75 (2016) 174–182.
[28] L. Du, X. Wang, T. Liu, J. Li, J. Wang, M. Gao, et al., Magnetic solid-phase extraction of organophosphorus pesticides from fruit juices using $NiFe_2O_4$@polydopamine@Mg/Al-layered double hydroxides nanocomposites as an adsorbent, Microchem. J. 150 (2019) 104128.
[29] S. Tang, G.H. Chia, Y. Chang, H.K. Lee, Automated dispersive solid-phase extraction using dissolvable Fe_3O_4-layered double hydroxide core-shell microspheres as sorbent, Anal. Chem. 86 (2014) 11070–11076.
[30] L. Adlnasab, M. Ezoddin, M. Shabanian, B. Mahjoo, Development of ferrofluid mediated CLDH@Fe3O4@Tanic acid-based supramolecular solvent: application in air-assisted dispersive micro solid phase extraction for preconcentration of diazinon and metalaxyl from various fruit juice samples, Microchem. J. 146 (2019) 1–11.
[31] A. Wen, G. Li, D. Wu, Y. Yu, Y. Yang, N. Hu, et al., Sulphonate functionalized covalent organic framework-based magnetic sorbent for effective solid phase extraction and determination of fluoroquinolones, J. Chromatogr. A 1612 (2020) 460651.
[32] K.R. Molaei, H. Tajik, M. Moradi, Magnetic solid phase extraction based on mesoporous silica-coated iron oxide nanoparticles for simultaneous determination of biogenic amines in an Iranian traditional dairy product, Food Control. 101 (2019) 1–8.
[33] J. Liu, X. Ma, S. Zhang, T. Wu, H. Liu, M. Xia, et al., Cationic gemini surfactant templated magnetic cubic mesoporous silica and its application in the magnetic dispersive solid phase extraction of endocrine-disrupting compounds from the migrants of food contact materials, Microchem. J. 145 (2019) 606–613.
[34] W.H. Ji, Y.S. Guo, X. Wang, X.F. Lu, D.S. Guo, Amino-modified covalent organic framework as solid phase extraction absorbent for determination of carboxylic acid pesticides in environmental water samples, J. Chromatogr. A 1595 (2019) 11–18.
[35] N. Li, D. Wu, N. Hu, G. Fan, X. Li, J. Sun, et al., Effective enrichment and detection of trace polycyclic aromatic hydrocarbons in food samples based on magnetic covalent organic framework hybrid microspheres, J. Agric. Food Chem. 66 (2018) 3572–3580.
[36] H.D. de Faria, L. de CarvalhoAbrão, M.G. Santos, A.F. Barbosa, E.C. Figueiredo, New advances in restricted access materials for sample preparation: a review, Anal. Chim. Acta 959 (2017) 43–65.
[37] D.H.A. Florez, F.V.A. Dutra, K.B. Borges, Magnetic solid phase extraction employing a novel restricted access material based on mesoporous polyaniline coated with hydrophilic monomers and casein for determination of antibiotics in milk samples, Microchem. J. 150 (2019) 104145.
[38] Z. Wang, J.C. Zhao, H.Z. Lian, H.Y. Chen, Aptamer-based organic-silica hybrid affinity monolith prepared via "thiol-ene" click reaction for extraction of thrombin, Talanta 138 (2015) 52–58.

[39] V. Pichon, N. Delaunay-Bertoncini, M.C. Hennion, Immunosorbents in sample preparation, Compr. Anal. Chem. 37 (2002) 1081–1100.
[40] C. Tu, Y. Guo, Y. Dai, W. Wei, W. Wang, L. Wu, et al., Determination of chloramphenicol in honey and milk by HPLC coupled with aptamer functionalized Fe_3O_4/graphene oxide magnetic solid-phase extraction, J. Food Sci. 84 (2019) 3624–3633.
[41] M. Faraji, R. Noorbakhsh, H. Shafieyan, M. Ramezani, Determination of acetamiprid, imidacloprid, and spirotetramat and their relevant metabolites in pistachio using modified QuEChERS combined with liquid chromatography-tandem mass spectrometry, Food Chem. 240 (2018) 634–641.
[42] M. Shamsipur, N. Yazdanfar, M. Ghambarian, Combination of solid-phase extraction with dispersive liquid–liquid microextraction followed by GC–MS for determination of pesticide residues from water, milk, honey and fruit juice, Food Chem. 204 (2016) 289–297.
[43] S. Moradi Shahrebabak, M. Saber-Tehrani, M. Faraji, M. Shabanian, P. Aberoomand-Azar, Simultaneous magnetic solid phase extraction of acidic and basic pesticides using triazine-based polymeric network modified magnetic nanoparticles/graphene oxide nanocomposite in water and food samples, Microchem. J. 146 (2019) 630–639.
[44] N. Bakaraki Turan, B. TuğbaZaman, S. Bakırdere, Application of oleic acid functionalized magnetic nanoparticles for a highly sensitive and efficient dispersive magnetic solid phase extraction of fenazaquin in almond samples for determination by gas chromatrography mass spectrometry, Microchem. J. 153 (2020) 104329.
[45] A.C. do Lago, M.H. da Silva Cavalcanti, M.A. Rosa, A.T. Silveira, C.R.T. Tarley, E.C. Figueiredo, Magnetic restricted-access carbon nanotubes for dispersive solid phase extraction of organophosphates pesticides from bovine milk samples, Anal. Chim. Acta 1102 (2020) 11–23.
[46] N. Li, J. Chen, Y.P. Shi, Magnetic polyethyleneimine functionalized reduced graphene oxide as a novel magnetic solid-phase extraction adsorbent for the determination of polar acidic herbicides in rice, Anal. Chim. Acta 949 (2017) 23–34.
[47] Y. Jia, Y. Wang, M. Yan, Q. Wang, H. Xu, X. Wang, et al., Fabrication of iron oxide@MOF-808 as a sorbent for magnetic solid phase extraction of benzoylurea insecticides in tea beverages and juice samples, J. Chromatogr. A 1615 (2020) 460766.
[48] X. Yang, K. Qiao, F. Liu, X. Wu, M. Yang, J. Li, et al., Magnetic mixed hemimicelles dispersive solid-phase extraction based on ionic liquid-coated attapulgite/polyaniline-polypyrrole/Fe_3O_4 nanocomposites for determination of acaricides in fruit juice prior to highperformance liquid chromatography-diode array detection, Talanta 166 (2017) 93–100.
[49] E. Yilmaz, M. Soylak, Preparation and characterization of magnetic carboxylated nanodiamonds for vortex-assisted magnetic solid-phase extraction of ziram in food and water samples, Talanta 158 (2016) 152–158.
[50] F. Liu, X. Yang, X. Wu, X. Xi, H. Gao, S.Z.W. Zhou, et al., A dispersive magnetic solid phase microextraction based on ionic liquid coated and cyclodextrin-functionalized magnetic core dendrimer nanocomposites for the determination of pyrethroids in juice samples, Food Chem. 268 (2018) 485–491.
[51] J. Zhao, Z. Meng, Z. Zhao, L. Zhao, Ultrasound-assisted deep eutectic solvent as green and efficient media combined with functionalized magnetic multi-walled carbon nanotubes as solid-phase extraction to determine pesticide residues in food products, Food Chem. 310 (2020) 125863.
[52] P. Zhao, S. Li, X. Chen, X. Guo, L. Zhao, Simultaneous enantiomeric analysis of six chiral pesticides in functional foods using magnetic solid-phase extraction based on

carbon nanospheres as adsorbent and chiral liquid chromatography coupled with tandem mass spectrometry, J. Pharm. Biomed. Anal. 175 (2019) 112784.
[53] H. Sahebi, E. Konoz, A. Ezabadi, A. Niazi, S.H. Ahmadi, Simultaneous determination of five penicillins in milk using a new ionic liquid-modified magnetic nanoparticle based dispersive micro-solid phase extraction followed by ultra-performance liquid chromatography-tandem mass spectrometry, Microchem. J. 154 (2020) 104605.
[54] M.D. Contin, J.E. Quinsaat, R.M. Negri, V.P. Tripodi, D. Opris, N.B.D. Accorso, Development of carbohydrate functionalized magnetic nanoparticles for aminoglycosides magnetic solid phase extraction, Anal. Chim. Acta 1082 (2019) 37–48.
[55] C. Vakh, M. Alaboud, S. Lebedinets, D. Korolev, V. Postnov, L. Moskvin, et al., An automated magnetic dispersive micro-solid phase extraction in a fluidized reactor for the determination of fluoroquinolones in baby food samples, Anal. Chim. Acta 1001 (2018) 59–69.
[56] H. Liu, N. Li, X. Liu, Y. Qian, J. Qiu, X. Wang, Poly(*N*-acryloyl-glucosamine-*co*-methylenebisacrylamide)-based hydrophilic magnetic nanoparticles for the extraction of aminoglycosides in meat samples, J. Chromatogr. A 2020 (1609) 460517.
[57] D. Li, T. Li, L. Wang, S. Ji, A polyvinyl alcohol-coated core-shell magnetic nanoparticle for the extraction of aminoglycoside antibiotics residues from honey samples, J. Chromatogr. A 1581–1582 (2018) 1–7.
[58] Y. Zhao, R. Wu, H. Yu, J. Li, L. Liu, S. Wang, et al., Chan Magnetic solid-phase extraction of sulfonamide antibiotics in water and animal-derived food samples using core-shell magnetite and molybdenum disulfide nanocomposite adsorbent, J. Chromatogr. A 1610 (2020) 460543.
[59] L. Xia, L. Liu, X. Lv, F. Qu, G. Li, J. You, Towards the determination of sulfonamides in meat samples: a magnetic and mesoporous metal-organic framework as an efficient sorbent for magnetic solid phase extraction combined with high-performance liquid chromatography, J. Chromatogr. A 1500 (2017) 24–31.
[60] S. Che, J. Fu, Z. Li, Y. Zeng, Y. Li, X. Su, et al., Preparation and application of magnetic molecular imprinted polymers for extraction of cephalexin from pork and milk samples, J. Chromatogr. A 1602 (2019) 124–134.
[61] W. Zhu, H. Peng, M. Luo, N. Yu, H. Xiong, R. Wang, et al., Zipper-like magnetic molecularly imprinted microspheres for on/off switchable recognition and extraction of 17β-estradiol from food samples, Food Chem. 261 (2018) 87–95.
[62] A.N.M. Nasir, N. Yahaya, N.N.M. Zain, V. Lim, S. Kamaruzaman, B. Saad, et al., Thiol-functionalized magnetic carbon nanotubes for magnetic micro-solid phase extraction of sulfonamide antibiotics from milks and commercial chicken meat products, Food Chem. 276 (2019) 458–466.
[63] N. Al-Afy, H. Sereshti, A. Hijazi, H. Rashidi Nodeh, Determination of three tetracyclines in bovine milk using magnetic solid phase extraction in tandem with dispersive liquid-liquid microextraction coupled with HPLC, J. Chromatogr. B 1092 (2018) 480–488.
[64] Z. Li, C. Lei, N. Wang, X. Jiang, Y. Zeng, Z. Fu, et al., Preparation of magnetic molecularly imprinted polymers with double functional monomers for the extraction and detection of chloramphenicol in food, J. Chromatogr. B 1100–1101 (2018) 113–121.
[65] E. Yavuz, Ş. Tokalıoğlu, Ş. Patat, Magnetic dispersive solid phase extraction with graphene/$ZnFe_2O_4$ nanocomposite adsorbent for the sensitive determination of mercury in water and fish samples by cold vapor atomic absorption spectrometry, Microchem. J. 142 (2018) 85–93.

[66] F. Shakerian, Y. Chelongar, A.M.H. Shabani, S. Dadfarnia, Mixed hemimicelles solid phase extraction based on sodium dodecyl sulphate-coated nano-magnets Fe_3O_4 for the simultaneous separation and preconcentration of cobalt and nickel, Microchem. J. 146 (2019) 234–238.
[67] I.S. Barreto, S.I.E. Andrade, F.A.S. Cunha, M.B. Lima, M.C.U. Araujo, L.F. Almedia, A robotic magnetic nanoparticle solid phase extraction system coupled to flow-batch analyzer and GFAAS for determination of trace cadmium in edible oils without external pretreatment, Talanta 178 (2018) 384–391.
[68] E. Kazemi, S. Dadfarnia, A.M.H. Shabani, M. Ranjbar, Synthesis, characterization, and application of a Zn (II)-imprinted polymer grafted on graphene oxide/magnetic chitosan nanocomposite for selective extraction of zinc ions from different food samples, Food Chem. 237 (2017) 921–928.
[69] S. Pan, C. Jin, X. Yang, W. Zhang, Ultrasound enhanced solid-phase extraction of ultra-trace arsenic on Fe_3O_4@AuNPs magnetic particles, Talanta 209 (2020) 120553.
[70] S. Zhu, B. Chen, M. He, T. Huang, B. Hu, Speciation of mercury in water and fish samples by HPLC-ICP-MS after magnetic solid phase extraction, Talanta 171 (2017) 213–219.
[71] Y. Yamini, M. Safari, Modified magnetic nanoparticles with catechol as a selective sorbent for magnetic solid phase extraction of ultra-trace amounts of heavy metals in water and fruit samples followed by flow injection ICP-OES, Microchem. J. 143 (2018) 503–511.
[72] M. Azizi, S. Seidi, A. Rouhollahi, A novel N,N′-bis(acetylacetone)ethylenediimine functionalized silica-core shell magnetic nanosorbent for manetic dispersive solid phase extraction of copper in cereal and water samples, Food Chem. 249 (2018) 30–37.
[73] B.T. Zaman, A.F. Erulaş, D.S. Chormey, S. Bakirdere, Combination of stearic acid coated magnetic nanoparticle based sonication assisted dispersive solid phase extraction and slotted quartz tube-flame atomic absorption spectrophotometry for the accurate and sensitive determination of lead in red pepper samples and assessment of green profile, Food Chem. 303 (2020) 125396.
[74] E. Akkaya, F. At Erulas, Ç. Büyükpinar, S. Bakirdere, Accurate and sensitive determination of lead in black tea samples using cobalt magnetic particles based dispersive solid-phase microextraction prior to slotted quartz tube-flame atomic absorption spectrometry, Food Chem. 297 (2019) 124947.
[75] A.N. Acikkapi, M. Tuzen, B. Hazer, A newly synthesized graft copolymer for magnetic solid phase microextraction of total selenium and its electrothermal atomic absorption spectrometric determination in food and water samples, Food Chem. 284 (2019) 1–7.
[76] H. Filik, A.A. Avan, Dextran modified magnetic nanoparticles based solid phase extraction coupled with linear sweep voltammetry for the speciation of Cr(VI) and Cr (III) in tea, coffee, and mineral water samples, Food Chem. 292 (2019) 151–159.
[77] A. Keramat, R. Zare-Dorabei, Ultrasound-assisted dispersive magnetic solid phase extraction for preconcentration and determination of trace amount of Hg (II) ions from food samples and aqueous solution by magnetic graphene oxide (Fe_3O_4@GO/2-PTSC): central composite design optimization, Ultrason. Sonochem. 38 (2017) 421–429.
[78] X. Yang, Y. Yin, Y. Zong, T. Wan, X. Liao, Magnetic nanocomposite as sorbent for magnetic solid phase extraction coupled with high performance liquid chromatography for determination of polycyclic aromatic hydrocarbons, Microchem. J. 145 (2019) 26–34.
[79] Y.H. Boon, N.N.M. Zain, S. Mohamad, H. Osman, M. Raoov, Magnetic poly (β-cyclodextrin-ionic liquid) nanocomposites for micro-solid phase extraction of selected

polycyclic aromatic hydrocarbons in rice samples prior to GC-FID analysis, Food Chem. 278 (2019) 322–332.

[80] C. Bignardi, A. Cavazza, C. Lagana, P. Salvadeo, C. Corradini, Optimization of mass spectrometry acquisition parameters for determination of polycarbonate additives, degradation products, and colorants migrating from food contact materials to chocolate, J. Mass Spectrom. 53 (2018) 83–90.

[81] N. Li, D. Wu, J. Liu, N. Hu, X. Shi, C. Dai, et al., Magnetic covalent organic frameworks based on magnetic solid phase extraction for determination of six steroidal and phenolic endocrine disrupting chemicals in food samples, Microchem. J. 143 (2018) 350–358.

[82] W. Ding, X. Wang, T. Liu, M. Gao, F. Qian, H. Gu, et al., Preconcentration/extraction of trace bisphenols in milks using a novel effervescent reaction-assisted dispersive solid-phase extraction based on magnetic nickel-based N-doped graphene tubes, Microchem. J. 150 (2019) 104109.

[83] N. Li, J. Chen, Y.P. Shi, Magnetic nitrogen-doped reduced graphene oxide as a novel magnetic solid-phase extraction adsorbent for the separation of bisphenol endocrine disruptors in carbonated beverages, Talanta 201 (2019) 194–203.

[84] Z.D. Du, Y.Y. Cui, C.X. Yang, X.P. Yan, Synthesis of magnetic amino-functionalized microporous organic network composites for magnetic solid phase extraction of endocrine disrupting chemicals from water, beverage bottle and juice samples, Talanta 206 (2020) 120179.

[85] M. Pastor-Belda, P. Viñas, N. Campillo, M. Hernández-Córdoba, Magnetic solid phase extraction with $CoFe_2O_4$/oleic acid nanoparticles coupled to gas chromatography-mass spectrometry for the determination of alkylphenols in baby foods, Food Chem. 221 (2017) 76–81.

[86] X. Wu, J. Hu, B. Zhu, L. Lu, X. Huang, D. Pang, Aptamer-targeted magnetic nanospheres as a solid-phase extraction sorbent for determination of ochratoxin A in food samples, J. Chromatogr. A 1218 (2011) 7341–7346.

[87] M. Zhao, H. Shao, J. Ma, H. Li, Y. He, M. Wang, et al., Preparation of core-shell magnetic molecularly imprinted polymers for extraction of patulin from juice samples, J. Chromatogr. A 1615 (2020) 460751.

[88] X. Yu, Z. Li, M. Zhao, S.C.S. Lau, H.R. Tan, W.J. Teh, et al., Quantification of aflatoxin B1 in vegetable oils using low temperature cleanup followed by immunomagnetic solid phase extraction, Food Chem. 275 (2019) 390–396.

[89] Z. Han, K. Jiang, Z. Fan, J.D.D. Mavungu, M. Dong, W. Guo, et al., Multi-walled carbon nanotubes-based magnetic solid-phase extraction for the determination of zearalenone and its derivatives in maize by ultra-high performance liquid chromatography-tandem mass spectrometry, Food Control. 79 (2017) 177–184.

[90] M. Zhang, J. Li, C. Zhang, Z. Wu, Y. Yang, J. Li, et al., In-situ synthesis of fluorinated magnetic covalent organic frameworks for fluorinated magnetic solid-phase extraction of ultra-trace perfluorinated compounds from milk, J. Chromatogr. A 2020 (1615) 460773.

[91] A.R. Bagheri, M. Arabi, M. Ghaedi, A. Ostovan, X. Wang, J. Li, et al., Dummy molecularly imprinted polymers based on a green synthesis strategy for magnetic solid-phase extraction of acrylamide in food samples, Talanta 195 (2019) 390–400.

[92] X. He, G.N. Wang, K. Yang, H.Z. Liu, X.J. Wu, J.P. Wang, Magnetic grapheme dispersive solid phase extraction combining high performance liquid chromatography for determination of fluoroquinolones in foods, Food Chem. 221 (2016) 1226–1231.

[93] M. Alvand, F. Shemirani, Fabrication of Fe_3O_4@graphene oxide core-shell nanospheres for ferrofluid-based dispersive solid phase extraction as exemplified for Cd(II) as a model analyte, Microchim. Acta 183 (2016) 1749–1757.
[94] Y. Zhang, S. Tao, Global atmospheric emission inventory of polycyclic aromatic hydrocarbons (PAHs) for 2004, Atmos. Environ. 43 (2009) 812–819.
[95] Z. Xia, X. Duan, S. Tao, W. Qiu, D. Liu, Y. Wang, et al., Pollution level, inhalation exposure and lung cancer risk of ambient atmospheric polycyclic aromatic hydrocarbons (PAHs) in Taiyuan, China, Environ. Pollut. 173 (2013) 150–156.
[96] P. Plaza-Bolanos, A.G. Frenich, J.L.M. Vidal, Polycyclic aromatic hydrocarbons in food and beverages: analytical methods and trends, J. Chromatogr. A 1217 (2010) 6303–6326.
[97] S.M. Wu, W.J. Yu, Liquid-liquid extraction of polycyclic aromatic hydrocarbons in four different edible oils from China, Food Chem. 134 (2012) 597–601.
[98] US Agency for Toxic Substances & Disease Registry (ATSDR). Available from http://www.atsdr.cdc.gov/spl/, cited November 8, 2013 [Google Scholar].
[99] Food Safety Authority of Ireland, Polycyclic Aromatic Hydrocarbons (PAHs) in Food, Food Safety Authority of Ireland 2, 2015, pp. 1–10.
[100] D.B. Zhou, X. Sheng, F. Han, Y.Y. Hu, L. Ding, Y.L. Lv, et al., Magnetic solid-phase extraction based on [60]fullerene functionalization of magnetic nanoparticles for the determination of sixteen polycyclic aromatic hydrocarbons in tea samples, J. Chromatogr. A 1578 (2018) 53–60.
[101] E.D. van Asselt, H.J. van der Fels-Klerx, H.J.P. Marvin, H. van Bokhorst-van De Veen, M.N. Groot, Overview of food safety hazards in the European dairy supply chain, Compr. Rev. Food Sci. Food Saf. 16 (2017) 59–75.
[102] E. Yıldırım, R. Gürkan, N. Altunay, A new ultrasonic thermostatic-assisted cloud point extraction/spectrophotometric method for the preconcentration and determination of bisphenol A in food, milk, and water samples in contact with plastic products, Food Anal. Methods 10 (2017) 1765–1776.
[103] C. Bignardi, A. Cavazza, C. Laganà, P. Salvadeo, C. Corradini, UHPLC-high-resolution mass spectrometry determination of bisphenol A and plastic additives released by polycarbonate tableware: influence of ageing and surface damage, Anal. Bioanal. Chem. 407 (2015) 7917–7924.
[104] S.M. Zimmers, E.P. Browne, P.W. O'Keefe, D.L. Anderton, L. Kramer, D.A. Reckhow, et al., Determination of free Bisphenol A (BPA) concentrations in breast milk of U.S. women using a sensitive LC/MS/MS method, Chemosphere 104 (2014) 237–243.
[105] D. Pérez-Palacios, M.Á. Fernández-Recio, C. Moreta, M.T. Tena, Determination of bisphenol-type endocrine disrupting compounds in food-contact recycled-paper materials by focused ultrasonic solid–liquid extraction and ultra-performance liquid chromatography-high resolution mass spectrometry, Talanta 99 (2012) 167–174.
[106] A. Chisvert, J.L. Benedé, A. Salvador, Current trends on the determination of organic UV filters in environmental water samples based on microextraction techniques - a review, Anal. Chim. Acta 1034 (2018) 22–38.
[107] Y. Yanga, G. Lia, D. Wuf, J. Liua, X. Lie, P. Luob, et al., Recent advances on toxicity and determination methods of mycotoxins in foodstuffs, Trends Food Sci. Technol. 96 (2020) 233–252.
[108] M. Peraica, B. Radić, A. Lucić, M. Pavlović, Toxic effects of mycotoxins in humans, Bull. World Health Organ. 77 (1999) 754–766.
[109] European Commission, Regulation (EC) No. 1881/2006 setting maximum levels for certain contaminants in foodstuffs, Off. J. Eur. Union L364 (2006) 5–24.

[110] R. Xiao, X. Zhang, X. Zhang, J. Niu, M. Lu, X. Liu, et al., Analysis of flavors and fragrances by HPLC with Fe_3O_4@GO magnetic nanocomposite as the adsorbent, Talanta 166 (2017) 262–267.

[111] G. Zhu, F. Liu, P. Li, S. He, S. Zhu, Q. Gao, et al., Profiling free fatty acids in edible oils via magnetic dispersive extraction and comprehensive two-dimensional gas chromatography-mass spectrometry, Food Chem. 297 (2019) 124998.

[112] R. Wu, F. Ma, L. Zhang, P. Li, G. Li, Q. Zhang, et al., Simultaneous determination of phenolic compounds in sesame oil using LC–MS/MS combined with magnetic carboxylated multi-walled carbon nanotubes, Food Chem. 204 (2016) 334–342.

[113] H. Lang, R. Yang, X. Dou, D. Wang, L. Zhang, J. Li, et al., Simultaneous determination of 19 phenolic compounds in oilseeds using magnetic solid phase extraction and LC-MS/MS, LWT-Food Sci. Technol. 107 (2019) 221–227.

[114] K. Hu, J. Qiao, W. Zhu, X. Chen, C. Dong, Magnetic naphthalene-based polyimide polymer for extraction of Sudan dyes in chili sauce, Microchem. J. 149 (2019) 104073.

[115] H.R. Noormohamadi, M.R. Fat̆hi, M. Ghaedi, Fabrication of polyethyleneimine modified cobalt ferrite as a new magnetic sorbent for the micro-solid phase extraction of tartrazine from food and water samples, J. Colloid Interface Sci. 531 (2018) 343–351.

[116] A. Rahimi, M.A. Zanjanchi, S. Bakhtiari, M. Dehsaraei, Selective determination of caffeine in foods with 3D-graphene based ultrasound-assisted magnetic solid phase extraction, Food Chem. 262 (2018) 206–214.

[117] M. Musa, W.A.W. Ibrahim, F.M. Marsin, A.S. Abdul Keyon, H. Rashidi Nodeh, Graphene-magnetite as adsorbent for magnetic solid phase extraction of 4-hydroxybenzoic acid and 3,4-dihydroxybenzoic acid in stingless bee honey, Food Chem. 265 (2018) 165–172.

[118] E. Kíl̇ınç, K.S. Çelik, H. Bilgetekin, γ-Fe_2O_3 magnetic nanoparticle functionalized with carboxylated multi walled carbon nanotube for magnetic solid phase extractions and determinations of Sudan dyes and Para Red in food samples, Food Chem. 242 (2018) 533–537.

[119] L. Wu, L. Yu, X. Ding, P. Li, X. Dai, X. Chen, et al., Magnetic solid-phase extraction based on graphene oxide for the determination of lignans in sesame oil, Food Chem. 217 (2017) 320–325.

[120] H. Chen, X. Deng, G. Ding, Y. Qiao, The synthesis, adsorption mechanism and application of polyethyleneimine functionalized magnetic nanoparticles for the analysis of synthetic colorants in candies and beverages, Food Chem. 293 (2019) 340–347.

[121] Q. Zhao, D.Q. Cheng, M. Tao, W.J. Ning, Y.J. Yang, K.Y. Meng, et al., Rapid magnetic solid-phase extraction based on alendronate sodium grafted mesoporous magnetic nanoparticle for the determination of trans-resveratrol in peanut oils, Food Chem. 279 (2019) 187–193.

[122] X.R. Shi, X.L. Chen, Y.L. Hao, L. Li, H.J. Xu, M.M. Wang, Magnetic metal-organic frameworks for fast and efficient solid-phase extraction of six Sudan dyes in tomato sauce, J. Chromatogr. B 1086 (2018) 146–152.

[123] Y. Cheng, J. Nie, J. Li, H. Liu, Z. Yan, L. Kuang, Synthesis and characterization of core–shell magnetic molecularly imprinted polymers for selective recognition and determination of quercetin in apple samples, Food Chem. 287 (2019) 100–106.

[124] V.C. Fernandes, M. Freitas, J.G. Pacheco, V.F. Domingues, C. Delerue-Matos, Evaluation of the QuEChERS and magnetic micro dispersive solid-phaseextraction of brominated flame retardants in red fruits with determination by GC/MS, Food Chem. 309 (2020) 125572.

[125] H.B. Zheng, J. Ding, S.J. Zheng, Q.W. Yu, B.F. Yuan, Y.Q. Feng, Magnetic "one-step" quick, easy, cheap, effective, rugged and safe method for the fast determination of pesticide residues in freshly squeezed juice, J. Chromatogr. A 1398 (2015) 1–10.

[126] H. Rashidi Nodeh, W.A.W. Ibrahim, M.A. Kamboh, M.M. Sanagi, Magnetic graphene sol–gel hybrid as clean-up adsorbent for acrylamide analysis in food samples prior to GC–MS, Food Chem. 239 (2018) 208–216.

[127] V.C. Fernandes, M. Freitas, J.P.G. Pacheco, J.M. Oliveira, V.F. Domingues, C. Delerue-Matos, Magnetic dispersive micro solid-phase extraction and gas chromatography determination of organophosphorus pesticides in strawberries, J. Chromatogr. A 1566 (2018) 1–12.

[128] Y.F. Li, L.Q. Qiao, F.W. Li, Y. Ding, Z.J. Yang, M.L. Wang, Determination of multiple pesticides in fruits and vegetables using a modified quick, easy, cheap, effective, rugged and safe method with magnetic nanoparticles and gas chromatography tandem mass spectrometry, J. Chromatogr. A 1361 (2014) 77–87.

[129] X.T. Peng, L. Jiang, Y. Gong, X.Z. Hu, L.J. Peng, Y.Q. Feng, Preparation of mesoporous ZrO_2-coated magnetic microsphere and its application in the multi-residue analysis of pesticides and PCBs in fish by GC–MS/MS, Talanta 132 (2015) 118–125.

Application of magnetic nanomaterials in bioanalysis

6

Saeed Nojavan, Parisa Mahdavi, Mohammad Reza Bayatloo and Soheila Chalavi
Department of Analytical Chemistry and Pollutants, Shahid Beheshti University, Tehran, Iran

6.1 Introduction

Bioanalysis is a subdiscipline of analytical chemistry covering the chemical analysis of the xenobiotics (e.g., drugs and their metabolites and biological molecules) and biotics (e.g., macromolecules, proteins, DNA, and peptide) in biological samples [1]. The purpose of chemical analysis is normally both to identify and to quantify the substances of interest in a given biological sample [2]. Bioanalysis is applied in many important fields such as medical diagnostics and treatment, medical studies, pharmaceutical research, biochemical research, and forensic. For example, hospital laboratories measure drug concentration in blood samples of patients to check that they are properly medicated. Also, bioanalysis is used in forensic toxicology laboratories, where a large number of samples such as blood, saliva, and hair are analyzed to prove the events of a crime [2,3]. It is noteworthy that bioanalysis is performed both on human samples and on animal samples [2]. Biological samples based on the physical state are mainly distinguished into three types: gas, liquid, and solid. Exhaled breath condensate has been utilized as a potential and auxiliary diagnostic way to recognize lung cancer and infectious illnesses in the early stage [4]. In this sense, aldehyde metabolites were measured in exhaled breath condensates of lung cancer patients and healthy people [4]. Biological liquid samples are used more frequently than other types in bioanalysis researches. This category includes whole blood, plasma, serum, urine, saliva, cerebrospinal fluid (CSF), synovial fluid, and breast milk. Whole blood from the human is an inhomogeneous suspension of erythrocytes, leukocytes, thrombocytes, and other minor cell fractions in a matrix called plasma [2]. In volume, the cell fraction in total constitutes about 45% and the plasma volume about 55%. As an alternative, blood may also be collected in the form of dried blood spots. A drop of blood is placed on a piece of paper or filter paper, and then dried. Plasma is obtained from whole blood collected in vials containing anticoagulant and then centrifuged typically for 10 minutes at 1000*g*. Plasma is normally a clear, homogeneous yellow liquid, and it consists of 90% water and 10% inorganic (primarily Na^+ and Cl^-) and organic solutes, primarily proteins (main part of proteins is albumin) [2]. Serum, as a clear homogeneous liquid, is obtained from whole blood after coagulation and by centrifugation, typically for

Magnetic Nanomaterials in Analytical Chemistry. DOI: https://doi.org/10.1016/B978-0-12-822131-0.00001-7

10 minutes at 1000*g*. The constituents in the serum are very similar to those of plasma. The only difference is the lack of fibrinogen proteins that act as the coagulation agents. Urine from humans is normally a yellow, clear, and sterile liquid. The contents of solutes in urine may vary a lot, depending on the intake of food and drink; most of the major solutes are anionic (carboxylic acids) or amphoteric (amino acids) in nature. Still, the urine contains thousands of different solutes in very different concentrations [2]. It is normally more concentrated in the morning but maybe diluted due to increased intake of beverages and alcohol. The pH of urine is about 6 but may vary between 4.6 and 8 depending on food intake. Saliva is an exocrine fluid secretion. It consists of approximately 99% water and contains a variety of electrolytes (e.g., calcium, magnesium, sodium, and potassium). Small organic molecules are also found in blood, as well as small amounts of proteins and polysaccharides. The advantages of saliva over blood and plasma are noninvasive and ease collection [5]. CSF is a clear colorless liquid containing a lot of inorganic and organic molecules but in a relatively small amount. CSF is collected from the spinal cord and can be used to obtain information on brain status. Synovial fluid is a clear, colorless, viscous liquid with a protein content between that of CSF and plasma [2]. It is redrawn from joints (e.g., the knee or elbow) using a syringe. Human breast milk is a complex matrix, its complexity lies not only in the high protein and lipid content but also in the changing composition along with the postpartum period [6]. Almost all medicines pass into breast milk in small amounts by a concentration gradient [7]. For this reason, it can be used to evaluate infant exposure to medicine substances during the breastfeeding period [6]. In another category, solid biological samples include feces, hair, and nails, tissue sections (biopsies). Human feces are an inhomogeneous, bad-smelling substance that may vary a lot in consistency depending on diet and general wellbeing. Bioanalysis of feces is most often performed in relation to recovery studies of drug substances for the diagnosis of digestive disorders [2]. The main constituent of hair and nails is keratin, a protein with a high content of the cysteine, which forms disulfide bridges, making the protein very dense and thorough. The proteins fibrillate into fibers. This makes hair and nails difficult to degrade. Hair and nails have attracted attention in forensic toxicology because it is possible to track some history due to storage of analytes in the hair and nails over time. Small amount of compounds present in blood can be deposited in hair and nails, and make these materials useful for retrospective measurements of special analytes in samples from a person. Tissue sections (biopsies) can be drawn from nearly any tissue such as liver, kidney, muscle, brain, colon, and prostate [2].

Biological samples have complex matrices and contain many components such as acids, salts, bases, proteins, and endogenous and exogenous substances [8]. To quantify chemicals and biochemicals in complex biological samples by analytical instruments a sample preparation step before chemical analysis is necessary for several reasons [9]:

1. The concentration of analytes is too low for detection in the analytical instruments.
2. Most of the biological samples are incompatible with the analytical instruments (device destruction).
3. The biological samples are too complicated for direct analysis, due to the probable interferences of other substances in the sample during measurement [3].

Traditional liquid–liquid and solid-phase extraction (SPE) procedures, and a variety of microextraction procedures are available for the extraction of analytes from biological samples [10]. In recent years, the application of magnetic nanomaterials (MNMs) as a sorbent in sample preparation has attracted the attention of several research groups. In recent years, MNMs have been emerging as a new type of important functional SPE material because they simplify the extraction process and save time as a result of their isolation from the sample matrix by an external magnetic field [11].

6.2 Magnetic nanomaterials

The magnetic materials based on composition can be divided into three classes, including metals, alloys, and oxides. The only metallic elements showing magnetism at room temperature are iron (Fe), cobalt (Co), and nickel (Ni) [12]. Some examples of magnetic alloys include FePt, CoPt, FeNi, and Fe_5C_2 [13]. Common magnetic oxides can be distinguished to iron oxides (Fe_xO_y) and various types of ferrites ($MO \cdot Fe_2O_3$ where M = Ni, Co, Mg, Zn, or Mn) [14]. Among these magnetic materials, iron oxide nanomaterials (NMs) such as magnetite (Fe_3O_4) and maghemite (Fe_2O_3) are the most commonly employed probe for bioanalytical applications due to their favorable properties including ease of synthesis, low cost, biocompatibility, easiness of surface modification, good recoverability, and excellent diffuseness in aqueous solution [15]. The bare MNMs are not usually employed for complex analysis due to selectivity problems, the formation of agglomerates, and oxidation [10]. Thus, to overcome the mentioned constraints, the magnetic core should be modified using the specific active groups. This coating of MNMs' surface considering as a shell is obtained by attaching inorganic components (e.g., silica or alumina) or organic molecules (e.g., polymer or surfactant), enhancing their chemical stability, preventing their oxidation, and improving desirable properties in them such as selectivity for analyte adsorption. Furthermore, inorganic coating can be modified or functionalized by utilizing organic or inorganic functional groups to improve the sorption features [16]. The following subsections briefly describe the main types of MNMs used as a sorbent for bioanalysis.

6.2.1 Inorganic magnetic nanomaterials

Bare MNMs are not good choices for bioanalysis applications. Surface modifications on magnetic cores can be accomplished with inorganic compounds such as silica or metals. This modification significantly increases antioxidant properties of naked magnetic cores for wider applications [17]. The main types of inorganic coatings such as metals, metal oxides, and silica are used in the preparation of inorganic MNMs. The modified magnetic core retains the properties of both inorganic material and core magnetism. This type of MNMs is simply investigated in the next subsections.

6.2.1.1 Silica-coated magnetic nanomaterials

Silica is widely used in the preparation of magnetic sorbents. Generally, three methods are used for coating silica on the surface of MNMs, namely Stöber reaction/sol–gel [18], embedding [19], and microemulsion [20] procedures. The simple sol–gel technique can be rather useful for coating the silica shell on the magnetic phase/core [21]. Silica is known as an adaptable coating agent owing to its accessibility, low toxicity, cheapness, favorable heat endurance, chemical inertness, easy grafting onto the magnetic core, and straightforward surface functionalization. Silane coupling agents have different functional groups such as thiol [22], amine [23], and carboxylic group [24]. Hence, the silanization process by using these agents can ameliorate the selectivity of hybrid NMs considerably [21]. Boojaria et al. synthesized magnetic nanoparticles (MNPs) by chemical coprecipitation technique and directly coated them with 3-(trimethoxysilyl)propyl methacrylate that is chemically adsorbed on the surface of Fe_3O_4 MNPs [25]. Moreover, to improve the selectivity of silica-coated MNPs, these materials can be further functionalized by organic or inorganic molecules, functionalization takes place via chemical immobilization or physical adsorption [16]. Octadecyl-functionalized magnetic silica ($Fe_3O_4@SiO_2@C_{18}$) was used as a sorbent to extract urinary volatile organic metabolites in a research study [26]. MNPs were obtained by the coprecipitation method. To the surface silica coating, $Fe_3O_4@SiO_2$ nanoparticles (NPs) were obtained by simply dispersing as-prepared Fe_3O_4 NPs in ethanol/water solution in the presence of tetraethyl orthosilicate (TEOS) under a nitrogen atmosphere. Finally, by dispersing in anhydrous toluene containing 1% (v/v) of octadecyltriethoxysilane and subsequently refluxed for 12 hours, the surface modification of the as-prepared $Fe_3O_4@SiO_2$ NPs was achieved. The process of preparation of $Fe_3O_4@SiO_2@C_{18}$ NPs is illustrated in Fig. 6.1.

6.2.1.2 Metal and metal oxide–coated magnetic nanomaterials

Another interesting MNM is the core–shell configuration, where either a magnetic core can be coated with metallic (metal or metal oxides) shells. Metal oxides not only can be served as effectively individual shells, whether onto pure MNP or MNP enclosed by SiO_2 NP, but also can be a proper potential for acquiring further organic/inorganic functionalizations [27]. They have unique surface active sites and physicochemical properties. A variety of nanometer-sized metal/metal oxides have been introduced into magnetic functional sorbents, including silver, gold, zirconia (ZrO_2), titanium oxide (TiO_2), alumina (Al_2O_3), manganese oxide (MnO_2), copper oxide (CuO), zinc oxide (ZnO), and magnesium oxide (MgO) [27]. Hsu and coworkers prepared L-cysteine-$Fe_3O_4@Au$ NPs [28]. Magnetic cores (Fe_3O_4) were synthesized by the coprecipitation method. Au-coated MNPs were obtained by reduction of Au salt. Finally, core/shell NPs were added to L-cysteine solution and L-cysteine-modified $Fe_3O_4@Au$ NPs were prepared [28]. In an interesting study, $Fe_3O_4@ZrO_2$ NPs were established by a sol–gel method as a sorbent for speciation of chromium(III) in human serum and urine samples [29]. Alumina is one of the most important metal oxides used in the preparation of MNMs. It assists sorbents

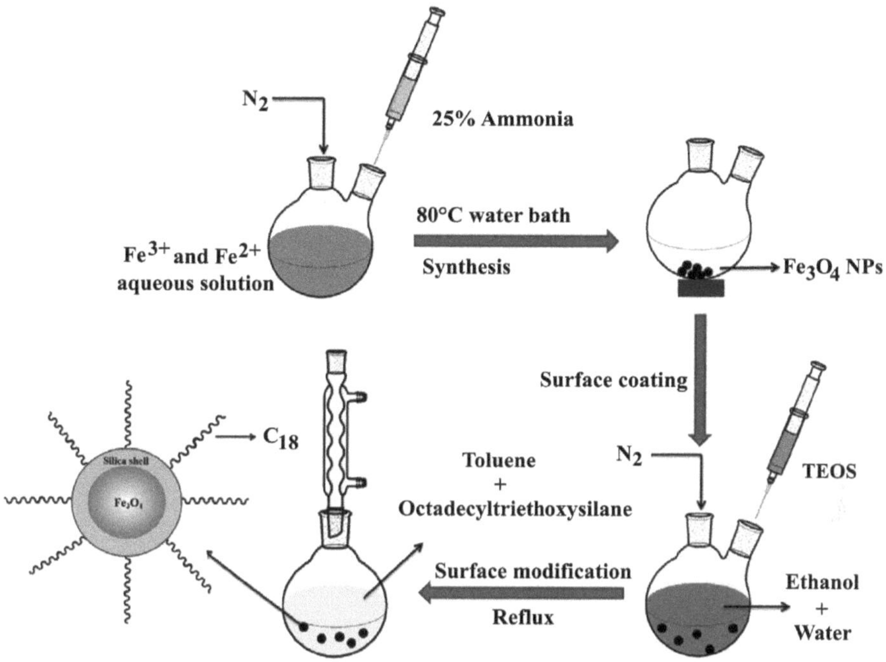

Figure 6.1 Schematic illustration of the preparation of $Fe_3O_4@SiO_2$-C_{18} NPs. *TEOS*, Tetraethyl orthosilicate.
Source: Reprinted with permission from Z. Qiao, R. Perestrelo, E.M. Reyes-Gallardo, R. Lucena, S. Cárdenas, J. Rodrigues, et al., Octadecyl functionalized core-shell magnetic silica nanoparticle as a powerful nanocomposite sorbent to extract urinary volatile organic metabolites, J. Chromatogr. A 1393 (2015) 18–25, Elsevier.

not to solve in the acidic solutions and increases their lifetime without changing in their magnetic characterizations [16]. Alumina can be functionalized with organic or inorganic compounds. Karimi et al. prepared dithizone/sodium dodecyl sulfate immobilized on alumina-coated MNMs by the ion-pair formation and simple physical adsorption [30].

6.2.1.3 Magnetic layered double hydroxide nanomaterials

Layered double hydroxides (LDHs), also known as "anionic clays" and hydrotalcite-like compounds, are a class of synthetic two-dimensional nanostructured inorganic materials that have positively charged layers of metal hydroxides separated by charge balancing anions and water molecules [31]. LDHs can be represented by the general formula $[M^{2+}{}_{1-x}\ M^{3+}{}_{x}\ (OH)_2]^{x+}\ (A^{n-})_{x/n}\cdot mH_2O$, where M^{2+} (M = Mg, Fe, Co, Cu, Ni, or Zn) and M^{3+} (Al, Cr, Ga, Mn, or Fe) are di- and trivalent cations, respectively; x is equal to the molar ratio of $M^{2+}/(M^{2+}+M^{3+})$ in the range 0.2–0.33; and A^{n-} is interlayer anions [21]. LDHs have many advantageous attributes as adsorbents, large specific surface area, good anion-exchange capacity, and easy synthesis [32]. A combination of LDHs and MNPs were evolved recently to improve the separation and reproduction of adsorbents/LDHs from sample media [32]. Li and coworkers reported the synthesis of Fe_3O_4@polydopamin@Mg-Al LDH [33]. First, water-dispersible Fe_3O_4 NPs were prepared through the hydrothermal method, then the as-prepared Fe_3O_4 NPs were dispersed into tris buffer solution (pH = 8.5) containing dopamine. Under this buffer solution, spontaneous polymerization of dopamine occurred, and Fe_3O_4 NPs were embedded into polydopamine (PDA) to form Fe_3O_4@PDA. Well-dispersed Mg-Al LDHs were synthesized in pure methanol solvent by the coprecipitation method under an inert atmosphere. For preparing Fe_3O_4@PDA@Mg-Al LDH, the desired amount of LDH and PDA-coated Fe_3O_4 was dispersed in water under the assistance of ultrasonication. The obtained homogenous solution was transferred to a Teflon-lined stainless-steel autoclave and heated. After the reaction for 24 hours, magnetic composite adsorbents were obtained [33]. Zhou et al. developed Fe@Mg-Al LDHs nanocomposite [34]. Fe nanocubes were synthesized by reduction of ferric chloride through hydrothermal reaction, and then Fe@Mg-Al LDHs were obtained by coating Mg-Al LDHs on the surface of the dispersed nanoscale zero-valent irons with the coprecipitation method [34].

6.2.2 Carbonaceous magnetic nanomaterials

There are many types of carbonaceous NMs, including carbon (C), carbon nanotube (CNT), graphene (G), graphene oxide (GO), and reduced graphene oxide (rGO). They have found a special position in the SPE/sample preparation due to intrinsic characteristics such as high surface-to-volume ratio, large sorption capacity, and excellent chemical and thermal stabilities. The integration of MNPs with carbonaceous NMs provides a source of carbonaceous NMs easy to separate from aqueous solutions. This section attempts to make an overview of magnetic carbon compounds.

6.2.2.1 Carbon-coated magnetic nanomaterials

The biocompatibility and chemical stability of carbon make it an excellent substance for coating on the MNMs [35]. The new carbon materials were prepared by high-temperature carbonization and template etching using different carbon-based substrates [36]. Generally, magnetic carbon can be obtained by mixing and calcining a magnetic source (e.g., Co, Fe_3O_4, and Fe_3O_4@SiO_2) and carbon source [37]. The source of carbon precursors used in calcination is extensive, for example, biological matrices, polymers, CNTs, and a combination of these with various templates [e.g., metal oxide and silica NPs, polymers, and metal-organic frameworks (MOFs), etc.] [36]. In a study, carbon-coated magnetite (Fe_3O_4@C) was synthesized for the simultaneous extraction of methamphetamine and ephedrine from urine samples [38]. For the synthesis of (Fe_3O_4@C) composite, Fe_3O_4 NMs were mixed with a glucose solution, followed by a hydrothermal reaction. Peng and coworkers prepared sulfur-doped magnetic carbon by carbonization at high temperature (about 600ºC) with $Fe_3O_4@SiO_2$ as magnetic and carbon source and basic magnesium sulfate as the template and sulfur source [39]. Another strategy is to mix magnetic metal salts and carbon precursors to produce magnetic carbon materials after calcination [40]. Wen et al. used ferric chloride as a magnetic source and tea residue as a carbon source, which was mixed and dried, followed by calcination. Tea residue was carbonized to produce porous carbon while iron salt was pyrolyzed to form γ-Fe_2O_3 on the porous carbon, producing magnetic carbon [40].

6.2.2.2 Magnetic carbon nanotube

CNTs are allotropes of carbon, which are composed of G sheets [41]. Generally, the monolayer or multilayer G sheets are rolled up to make one-dimensional hollow tubular nanocomposition-entitled CNTs. CNTs are classified based on the number of G layers. This classification of CNTs contains the single-walled CNTs and the multiwalled CNTs (MWCNTs) [41]. CNTs have attracted a myriad of attention in the field of SPE/sample preparation methods owing to their properties such as their high surface area and inner volume, stability, mechanical strength, the possibility of establishing $\pi-\pi$ interactions, and their capacity for functionalization. However, CNTs show two serious drawbacks that limit the use of these substances. First, their solubility is low in aqueous solutions, and second, their recovery from dispersing media is rather difficult [42]. There are two effective approaches to improve these drawbacks: (1) particular chemical functional groups can be anchored on the sidewall or bottom of CNTs, such as hydroxyl, carbonyl, and carboxyl groups. These mentioned groups can change the features of CNTs' surfaces; besides, they will be more hydrophilic. Also, these groups endow the ability to retain a variety of analytes and provide abundant active sites for further modification. (2) The introduction of magnetic compounds to prepare magnetic CNTs can enhance the separation efficiency throughout the analytical applications [41]. Yang et al. established amine-terminated poly(amidoamine) dendrimer (PAMAM)-functionalized magnetic MWCNT as a sorbent for the separation of flavonoids from plant extraction. The reaction between PAMAM and carboxylated MWCNTs was accelerated by *N*-hydroxysuccinimide (NHS) and *N*-ethyl-3-(3-dimethylaminopropyl)carbodiimide (EDC). Finally, magnetic PAMAM-functionalized MWCNTs were prepared by the introduction of MNPs on the surface of CNTs through the coprecipitation method (Fig. 6.2A) [43]. Shen and coworkers synthesized magnetic MWCNTs via a simple hydrothermal process and it was used as an adsorbent for the determination of tissue distribution of doxorubicin liposome inject in rats [45]. In another study, Fan et al. developed magnetic MWCNTs composites modified with thiol groups (MWCNTs-SH@Fe_3O_4) as an adsorbent. The thiol group was first decorated onto the MWCNTs, and the functionalized MWCNTs-Fe_3O_4 composites were subsequently prepared through a chemical coprecipitation method [46].

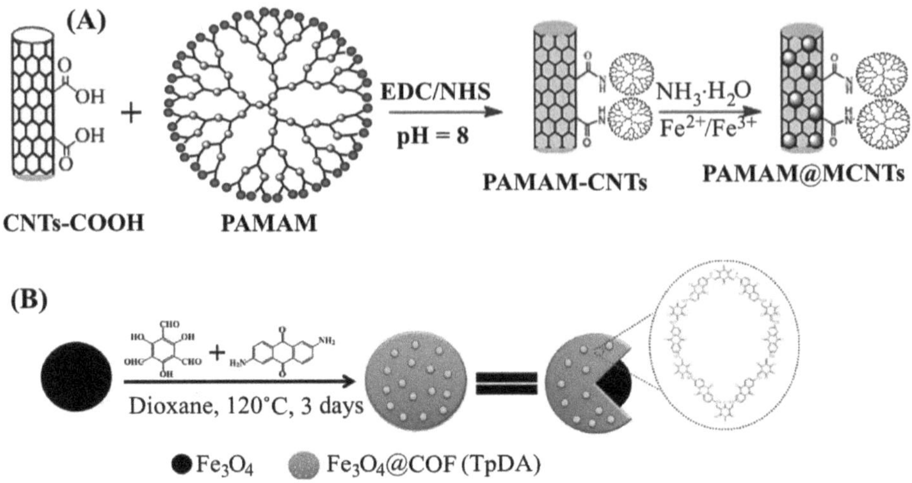

Figure 6.2 Schematic illustration of (A) proposed preparation route of PAMAM@MCNTs and (B) schematic fabrication process of Fe_3O_4@COF(TpDA). *COF*, Covalent organic framework; *EDC*, *N*-ethyl-*N*-3-(dimethylaminopropyl) carbodiimide; *MCNTs*, magnetic carbon nanotubes; *NHS*, *N*-hydroxysuccinimide; *PAMAM*, polyamidoamine dendrimer; *TpDA*, 1,3,5-triformylphloroglucinol (Tp) and 2,6-diaminoanthraquinone (DA).
Source: (A) Reprinted with permission from J. Yang, Z. Zhang, W. Pang, H. Chen, G. Yan, Polyamidoamine dendrimers functionalized magnetic carbon nanotubes as an efficient adsorbent for the separation of flavonoids from plant extraction. Sep. Purif. Technol. 227 (2019) 115710 [43], Elsevier. (B) reprinted with permission from N. Li, D. Wu, X. Li, X. Zhou, G. Fan, G. Li, et al., Effective enrichment and detection of plant growth regulators in fruits and vegetables using a novel magnetic covalent organic framework material as the adsorbents, Food Chem. 306 (2020) 125455 [44], Elsevier.

6.2.2.3 Magnetic graphene, graphene oxide, and reduced graphene oxide

G is a common carbon-based nanocomposite. This substance comprises of a single-atom layer, two-dimensional (2D) sheets of sp^2 hybridized carbon atoms with the structural arrangement in a hexagonal lattice [41]. G possesses a large surface area along with a myriad of delocalized π-electrons, so this material is highly practical in the extraction procedure of organic compounds, which contains a benzene ring structure based on $\pi-\pi$ stacking. G is ineffective for the adsorption of numerous polar compounds containing hydrophilic chemical groups [41]. Thus, to achieve particular requirements of practical applications, a suitable modification seems necessary. GO is the production of a reaction that is attained by the chemical oxidation of G. As a result, GO possesses more functional groups such as carboxyl and epoxy. Not only can these functional groups ameliorate the selective affinity for different analytes, but also they can simplify further modification depending on the active interaction sites relative to pristine G. rGO can be prepared via an appropriate chemical reduction method; as a result, this material will have fewer residual oxygenated groups and vacancy defect than GO [41]. These G-based NMs suffer from some drawbacks: uniform dispersion and satisfying recycling of pure G-based NMs from sample solutions are difficult. The introduction of magnetic components onto the G-based NMs combines the high adsorption capacity of G-based NMs and the convenience separation of the magnetic compounds [35]. Sha et al. synthesized magnetic G through an easy hydrothermal process [47]. These magnetic sorbents were employed as adsorbent and matrix for matrix-assisted laser desorption ionization-time of flight-mass spectrometry for extraction and determination of crotonaldehyde from saliva samples, respectively [47]. Fe_3O_4/GO composite as a magnetic solid-phase extraction (MSPE) adsorbent coupled with liquid chromatography was applied for the extraction of pseudoephedrine from urine samples [48]. Li et al. reported the synthesis of $ZnFe_2O_4$/rGO, then applied as a magnetic sorbent for the extraction of estrogens from water, soil, and fish samples [49].

6.2.3 Magnetic porous nanomaterials

Among the different materials used as SPE sorbents, the use of MOFs and covalent organic frameworks (COFs) for analytical extractions are gaining interest [50]. These novel substances have drawn a great deal of attention from analytical chemists due to their ability to function as hosts [50]. In other words, the available pores of these NMs can perform as a reactor in which analytes can be adsorbed and separated from the sample solution. These porous materials are well known for several extraordinary properties, for example, high porosity, large surface area, uniformly structured cavities, and chemical/thermal stability. The recovery of these nanosorbents based on very small particles dispersed in a sample solution can be a very hard task when direct use of these NMs for SPE. An interesting alternative to solve this problem is the magnetization of sorbents that can accelerate the separation of sorbents from the sample solution by simply applying an external magnetic field.

6.2.3.1 Magnetic metal-organic frameworks

MOFs as a multidimensional periodic network structure contain inorganic subunits such as ions or clusters, and organic linkers that are joined by coordination bonds. Magnetic MOFs are novel applicable components that have emerged recently. This functional material is a combination of porous MOFs and MNMs. It not only retains the structure and performance of the MOFs material but also possesses magnetic properties. These materials are selective, reusable, and have desirable dispersion in the target area in addition to their easy separation from aqueous solutions by a simple magnet [51]. Various synthetic strategies have been used to prepare magnetic MOFs and can be divided into the following categories: embedding method [52], encapsulation method [53], layer-by-layer self-assembly [54], physical mixing [55], postmagnetization [56], and other methods [57]. In a study, Fe_3O_4@SiO_2@Ti-MOF nanocomposite was used as a viable and efficient sorbent for the preconcentration of caffeic acid in the medical extracts of plants and water sample [58]. The prepared Fe_3O_4@SiO_2 was added into the solution of MOF precursor containing inorganic metal salts and organic linkers with stirring. Ti-MOFs are grown on the surface of the MNPs by the hydrothermal method [58]. Yan and coworkers proposed a physical mixing method strategy for the first time. They mixed Fe_3O_4@SiO_2 NPs and MIL-101 in aqueous solution under ultrasound, based on strong electrostatic interaction between negatively charged Fe_3O_4@SiO_2 NPs and MIL-101 with positive surface charge, Fe_3O_4@SiO_2-MIL-101 composites were obtained [55]. Xia et al. developed a magnetic, mesoporous core/shell structured Fe_3O_4@JUC-48 nanocomposite by encapsulation strategy and employed as an MSPE sorbent for the determination of trace sulfonamides in meat samples. Fe_3O_4 NPs were synthesized by the solvothermal method. Subsequently, the obtained Fe_3O_4 NPs were modified with mercaptoacetic acid (MAA). Finally, MMA-functionalized Fe_3O_4 NPs were dispersed in a precursor solution for JUC-48 synthesis [59].

6.2.3.2 Magnetic covalent organic frameworks

The COFs are a new type of crystalline porous polymeric materials in which the covalent bonds between organic building units form their structures. COFs have drawn intense interest due to low density, large specific surface area, excellent stability, and permanent porosity [44]. COFs can be coated on the surface of iron oxide particles by physical and chemical adsorption [41]. In a research study, a novel magnetic COF (Fe_3O_4@COF (TpDA)) was prepared via a rapid strategy [44]. Fe_3O_4 NPs were synthesized by the solvothermal method. COF(TpDA) shell was created through a Schiff-base condensation reaction of 2,6-diaminoanthraquinone (DA) and 1,3,5-triformylphloroglucinol (Tp) on the surface of Fe_3O_4 NPs under the typically solvothermal conditions. These magnetic sorbents were utilized in the MSPE procedure for extracting the plant growth regulators from fruits and vegetables (Fig. 6.2B) [44]. He et al. synthesized novel bouquet-shaped magnetic COFs, the surface of amino-functionalized Fe_3O_4 NPs was treated with Tp monomers, which was necessary for the ensuring formation of the COFs and directed growth. Tp and *p*-phenylenediamine (Pa-1) were then added to the surface-modified MNPs to grow COFs (TpPa-1) nanofibers (NFs) through the Schiff-base reaction by a room temperature solution-phase approach [60].

6.2.4 Magnetic polymers

Magnetic polymers are another important functionalized magnetic sorbent. Polymers containing different functional groups are often employed to protect the surface of the MNPs to avoid agglomeration and form stable suspension [35]. In general, polymers can be chemically bonded or physically adsorbed on MNPs. These functionalized MNMs have been widely used for SPE/sample preparation due to their abundant adsorption sites, high pH stability, as well as flexibility for modification [35].

6.2.4.1 Magnetic nonimprinted polymers

Recently, polymers have become significant and attractive coating materials because of their unique features such as acceptable chemical stability, low skeleton density, and large surface area. [61]. Magnetic polymers are generally composed of spherical Fe_3O_4 cores with a polymer coating on the surface (Fe_3O_4 @ polymer) [35]. Rodriguez and coworkers prepared PDA-coated iron oxide magnetic NMs as the sorbent for the extraction of estrogenic compounds from different types of milk [62]. In another study, Asgharinezhad et al. established polypyrrole MNMs as an efficient adsorbent for the extraction of antidepressant drugs from biological fluids [63].

6.2.4.2 Magnetic molecularly imprinted polymers

Magnetic molecularly imprinted polymers (MMIPs) are synthesized by copolymerizing the suitable monomer, cross-linking agent, and MNPs in the presence of the target molecule to which the sorbent is intended to be selective for it [64]. Therefore this molecule is known as the template molecule. The high degree of selectivity arises because, during the synthesis of the MMIP, the molecule that the MMIP is intended to be selective is involved in the polymerization processes, leaving a fingerprint for further recognition. A new and effective magnetic sorbent was prepared by Peyrovi et al. through coating a molecularly imprinted polymer (MIP) on the surface of MNMs, which was modified by oleic acid [65]. This selective sorbent was applied for the extraction of ezetimibe from biological samples. Methacrylic acid as a functional monomer, ezetimibe as a template molecule, and ethylene glycol dimethacrylate as a cross-linker were used in the polymerization process of MIP on the surface of MNPs [65]. Sorribes-Soriano and coworkers developed an MMIP for selective extraction of cocaine by polymerization in the presence of MNPs. First, polyethylene glycol and 3-(trimethoxysilyl) propyl methacrylate were used for coating the surface of Fe_3O_4 NPs. For the preparation of MMIPs, a mixture of cocaine (template), methacrylic acid (monomer), ethylene glycol dimethacrylate (cross-linker), azobisisobutyronitrile (initiator), and functionalized MNPs was prepared in acetonitrile, and polymerization carried out at 60ºC for 24 hours with magnetic stirring. After polymerization, the MMIPs were crushed in an Agata mortar, dried at 80ºC overnight, and sieved to obtain a homogeneous particle size. MMIPs were treated with a mixture of acetic acid and methanol in Soxhlet extractor for 24 hours to remove the template molecule from the polymer [66]. A schematic representation of preparation for MMIPs is shown in Fig. 6.3A [66].

6.2.5 Magnetic ionic liquids

Ionic liquids (ILs) are a class of molten salts exhibiting melting points at or below 100ºC. They are environment-friendly solvents because they have low vapor pressure, great chemical and thermal stability, variable viscosity, and favorable solubility for organic and inorganic compounds. Besides, if the type of anion or cation of ILs changes, the physical and chemical properties of ILs will change [68]. Physical or chemical immobilization of ILs on the surface of Fe_3O_4 NMs can merge the benefits of ILs and magnetic MNMs. Badragheh et al. fabricated silica-coated magnetic iron oxide functionalized with a polymeric IL as a sorbent for the simultaneous extraction of drugs from human plasma [69]. In another research, a novel magnetic

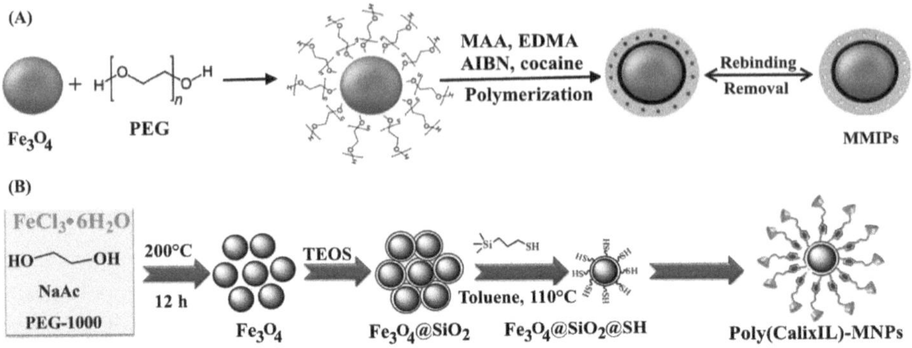

Figure 6.3 (A) The schematic diagram for the preparation of MMIPs using MNPs modified with PEG and (B) schematics of the preparation of poly(CalixIL)-MNPs. *AIBN*, Azobisisobutyronitrile; *EDMA*, ethylene glycol dimethacrylate; *MAA*, methacrylic acid; *MMIPs*, magnetic molecularly imprinted polymers; *MNPs*, magnetic nanoparticles; *PEG*, polyethylene glycol; *poly(CalixIL)*, poly(calixarene ionic liquid); *TEOS*, tetraethyl orthosilicate.
Source: (A) Reprinted with permission from A. Sorribes-Soriano, F.A. Esteve-Turrillas, S. Armenta, A. Montoya, J.M. Herrero-Martínez, M. de la Guardia, Magnetic molecularly imprinted polymers for the selective determination of cocaine by ion mobility spectrometry, J. Chromatogr. A 1545 (2018) 22–31 [66], Elsevier. (B) Reprinted with permission from K. Hu, J. Qiao, X. Wu, H. Yang, Y. Huang, S. Zhang, Poly(calixarene ionic liquid) modified Fe3O4 nanoparticles as new sorbent for extraction of flavonoids in fruit juice and green tea, Microchem. J. 143 (2018) 39–46 [67], Elsevier.

sorbent was prepared by the copolymerization of calixarene IL and mercaptan-modified Fe_3O_4 NPs [67]. A schematic of the preparation of poly(CalixIL)-MNPs is shown in Fig. 6.3B [67].

Recently, a new type of magnetic ILs (MILs) has been the subject of intensified interest in numerous analytical applications. These MILs are produced by the incorporation of a paramagnetic component in either the anion or cation of the IL structure [70]. They possess similar physicochemical properties to conventional ILs while also exhibiting a strong response to external magnetic fields. The paramagnetic behavior of the 1-butyl-3-methylimidazolium tetrachloroferrate(III) ($[BMIM^+][FeCl_4^-]$) MIL was first demonstrated by Hayashi et al. [70]. Subsequently, MILs based on other transition metals including manganese and cobalt were prepared and similarly shown to be paramagnetic liquids. In an interesting study, Clark et al. described a simple and rapid method for MIL-based plasmid DNA extraction from crude bacterial cell lysate in which DNA-enriched MIL is transferred directly to a polymerase chain reaction (PCR) tube for analysis [71]. The results demonstrate the feasibility of interfacing MIL extraction solvents with biochemical assays to achieve rapid enrichment and analysis of DNA. In this study the reaction mixture was spiked with 0.5 mL of either the trihexyl(tetradecyl)phosphonium tetrachloroferrate(III) ($[P_{6,6,6,14}{}^+][FeCl_4^-]$) MIL or the trioctylbenzylammonium bromotrichloroferrate(III) ($[(C_8)_3BnN^+][FeCl_3Br^-]$) MIL.

6.3 Applications of magnetic nanomaterials in bioanalysis

In comparison with microscale and bulk materials, MNMs have specific physical and chemical properties. They have various types, including spheres, nanotubes, nanocages, nanohorns, and nanolayers. Besides, their high surface area, which causes a considerable sorption capacity along with a magnetic field, enables them to adsorb the analytes strongly. These features and its straightforward operation steps make them robust, applicable, and efficient materials for bioanalysis. Nowadays, one of the most attractive MNMs is magnetic iron oxide in bare, core–shell, composite, and organic and inorganic ligand-modified forms [72]. These particles are invaluable in the bioanalysis studies because of their stability, shape controllability, and narrow size distribution. Therefore in recent years a myriad of studies are allocated to these kinds of MNMs [73]. Biologics can be purified or separated from their media with little damage via MNMs. In the following, a robust evaluation can be conducted on these biologics with minimal purification steps [13]. Intrinsically negligible background noises in biological media and the controllable magnetic field of MNMs are some of their most critical benefits in the biological analysis [13]. They also can be fruitful in the extraction and separation of drugs from biological media [74]. Analysis of drugs is essential in some areas such as therapeutic drug monitoring, toxicological analysis, and pharmacokinetic studies. However, direct determination of them in the biological systems is challenging because of the presence of endogenous bothering substances and low concentration of target drugs. As a result, employing the MNMs as a reliable sample pretreatment step is very helpful. Also, there is a myriad of roles that MNMs can play in analytical studies about the biological samples. Utilizing MNMs as extracting agent in the biological media instead of NMs or conventional microscale sorbents enhances the efficiency of techniques [72]. These magnetic substances can clean up the complicated samples before the injection to the chromatographic systems or even they can be used in online and automated analysis of biological samples [72]. Hence, a precise study on the application of various NMs in analyzing the biological samples seems essential. In the following subsections, recent researches and discoveries about the analytical applications of MNMs in some biological samples, including blood, plasma, serum, urine, milk, saliva, and hair will be reviewed.

6.3.1 Analysis of plasma and serum samples

In general, the most pioneering works in the field of magnetic materials were conducted on biological media. For the first time, the application of MNMs in bioanalysis was reported in 1972 [75]. This study indicated the potentiality of the MNMs in separating the biologically active proteins. Besides, two enzymes were recovered by using the magnetic supports in 1973 [76]. By improving the designs and structures of the MNMs throughout the years, the importance of their application in analyzing the biological samples increased. Some of the essential biological fluids are blood, serum, and plasma. Sample preparation of these fluids is so vital because

(1) their too complicated media is a severe problem for the analysis of a specific analyte existed in them, (2) a lot of valuable components such as drugs, nucleic acids, polypeptides or proteins, and even cells can be found in these areas for analysis, and (3) biologically active species can clog the columns and capillaries in the separation systems if they are injected directly to these devices [77]. Therefore the analysis of these components is fundamental in fields such as medicine, biotechnology (nanomedicine), and biomedicine. MNMs (especially iron oxide NMs) are suitable substances for separating and extracting the components from complex matrices such as plasma, blood, and serum [78]. Analytes in the complex samples can easily bind to the MNMs and the usage of a simple magnet separates them. However, naked MNMs cannot be selective material for analytical applications, so a functionalization of affinity ligands such as proteins, enzymes, receptors, and DNA will be indispensable [79]. Therefore, in the following paragraphs, some new and impressive applications of MNMs in the extraction and separation of target analytes from blood, serum, and plasma samples will be mentioned.

As it was mentioned, extracting and separating the drugs from biological samples by MNMs have a long history. A myriad of articles have studied the extraction of a wide range of drugs via various MNMs that had been synthesized in different ways. MNMs in various SPE methods are used as a preconcentration and separation step before analyzing the drugs and medicines by the sensitive analytical device. Not only the serum or plasma samples are too complicated to analyze the medications in their media, but also sometimes the concentration of analytes is below the detection limit of an analytical system [80]. Lotfi et al. synthesized useful nanosheets by using PAMAM and magnetic GO. They aimed to preconcentrate four antidepressant drugs, including citalopram, sertraline, fluvoxamine, and fluoxetine [81]. In this study, the novel magnetic sorbents were synthesized by covalently binding the magnetic GO to the PAMAM [dendrimer-functionalized magnetic graphene oxide (DMGO)]. The large surface area, accomplished biocompatibility, and many functional groups made this nanosorbent a suitable substance for extracting the drugs from human plasma. DMGO nanosorbents were rather reusable, so could be used 20 times. Besides, using low sample volume, few matrix effects, and short extraction time were other advantages of this technique. The extraction method was ultrasound-assisted MSPE, and the plasma sample was prepared before the procedure. In this way, the proteins of plasma were precipitated by acetonitrile. After spiking a suitable amount of the mentioned drugs' solution into the 1.5 mL of plasma, 2.5 mL of acetonitrile was added to this solution. Then, this solution was centrifuged for 5 minutes at 5000 rpm. The supernatant fluid was heated at 40°C in the protection of nitrogen.

One of the most attractive and useful strategies is attaching MIPs to NMs. This approach can improve the properties of a magnetic nanosorbent for sample clean up in biological media [82]. When MIPs are attached to the MNMs, the specific cavity design of these polymers, which is contemporary for the template molecules, their thermostability, and physical robustness, can cause high recovery in the extraction procedures. Besides, MIPs improve the selectivity of MNMs considerably, and it is a decent feature for sample preparation of the complex biological fluids [83]. In 2015 Madrakian et al. extracted the venlafaxine (VEN) from human blood serum by a novel MNM in which a layer of VEN-imprinted polymer was coated on the NMs' surface (VENMNPs) [84]. VEN is an antidepressant drug that inhibits the reuptake of norepinephrine and serotonin. In this study, 1.0 mL of acetonitrile was added to the mixture solution of 2.0 mL serum and 0.5 mL VEN solution to precipitate the proteins. Then, the whole solution was centrifuged at 4000 rpm, and proteins were separated from the sample. This technique was cost-effective and straightforward with low LOD values and an acceptable linearity range. Besides, the adsorbent capacity of VENMNPs was high, and the separation of the analyte was easy. Another interesting work introduced boronate-affinity magnetic hollow molecularly imprinted polymers (B-MhMIPs) as sorbents that were synthesized based on Fe_3O_4 NMs in 2019 [85]. The hollow structure of this magnetic sorbent brings more binding sites for adsorbing template molecules, both inside and outside surfaces. In addition, because of the presence of Fe_3O_4 as a magnetic base, these sorbents have a great magnetism and rapid adsorption rate. It has worth to say that the enhanced selectivity of B-MhMIPs makes them suitable materials for using in complex biological samples. Sialic acid (SA) was considered as a template and was extracted from three different rabbit serums. SA is a kind of monosaccharide and plays a critical role in physiological and pathological procedures. Dispersive SPE was employed as an extraction method in the prepared rabbit serum samples. In this way, 8.0 mL acetonitrile was added to 2.0 mL of the serum sample, and the mixture was centrifuged at 10,000 rpm to separate the proteins. It should be mentioned that ethylenediaminetetraacetic acid was added to a tube that contained the rabbit serum as an anticoagulant additive before the preparation. Another attractive way for improving the application of MNMs in biological samples is by using carbon dots (CDs) in the structure of sorbents [86]. CDs can be obtained from various sources such as natural, biological, or synthetic procedures. They have some physical and chemical features such as photoluminescence and cytotoxicity that make them suitable materials for bioapplications. Besides, some other properties, including chemically inert, biocompatibility, low toxicity, and simple operation, make them an appealing candidate for utilizing in the sample preparation of biological fluids [86]. As a piece of evidence in 2020, Yuvali et al. synthesized magnetic carbon nanodot/GO hybrid material (Fe_3O_4@CD@GO) through a green production process for extracting ibuprofen [80]. The extraction method was MSPE, and ibuprofen was extracted from human plasma. The human plasma sample was obtained from a blood sample by centrifuging it at 4000 rpm for 10 minutes, and before the extraction procedure, 1.0 mL acetonitrile was added to the 0.5 mL of sample for protein precipitation. This method is rather rapid because reaching adsorption and desorption equilibrium happens just in 2 minutes. Furthermore, results show a high extraction recovery, low limit of detections (LODs) and limit of quantifications, and reasonable repeatability.

Another essential role of modified MNMs is in the SPE of biological macromolecules before analyzing [87]. These materials can adsorb a large amount of these macromolecules owing to their high specific surface area. MNMs show desirable features such as stability under physiological environment, biocompatibility, exceptional magnetic susceptibility, and easy preparation steps. Therefore they have considerable attraction for biological macromolecules adsorption and recognition [88]. It is worth to be mentioned that captured macromolecules can be separated swiftly by a magnetic force without any centrifugation or filtration step [89]. For instance, a novel type of magnetic fluorescent molecularly imprinted nanoparticles (FMINPs) was synthesized by Yi-Da et al. (Fig. 6.4) in 2018 with the target of extraction and fluorescence detection of transferrin (TrF) [90]. This protein is one of the most famous glycoproteins, its role is controlling the iron ion balance in human blood, and so it was extracted from human serum in this study. As is indicated in Fig. 6.4, at first, the surface of Fe_3O_4 NPs was modified by utilizing 3-carboxybenzeneboronic acid, and then, an imprinted material of TrF was attained by a particular method. After this step, a near-infrared semicyanine fluorescent compound has been introduced to the surface of the mentioned imprinted material. As a result, in this technique, the reeluent of TrF from FMINPs is not indispensable, so the method is so fast, and there is no risk of deactivity of TrF. However, the problem of this method was using fluorescence quenching signals to quantify the protein with nonlinear curve. Hence, some materials for increasing the fluorescence signal should be used. Three different human serum samples were just diluted with proportion of 1:50, 1:75, and 1:100 by phosphate buffer saline solution (50 mM, pH = 7.4).

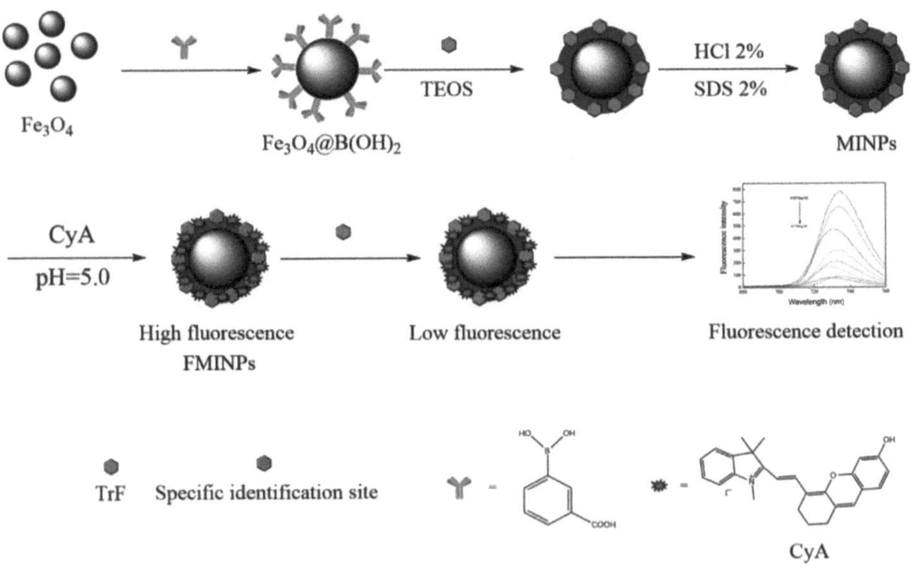

Figure 6.4 Synthesis scheme of the fluorescent molecularly imprinted nanoparticles. *CyA*, Cyclosporin A; *FMINPs*, fluorescent molecularly imprinted nanoparticles; *MINPs*, molecularly imprinted nanoparticles; *TEOS*, tetraethyl orthosilicate; *TrF*, transferrin. *Source*: Reprinted with permission from Y. Da Zhang, Q.W. Huang, C. Ma, X.Y. Liu, H.X. Zhang, Magnetic fluorescent molecularly imprinted nanoparticles for detection and separation of transferrin in human serum, Talanta 188 (2018) 540–545.

In 2019 Fe_3O_4@GO nanoadsorbents were introduced as a novel magnetic sorbent for DNA adduct analysis [91]. This is the first application of a Fe_3O_4@GO nanoadsorbent for chromatographic analysis of DNA adducts. In this study, human blood DNA was isolated by a specific method using a Qiagen kit from the manufacturer. In this method, four different human blood samples were assayed and DNA adducts were considered as existing analytes in two samples. However, the technique could not recognize DNA adducts in other samples. Their amounts might be below the detection limit of the technique. Several adducts were identified in a small volume of the blood sample with an acceptable extraction recovery. Furthermore, the kinds of other MNMs that were used in biological fluid include blood, serum, and plasma along with their figures of merits were collected in Table 6.1 [69,92–100]. As it is presented in Table 6.1, a myriad of essential compounds can be extracted from plasma and serum media. Based on the criteria collected in this table, some recent works show remarkable benefits of MNMs for extracting the components, including various kinds of drugs [92,93], proteins [94], peptides [95], agents [96], and even biphenyls [97]. For instance, improving the structure of NMs and using three-dimensional magnetic MNMs bring a robust and reliable sorbent that has extracted seven various biphenyls from both biological and environmental samples [97]. The highest number of extracted analytes among other works and the applicability of the proposed sorbent in different real samples show that magnetic metal-organic nanotubes are promising adsorbent materials for the MSPE. Therefore focusing on the developments of nanosorbents can be so helpful for analyzing critical components existing in plasma and serum samples. As a piece of evidence, using novel Fe_3O_4/cetyltrimethylammonium bromide MNMs increases the sorption capacity (three times) for extracting specific and important drugs entitled tetracyclines [98]. Besides, these NMs showed the lowest LODs, among other works in Table 6.1. It is worth to be mentioned that numerous practical approaches were introduced recently for extracting the endogenous substances of plasma and serum. In this way, four kinds of choline chloride (ChCl)–based deep eutectic solvents were evaluated as nontoxic solvents for synthesizing the best core–shell structure magnetic GO [94]. Hence, seven different proteins were extracted successfully from whole bovine blood with acceptable recoveries through an environment-friendly approach.

6.3.2 Analysis of urine samples

Urine is one of the most common biological samples, and it has numerous applications in analytical studies. Accessibility to this sample in large volumes is easy, plus its media is noninvasive [101]. Therefore, recently, this biological fluid is used for analyzing a wide range of analytes that exist in it. As an instance, some routine components that are extracted from urine efficiently are harmful materials such as heavy metals [102], drugs [103], volatile organic metabolites [104], and even phosphorous [105]. The same as other biological samples such as plasma, serum, and blood, urine is a complicated sample and causes some difficulties for analyzing the analytes in its media. Direct analysis of target analytes in urine media is not

Table 6.1 Application of magnetic nanomaterials in the analysis of plasma and serum samples.

Magnetic nanomaterial	Sample	Analyte	Method	Desorption solvent	LOD ($ng\ mL^{-1}$)	*R*%	Ref.
Fe_3O_4@SiO_2@PIL-PF_6	Plasma	Antidiabetic drugs	MSPE-HPLC-UV	Acetonitrile	0.8–6.0	86.5–92.5	[69]
C/Fe_3O_4	Plasma	Losartan	MSPE-HPLC-UV	Methanol and acetonitrile (1:1 v/v)	0.09–0.69	56.3–66.4	[92]
MWCNT-MMIPs	Plasma	Sotalol	UA-DSPME-HPLC-UV	Methanol/acetic acid (9:1, v/v)	0.31	97.4–101.6	[93]
Fe_3O_4–NH_2@GO@DES	Bovine serum	Proteins	MSPE-UV-Vis	Na_2HPO_4 containing 1 M NaCl	13.74	86.2–95.5	[94]
Fe_3O_4@MIL(Fe/Ti)	Serum	Phosphorylated peptides	MALDI-TOF-MS	$NH_3 \cdot H_2O$ (10 wt.%)	–	85.0	[95]
Fe_3O_4@PDA	Dog plasma	Glimepiride	MSPELC-MS/MS	Methanol-acetonitrile (1:1 v/v)	–	71.2–85.7	[96]
Fe_3O_4@Co-MONT	Dog plasma	PCBs	MSPE-GC-MS/MS	*n*-Hexane	0.31–0.49	66.7–82.0	[97]
Fe_3O_4/CTAB	Serum	Tetracyclines	MD-μ-SPE-HPLC-DAD	Etidronic acid (0.1 M)	0.03–0.08	90.0–115.0	[98]
Fe_3O_4@PDA@Cu^{2+}	Rat and rabbit plasma	Purines	MSPE-HPLC-DAD	Imidazole (30 mM) + acetic acid (10 mM)	0.42–2.15	70.0–102.4	[99]
Fe_3O_4@Ni_xSiO_y	Serum	Tyrosine kinase inhibitors	MUA-DμSPE-HPLC-UV	THF containing 10 mM imidazole	0.29–0.57	90.0–98.6	[100]

C, Carbon; *CTAB*, cetyltrimethylammonium bromide; *DAD*, diode array detection; *DESs*, deep eutectic solvents; *GO*, graphene oxide; *MALDI-TOF-MS*, matrix-assisted laser desorption ionization-time of flight-mass spectrometry; *MD-μ-SPE*, magnetic dispersive micro-solid-phase extraction; *MONT*, metal-organic nanotube; *MSPE*, magnetic solid-phase extraction; *MUA-DμSPE*, magnetic ultrasound-assisted dispersive micro-solid-phase extraction; *MWCNT-MMIPs*, multiwalled carbon nanotubes-magnetic molecularly imprinted polymer; *PCBs*, polychlorinated biphenyls; *PDA*, polydopamine; *PF6*, hexafluorophosphate; *PIL*, poly-1-vinyl-3-octylimidazoliumbromide; *UA-DSPME*, ultrasonic-assisted dispersive solid-phase microextraction.

possible because the amount of target analytes is usually low in this sample. Besides, the complex urine medium decreases the selectivity and sensitivity of the determination of analytes [106]. Hence, a feasible sample preparation method is required before analyzing the target analytes in urine samples. As it was mentioned, MNMs are compatible substances with biological samples, so they are rather useful for sample preparation of urine. Using modified MNMs in SPE methods is rapid and time consuming because a simple magnet separates analytes after adsorbing. Therefore no filtering or centrifuging step is required [107]. Besides, these materials are biocompatible and have high surface-area-to-volume that brings maximum extraction efficiency from a complex urine medium. In the following paragraph, some recent papers will be investigated in the field of analysis of analytes by MNMs in the urine samples.

One of the most notable elements that are used as a modifier for improvement of MNMs surface is polymer structures. It is not without grace to mention a study about this technique for extracting the target analytes from urine samples. In 2015 poly (*para*-phenylenediamine) modified with Fe_3O_4 NPs was prepared for extracting monocyclic aromatic amines from a human urine sample [108]. Polymers possess extended conjugated p-electron systems along with a polymer backbone that makes them efficient sorbents. In this paper, human urine samples that were obtained from the smoky and nonsmoky persons were just stored in polyethylene bottles in the dark at $-18^{\circ}C$. The advantage of this method was its sensitivity to aromatic amines. Also, the proposed method presented an efficient and cost-effective way with very low LOD, satisfactory recoveries, and good relative standard deviation.

MIPs are so applicable to modify the sorbents because of their specific characteristics, so the applications of magnetic nanosorbents modified by MIPs in plasma, serum, and blood samples were mentioned. Their application in urine also is attractive. Therefore many studies have been conducted on modifying MNMs by MIPs to sample pretreatment of urine samples in recent years. In this way, some new substances, namely ion-imprinted polymers (IIP), were introduced that are similar to MIPs. Due to the IIPs' specific properties, they are desirable candidates for improving the application of MNMs in bioanalysis [109]. IIPs have particular chemical structures so that they can be selective sorbents. The polymer memory effect toward the metal ion interaction with specific ligand in these materials, and their coordination geometry, metal ion coordination number, charge, and size are rational reasons for their excellent selectivity [110]. For example, Khoddami et al. have investigated the impact of IIPs on modifying the surface of MNMs for extracting cobalt(II) from the urine sample [111]. In this study, $Fe_3O_4@TiO_2@SiO_2$−IIP functionalized with −NH groups was prepared, and cobalt ion was considered as a template. Besides, TiO_2 was selected as a suitable choice for the imprinted coating material. To prepare the urine sample, a mixture of HNO_3 and $HClO_4$ (2:1) was added to 10 mL of urine. Then the solution was heated at 100°C and 150°C for 10 minutes. After adding 3 mL of $HClO_4$ to the residue, the outcome was heated at 160°C. Finally, a specific amount of H_2SO_4 was added, and the mixture was heated again. This novel magnetic nanosorbent brings swift sorption and desorption of cobalt, and its experimental data shows good fit into the pseudo-second-order kinetic model and the Langmuir

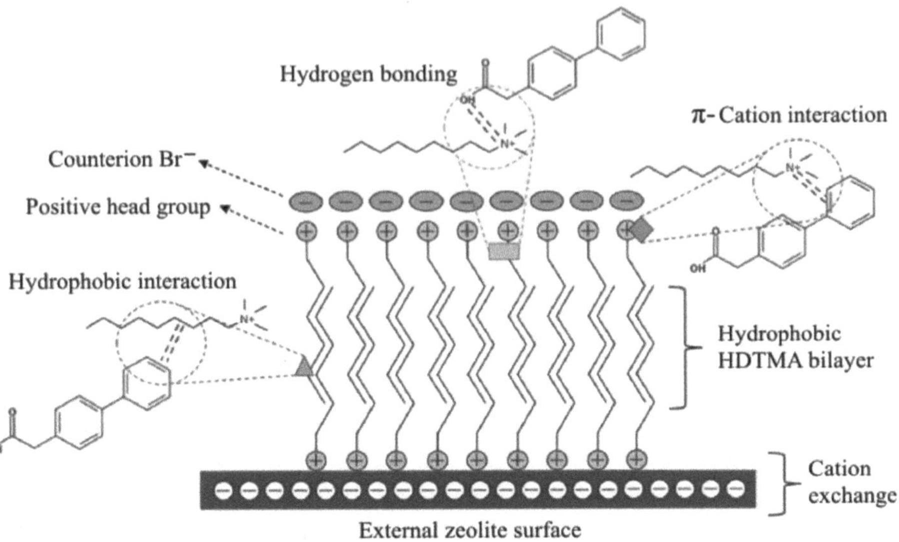

Figure 6.5 Scheme of zeolite surface modified by hexadecyltrimethylammonium bromide surfactant. *HDTMA*, Hexadecyltrimethylammonium.
Source: Reprinted with permission from P. Baile, L. Vidal, A. Canals, A modified zeolite/iron oxide composite as a sorbent for magnetic dispersive solid-phase extraction for the pre-concentration of non-steroidal anti-inflammatory drugs in water and urine samples, J. Chromatogr. A 1603 (2019) 33–43, Elsevier.

isotherm. Furthermore, the extraction recovery diminished just 5.8% after the seventh cycle of the adsorption–desorption process.

Other impressive materials that can be potential sorbents for SPE procedures are zeolite-based materials [112]. These materials have special features, so their combination with MNMs can result in producing a durable and influential sorbent. Zeolites have a high surface area and adsorption capacity, acceptable molecular selectivity, chemical and thermal stability, ion exchange capacity, and low expense. Hence, these materials can be so useful in analyzing the biological fluids, especially urine. In 2019 Baile et al. synthesized hexadecyltrimethylammonium bromide (HDTMABr) surfactant-ZSM-5 zeolite-based composite decorated with iron oxide MNMs in order to extract nonsteroidal antiinflammatory drugs (NSAIDs) from urine samples (Fig. 6.5) [113]. Generally, these drugs are utilized for curing chronic pains and are among the most widely used medications worldwide. It is shown in Fig. 6.5 that HDTMABr surfactant with a high concentration of 13.7 mM makes a bilayer on the external surface of the zeolite. This bilayer causes negative charges on the surface changes to the positive, and this positive charge has been balanced by adding bromide counter ions. The employed extraction method was magnetic dispersive solid-phase extraction (MdSPE) for extracting the target analytes from urine. Thus five urine samples were obtained from a healthy person, and several urine samples were attained from a person who was under treatment by ibuprofen. First, the buffer (2.5 mM H_3PO_4/2.9 mM KH_2PO_4) was added to the samples to adjust their pH values on 2.2. Finally, the urine samples were filtered by 0.45-μm pore size nylon filters. This technique was simple, rapid, economical, and user-friendly owing to the straightforward modification with a cationic surfactant.

As it was mentioned, naked MNMs have low selectivity and high dispersibility. Therefore one of the most feasible strategies for improving this nanosorbent in the extraction process is making carbon-based magnetic nanocomposites. These nanocomposites show acceptable biocompatibility and high chemical and thermal stability. Hence, they are suitable sorbents in the aqueous or biological mediums [114]. Taghvimi et al. used an IL on carbon-coated MNMs to improve the application of carbon-based MNMs in biological fluids such as urine [115]. ILs are molten salts that consist of organic cations and organic or inorganic anions, and they can form a micelle to perform as a surfactant. The ability to make micelle, high conductivity, and high thermal stability of ILs make them attractive materials for modifying the MNMs. In this paper, IL-C/MNPs were used in a mixed hemimicelle MdSPE method as nanosorbent for extracting tramadol from the urine sample. Tramadol is usually utilized as opioid pain medication in the treatment of mild pains, and approximately 10%–30% of the parent drug is observed in the urine. This technique is sensitive, inexpensive, and fast with a green synthesis pathway. Furthermore, this way can be compatible with clinical usages. The applications of some of the other MNMs in the urine samples along with their figures of merits were collected in Table 6.2 [48,93,97,116–123]. As it is observed in the table, there are also many invaluable analytes extracted from urine samples, the same as other biological media. Innovate developments in the structure of magnetic nanosorbents, and consequently, the extraction method brings more effective sample pretreatment ways for analyzing essential analytes in the complex urine medium. In this way, by introducing magnetic hydroxyapatite NMs, six various kinds of parabens were extracted simultaneously in a very short time from several real samples, including urine [116]. Also, in another study by employing GO-Fe_3O_4 nanocomposite MSPE method, eight psychoactive drugs were simultaneously extracted from urine samples [117]. This method delivered the best LOD values (0.02–0.2 ng mL^{-1}) among other data in the table and showed high accuracy and repeatability values. The extraction of several various analytes at the same time by an extraction method is not the only criterion to consider a technique as an applicable one. In many cases, developing a selective sorbent or a method to be sensitive to a specific component is also important. In this way, MIPs usually were used for forming unique templates on the magnetic nanosorbents, so some selective extraction methods were introduced. As it is observed in Table 6.2, the trace amount of some critical components in urine such as clenbuterol [118] and codeine [119] was extracted via a selective approach successfully. Also, it is worth to be mentioned that some developed NMs can be multifunctional. In an interesting study, multifunctional TiO_2@Fe_3O_4@C-NF nanocomposite was used for the preconcentration of ibuprofen, and as a photocatalyst for photocatalytic degradation of antibiotics, NSAIDs, and azo dye [120].

Table 6.2 Application of magnetic nanomaterials in the analysis of urine, hair, milk, and other samples.

Magnetic nanomaterial	Sample	Analyte	Method	Desorption solvent	LOD (ng mL^{-1})	*R*%	Ref.
MWCNT-MMIPs	Urine	Sotalol	UA-DSPME-HPLC-UV	Methanol/acetic acid (9:1, v/v)	0.31	94.6–102.5	[93]
Fe_3O_4@MIL(Fe/Ti)	Saliva	Phosphorylated peptides	MALDI-TOF-MS	$NH_3 \cdot H_2O$ (10 wt.%)	–	85.0	[95]
Fe_3O_4@Co-MONT	Urine	PCBs	MSPE-GC-MS/MS	*n*-Hexane	0.31–0.49	62.7–82.2	[97]
Fe_3O_4@Ni_xSiO_y	Cerebrospinal fluid	Tyrosine kinase inhibitors	MUA-DμSPE-HPLC-UV	THF containing 10 mM imidazole	0.29–0.57	94.0–98.2	[100]
γ-Fe_2O_3@HAP	Urine	Parabens	UAMSPE-GC-MS	Acetonitrile	5.0–10.0	–	[116]
GO-Fe_3O_4	Urine	Psychoactive drugs	MSPE-UHPLC-MS/MS	Methanol	0.02–0.2	80.4–105.5	[117]
MHIMs	Bovine urine	Clenbuterol	SM-dSPE-LC-UV	Acetonitrile-acetic acid (85:15 v/v)	0.085	91.4–105.3	[118]
COP-IP-MNSs	Urine	Codeine phosphate	SPE-FL	Methanol	0.67	97.3–98.1	[119]
TiO_2@Fe_3O_4 @C-NF	Urine	NSAIDs	MSPE-HPLC-DAD	Acetone	0.95	97.0–100.0	[120]
MNGO	Urine	Pseudoephedrine	MSPE-HPLC-UV	Methanol	25.0	95.1–97.7	[48]
Fe_3O_4@ZrGly	Urine	Ribonucleosides	MdSPE-HPLC-UV	TFA (0.1 wt.%)	1.7–19.0	90.6–115.0	[121]
Fe_3O_4@TCVA	Urine	NSAIDs	MSPE-HPLC-DAD	Acetonitrile/phosphoric acid (98.5/1.5, v/v)	11.0–39.0	83.1–120.0	[122]
MMWCNT-poly (STY-DVB)	Deer urine and hair	Catecholamines	MSPELC-MS	HFBA in methanol (0.01% v/v)	55.0–93.0	79.1–105.6	[123]

Tol-MCNT	Milk	Sulfonamides	MSPE-LC-HRMS	Acetonitrile	2.0−10.0	81.3−108.9	[124]
Fe_3O_4@[DABCO-DHP][Cl]	Milk	Cephalosporins	D-μ-SPE-UPLC-MS/MS	Methanol containing 1% (v/v) acetic acid	0.02−1.18	50.3−92.9	[125]
3D Net-like MCN	Milk	Sulfonamides	MSPE-UPLC-HRMS	Acetone containing 0.5% (v/v) ammonia	7.2−13.6	78.7−106.9	[126]

C-NF, Carbon nanofibers; *COP-IP-MNSs*, codeine phosphate−imprinted polymer magnetic nanospheres; *DABCO-DHP-Cl*, 1-(2,3-dihydroxypropyl)-1,4-diazabicyclo [2.2.2] octanylium chloride; *D-μ-SPE*, dispersive solid-phase microextraction; *HAP*, hydroxyapatite; *HFBA*, heptafluorobutyric acid; *HRMS*, high-resolution mass spectrometry; *MALDI-TOF-MS*, matrix-assisted laser desorption ionization-time of flight-mass spectrometry; *MCN*, mesoporous carbon nanocomposite; *MCNTs*, magnetic carbon nanotubes; *MdSPE*, magnetic dispersive solid-phase extraction; *MHIMs*, magnetic organic−inorganic hybrid composite molecularly imprinted microspheres; *MMWCNT-poly(STY-DVB)*, multiwalled carbon nanotube-poly(styrene-*co*-divinylbenzene); *MNGO*, magnetic nanographene oxide; *MSPE*, magnetic solid-phase extraction; *MUA-DμSPE*, magnetic ultrasound-assisted dispersive micro-solid-phase extraction; *NSAIDs*, nonsteroidal antiinflammatory drugs; *PCBs*, polychlorinated biphenyls; *SM-dSPE*, selective magnetic dispersive solid-phase extraction; *SPE-FL*, solid-phase extraction-spectrofluorometric; *TFA*, trifluoroacetic acid; *TOL*, *p*-tolyl isocyanate; *TVCA*, triethyl-(4-vinylbenzyl)aminium chloride and 4-vinylbenzeneboronic acid; *UA-DSPME*, ultrasonic-assisted dispersive solid-phase microextraction; *UAMSPE*, ultrasonic-assisted magnetic solid-phase extraction; *ZrGly*, zirconium glycerolate.

6.3.3 Analysis of other biological samples

Some conventional samples such as whole blood, serum, plasma, and urine are typically used as a biological medium that the target analytes can be extracted or separated from them by MNMs. However, some other papers reported the application of MNMs in different biological samples. These samples, such as hair, milk, and saliva, have their benefits for extracting the analytes for bioanalysis. Several analytes are recognized in these samples with a higher concentration than other biological samples. Besides, sometimes these mentioned samples are less invasive and more available [127]. Also, various kinds of analytes, including drugs, proteins, peptides [128], and hormones [129], are observed in this unconventional media. For instance, for analyzing the chronic drugs, the information that comes from analyzing the drugs in hair would be complementary to those obtained from whole blood [130]. Thus, in 2015, Boojaria et al. synthesized a novel and sensitive silane surface–modified MNMs for preconcentration of trace amount of morphine in hair samples by MSPE technique [25]. As it was mentioned, morphine is a usual therapeutic drug to control the chronic pains and abused as an illicit drug. Hair samples were washed by 20 mL dichloromethane, 20 mL acetone, and 15 mL methanol at the room temperature, and then dried. The prepared hair samples were cut into approximately 1 mm in the next step. Finally, 2.0 mL of menthol was added to the samples as a digestion step, and after 5 hours, the residual solid was filtered and rinsed with 0.5 mL ethanol. The target solution was diluted by 100 mL of deionized water. Silane-modified MNPs have a high surface area and they show desirable sensitivity, high enrichment factor, and large adsorption capacity, in the short extraction duration. Therefore this method was suitable for the fast extraction and determination of morphine in hair samples [25]. Another significant biological sample is a saliva sample that can contain an array of bacteria, cellular debris, salts, endogenous peptides, and proteins [131]. In 2020 Chu et al. provided hydrophilic citric acid (CA)–functionalized magnetic ferroferric oxide (Fe_3O_4) through a facile and straightforward one-pot solvothermal reaction. These NMs were used for the extraction of glycopeptides (glycosylation) from saliva samples [132]. Glycosylation is one of the most critical posttranslational modifications among other protein modifications. Human saliva was prepared through dispersing 4 μL of saliva in 16 μL of ammonium bicarbonate (25 mM) in the first step and then boiling the mixture for 10 minutes. The obtained solution was reduced by dithiothreitol for 30 minutes at 50°C and then alkylated by iodoacetamide in the dark situation for 1 hour at 37°C. Finally, 71 μL of a mixture consisting of ammonium bicarbonate and trypsin was added to the solution. After preparing the real sample, by adding proposed Fe_3O_4-CA NMs to the sample, glycopeptides were separated from nonglycopeptides. After elution, they were injected to the analyzing device. This method showed equal identification efficiency in comparison with previous hydrophilic materials with complex fabrication procedures.

As was mentioned, polymers are the most useful materials in improving the function of MNMs. One of these polymers that have been used in many studies is PDA. This substance is important because using it along with the MNMs can be so helpful in the extraction of target analytes from complicated biological samples such as milk. PDA has a high interaction capacity with organic components because it can establish $\pi-\pi$ interaction and hydrogen bonds [133]. Milk as a biological sample, which can be obtained from human or animals is a vital sample for the extraction and determination of compounds with estrogenic activities [129]. In this way, PDA-coated MNPs were synthesized by Socas-Rodrigues et al. for analyzing 21 estrogenic analytes (4 natural and 4 synthetic estrogens, 6 mycotoxins, and 7 phytoestrogens) in cow, goat, sheep, and human milk samples [62]. It is worth to be mentioned that milk samples (especially human breast milk) are suitable matrices for obtaining estrogens since the content of natural estrogens in this biological media is high. This phenomenon happens due to the augmentation of these hormone levels during the pregnancy and postbirth periods [134]. Therefore, to the preparation of milk samples, 1.5 mL of milk was filtered by polypropylene centrifuge tubes, and the internal standard (166 ng mL^{-1}) was added to each sample. Then, the sample's proteins were precipitated by adding 3 mL of acetonitrile and 75 μL of acetic acid. Then this solution was vortex-shaken at 2800 rpm for 1 minute. Finally, the mixture was stored in the darkness for 15 minutes and then was centrifuged at $3000 \times g$ for 15 minutes again. The supernatant was moved to an Erlenmeyer flask, and the remaining organic layer was evaporated at 40°C. Besides, 23.5 mL of Milli-Q water was added to the final solution. In this method, the amounts of used sorbents and the organic solvent were low, while the method was swift and simple. However, this technique could not recognize and determine glycitein and geniste in human breast milk. The applications of some of other MNMs in the other samples (such as hair) along with their figures of merits were collected in Table 6.2 [95,100,124–126]. In an interesting study, by a simple modification on the functionalization of amino-terminated CNTs, 13 various sulfonamides were extracted from whole milk with high accuracy and precision [124]. Also in another eco-friendly work, nine antibiotics were extracted from milk just by adding a trace amount of novel 1,4-diazobicyclo[2.2.2]octane (DABCO)–based IL-functionalized MNMs [126]. Therefore lower organic solvent was consumed and as a result, this method was less dangerous for environment. Accordingly, using MNMs in SPE of different analytes from various complicated samples seems so feasible.

6.4 Conclusions

Since the pretreatment/sample preparation procedure plays a significant role in the field of analysis in biological matrices (mostly urine, plasma, and serum), a simple, fast, and reliable method should be developed to overcome challenges in these complex matrices. As it possesses attractive properties such as small size, high surface-to-volume ratio, high magnetic susceptibility, and sufficient ability to bind, MNPs have recently received a lot of attention in the sample preparation and specially MSPE. In this procedure, a magnetic sorbent is added to the solution or suspension, which contains the target analytes under stirring condition. The desired analytes are adsorbed on the magnetic sorbent. Then, the separation of these NMs is performed easily with the application of an external magnetic field to the solution. In the following, the analytes are eluted and analyzed with a proper instrument. Therefore there is the advantage that the sorbent no longer needs to be filled and packed into the SPE cartridge like the traditional SPE. Furthermore, using the MSPE procedure minimizes the application of extra steps (precipitation, centrifugation, and filtration), which results in the reduction of sample manipulation.

By reviewing the related articles in this chapter, it is concluded that pretreatment with SPE using MNPs is a low-cost method in which the consumption of the extraction solvent is reduced compared to classical techniques, and it is, therefore, an eco-friendly method. Also, in all cases, this method has been proposed as a simple, sensitive, and fast method that provides good recoveries and preconcentration factors at room temperature. According to the studied articles, MNPs have been used for the separation and analysis of simple and complex analytes. Different drugs and inorganic analytes have been analyzed in biological samples directly by UV-Vis spectrometry, fluorimetry (FL), and MS applications or MSPE-analytical method combinations (-HPLC/UV, diode array detector, FL, MS, -GC/FID, MS, -capillary electrophoresis/UV, FL). In addition, one of the most essential recognized benefits of the MSPE method is the reusability of the MNPs.

References

[1] S. Dutta, Bioanalysis: an extensive application in pharmaceutical industries, World J. Pharmaceut. Med. Res. 2 (2016) 45–52.

[2] S.H. Hansen, S. Pedersen-Bjergaard (Eds.), Bioanalysis of Pharmaceuticals: Sample Preparation, Separation Techniques and Mass Spectrometry, John Wiley & Son, 2015.

[3] J.A. Ocaña-González, R. Fernández-Torres, M.Á. Bello-López, M. Ramos-Payán, New developments in microextraction techniques in bioanalysis. A review, Anal. Chim. Acta 905 (2016) 8–23.

[4] J. Huang, H. Deng, D. Song, H. Xu, Electrospun polystyrene/graphene nanofiber film as a novel adsorbent of thin film microextraction for extraction of aldehydes in human exhaled breath condensates, Anal. Chim. Acta 878 (2015) 102–108.

[5] H. Elmongy, M. Abdel-Rehim, Saliva as an alternative specimen to plasma for drug bioanalysis: a review, TrAC-Trends Anal. Chem. 83 (2016) 70–79.

[6] G.O. Silveira, A.M.F. Pego, J.P. E-Silva, M. Yonamine, Green sample preparations for the bioanalysis of drugs of abuse in complex matrices, Bioanalysis 11 (2019). 295–312.

[7] H.C. Atkinson, E.J. Begg, B.A. Darlow, Drugs in human milk, Clin. Pharmacokinet. 14 (1988) 217–240.

[8] G.N. Reddy, A.D. Zagade, P. Sengupta, Current direction and advances in analytical sample extraction techniques for drugs with special emphasis on bioanalysis, Bioanalysis 11 (2019) 313–332.

[9] Q.A. Xu, T.L. Madden, LC-MS in Drug Bioanalysis, Springer, 2012.

[10] G. Lasarte-Aragonés, R. Lucena, S. Cárdenas, M. Valcárcel, Nanoparticle-based microextraction techniques in bioanalysis, Bioanalysis 3 (2011) 2533–2548.

[11] M.M. Moein, A. Abdel-Rehim, M. Abdel-Rehim, Nanomaterials for Microextraction Techniques in Bioanalysis: Handbook of Nanomaterials in Analytical Chemistry, Elsevier, 2019, pp. 43–56.

[12] P. Biehl, M. Lühe, S. Dutz, F.H. Schacher, Synthesis, characterization, and applications of magnetic nanoparticles featuring polyzwitterionic coatings, Polymers 10 (2018) 91.
[13] G. Huang, C.H. Lu, H.H. Yang, Magnetic nanomaterials for magnetic bioanalysis, in: X. Wang, X. Chen (Eds.), Novel Nanomaterials for Biomedical, Environmental and Energy Applications, Elsevier, 2019, pp. 89–109.
[14] M.H. El-Dakdouki, K. El-Boubbou, J. Xia, H. Kavunja, X. Huang, Methods for magnetic nanoparticle synthesis and functionalization, in: R. Narain (Ed.), Chemistry of Bioconjugates: Synthesis, Characterization, and Biomedical Applications, Wiley, 2014, pp. 281–314.
[15] C.E.D. Nazario, B.H. Fumes, M.R. Silva, F.M. Lanças, New materials for sample preparation techniques in bioanalysis, J. Chromatogr. B 1043 (2017) 81–95.
[16] G. Giakisikli, A.N. Anthemidis, Magnetic materials as sorbents for metal/metalloid preconcentration and/or separation. A review, Anal. Chim. Acta 789 (2013) 1–16.
[17] S. Natarajan, K. Harini, G.P. Gajula, B. Sarmento, M.T. Neves-Petersen, V. Thiagarajan, Multifunctional magnetic iron oxide nanoparticles: diverse synthetic approaches, surface modifications, cytotoxicity towards biomedical and industrial applications, Biol. Biomed. Mater. 1 (2019) 2.
[18] Y. Lu, Y. Yin, B.T. Mayers, Y. Xia, Modifying the surface properties of superparamagnetic iron oxide nanoparticles through a sol-gel approach, Nano Lett. 2 (3) (2002) 183–186.
[19] J.H. Wu, X.S. Li, Y. Zhao, Q. Gao, L. Guo, Y.Q. Feng, Titania coated magnetic mesoporous hollow silica microspheres: fabrication and application to selective enrichment of phosphopeptides, Chem. Commun. 46 (2010) 9031–9033.
[20] Q. Ma, Y. Nakane, Y. Mori, M. Hasegawa, Y. Yoshioka, T.M. Watanabe, et al., Multilayered, core/shell nanoprobes based on magnetic ferric oxide particles and quantum dots for multimodality imaging of breast cancer tumors, Biomaterials 33 (2012) 8486–8494.
[21] M. Hemmati, M. Rajabi, A. Asghari, Magnetic nanoparticle based solid-phase extraction of heavy metal ions: a review on recent advances, Microchim. Acta 185 (2018) 160.
[22] S. Ma, M. He, B. Chen, W. Deng, Q. Zheng, B. Hu, Magnetic solid phase extraction coupled with inductively coupled plasma mass spectrometry for the speciation of mercury in environmental water and human hair samples, Talanta 146 (2016) 93–99.
[23] W. Jiahong, Z. Shourong, S. Yun, L. Jingliang, X. Zhaoyi, Z. Dongqiang, Amino-functionalized $Fe_3O_4@SiO_2$ core-shell magnetic nanomaterial as a novel adsorbent for aqueous heavy metals removal, J. Colloid Interface Sci. 349 (2010) 293–299.
[24] Z. Jiaqi, D. Yimin, L. Danyang, W. Shengyun, Z. Liling, Z. Yi, Synthesis of carboxyl-functionalized magnetic nanoparticle for the removal of methylene blue, Colloid Surf. A 572 (2019) 58–66.
[25] A. Boojaria, M. Masrournia, H. Ghorbani, A. Ebrahimitalab, M. Miandarhoie, Silane modified magnetic nanoparticles as a novel adsorbent for determination of morphine at trace levels in human hair samples by high-performance liquid chromatography with diode array detection, Forensic Sci. Med. Pat. 11 (2015) 497–503.
[26] Z. Qiao, R. Perestrelo, E.M. Reyes-Gallardo, R. Lucena, S. Cárdenas, J. Rodrigues, et al., Octadecyl functionalized core-shell magnetic silica nanoparticle as a powerful nanocomposite sorbent to extract urinary volatile organic metabolites, J. Chromatogr. A 1393 (2015) 18–25.
[27] B. Hu, M. He, B. Chen, Magnetic nanoparticle sorbents, Solid-Phase Extraction: Handbooks in Separation Science, Elsevier, 2020, pp. 235–284.
[28] K.C. Hsu, P.F. Hsu, C.C. Hung, C.H. Chiang, S.J. Jiang, C.C. Lin, et al., Microfluidic desorption-free magnetic solid phase extraction of Hg^{2+} from biological samples

using cysteine-coated gold-magnetite core-shell nanoparticles prior to its quantitation by ICP-MS, Talanta 162 (2017) 523–529.

[29] Y.W. Wu, J. Zhang, J.F. Liu, L. Chen, Z.L. Deng, M.X. Han, et al., Fe_3O_4@ZrO_2 nanoparticles magnetic solid phase extraction coupled with flame atomic absorption spectrometry for chromium(III) speciation in environmental and biological samples, Appl. Surf. Sci. 258 (2012) 6772–6776.

[30] M.A. Karimi, A. Hatefi-Mehrjardi, S.Z. Mohammadi, A. Mohadesi, M. Mazloum-Ardakani, A.A. Kabir, et al., Solid phase extraction of trace amounts of silver (I) using dithizone-immobilized alumina-coated magnetite nanoparticles prior to determination by flame atomic absorption spectrometry, Int. J. Environ. Anal. Chem. 92 (2012) 1325–1340.

[31] S. Tang, Y. Yao, T. Chen, D. Kong, W. Shen, H.K. Lee, Recent advances in the application of layered double hydroxides in analytical chemistry: a review, Anal. Chim. Acta 1103 (2020) 32–48.

[32] L. Du, X. Wang, T. Liu, J. Li, J. Wang, M. Gao, et al., Magnetic solid-phase extraction of organophosphorus pesticides from fruit juices using $NiFe_2O_4$@polydopamine@Mg/Al-layered double hydroxides nanocomposites as an adsorbent, Microchem. J. 150 (2019) 104128.

[33] J. Li, Q. Fan, Y. Wu, X. Wang, C. Chen, Z. Tang, et al., Magnetic polydopamine decorated with Mg–Al LDH nanoflakes as a novel bio-based adsorbent for simultaneous removal of potentially toxic metals and anionic dyes, J. Mater. Chem. A 4 (2016) 1737–1746.

[34] Q. Zhou, M. Lei, J. Li, K. Zhao, Y. Liu, Determination of 1-naphthol and 2-naphthol from environmental waters by magnetic solid phase extraction with Fe@ MgAl-layered double hydroxides nanoparticles as the adsorbents prior to high performance liquid chromatography, J. Chromatogr. A 1441 (2016) 1–7.

[35] D. Huang, C. Deng, X. Zhang, Functionalized magnetic nanomaterials as solid-phase extraction adsorbents for organic pollutants in environmental analysis, Anal. Methods 6 (2014) 7130–7141.

[36] J. Lee, J. Kim, T. Hyeon, Recent progress in the synthesis of porous carbon materials, Adv. Mater. 18 (2006) 2073–2094.

[37] A.H. Lu, W.C. Li, N. Matoussevitch, B. Spliethoff, H. Bönnemann, F. Schüth, Highly stable carbon-protected cobalt nanoparticles and graphite shells, Chem. Commun. 1 (2005) 98–100.

[38] A. Taghvimi, H. Hamishehkar, Carbon coated magnetic nanoparticles as a novel magnetic solid phase extraction adsorbent for simultaneous extraction of methamphetamine and ephedrine from urine samples, J. Chromatogr. B 1041 (2017) 113–119.

[39] C. Peng, M. He, B. Chen, L. Huang, B. Hu, Magnetic sulfur-doped porous carbon for preconcentration of trace mercury in environmental water prior to ICP-MS detection, Analyst 142 (2017) 4570–4579.

[40] T. Wen, J. Wang, S. Yu, Z. Chen, T. Hayat, X. Wang, Magnetic porous carbonaceous material produced from tea waste for efficient removal of As(V), Cr(VI), humic acid, and dyes, ACS Sustain. Chem. Eng. 5 (2017) 4371–4380.

[41] M. Yu, L. Wang, L. Hu, Y. Li, D. Luo, S. Mei, Recent applications of magnetic composites as extraction adsorbents for determination of environmental pollutants, TrAC-Trends Anal. Chem. 119 (2019) 115611.

[42] C. Herrero-Latorre, J. Barciela-García, S. García-Martín, R.M. Peña-Crecente, J. Otárola-Jiménez, Magnetic solid-phase extraction using carbon nanotubes as sorbents: a review, Anal. Chim. Acta 892 (2015) 10–26.

[43] J. Yang, Z. Zhang, W. Pang, H. Chen, G. Yan, Polyamidoamine dendrimers functionalized magnetic carbon nanotubes as an efficient adsorbent for the separation of flavonoids from plant extraction, Sep. Purif. Technol. 227 (2019) 115710.

[44] N. Li, D. Wu, X. Li, X. Zhou, G. Fan, G. Li, et al., Effective enrichment and detection of plant growth regulators in fruits and vegetables using a novel magnetic covalent organic framework material as the adsorbents, Food Chem. 306 (2020) 125455.

[45] S. Shen, J. Ren, J. Chen, X. Lu, C. Deng, X. Jiang, Development of magnetic multi-walled carbon nanotubes combined with near-infrared radiation-assisted desorption for the determination of tissue distribution of doxorubicin liposome injects in rats, J. Chromatogr. A 1218 (2011) 4619–4626.

[46] L. Fan, A. Zhou, L. Zhong, Z. Zhang, Y. Liu, Selective and effective adsorption of Hg (II) from aqueous solution over wide pH range by thiol functionalized magnetic carbon nanotubes, Chemosphere 226 (2019) 405–412.

[47] Y. Sha, D. Huang, S. Zheng, C. Deng, Development of magnetic graphene as an adsorbent and matrix for selective enrichment and detection of crotonaldehyde in saliva by MALDI-TOF-MS, Anal. Methods 5 (2013) 4585–4590.

[48] A. Taghvimi, H. Hamishehkar, M. Ebrahimi, Magnetic nano graphene oxide as solid phase extraction adsorbent coupled with liquid chromatography to determine pseudoephedrine in urine samples, J. Chromatogr. B. 1009–1010 (2016) 66–72.

[49] W. Li, J. Zhang, W. Zhu, P. Qin, Q. Zhou, M. Lu, et al., Facile preparation of reduced graphene oxide/$ZnFe_2O_4$ nanocomposite as magnetic sorbents for enrichment of estrogens, Talanta 208 (2020) 120440.

[50] G. Li, L. Xia, J. Dong, Y. Chen, Y. Li, Metal-organic frameworks, Solid-Phase Extraction: Handbooks in Separation Science, Elsevier, 2020, pp. 285–309.

[51] G. Zhao, N. Qin, A. Pan, X. Wu, C. Peng, F. Ke, et al., Magnetic nanoparticles@ metal-organic framework composites as sustainable environment adsorbents, J. Nanomater. 2019 (2019) 1454358.

[52] F. Ke, Y.P. Yuan, L.G. Qiu, Y.H. Shen, A.J. Xie, J.F. Zhu, et al., Facile fabrication of magnetic metal–organic framework nanocomposites for potential targeted drug delivery, J. Mater. Chem. 21 (2011) 3843–3848.

[53] T. Zhang, X. Zhang, X. Yan, L. Kong, G. Zhang, H. Liu, et al., Synthesis of Fe_3O_4@ZIF-8 magnetic core–shell microspheres and their potential application in a capillary microreactor, Chem. Eng. J. 228 (2013) 398–404.

[54] F. Ke, L.G. Qiu, Y.P. Yuan, X. Jiang, J.F. Zhu, Fe_3O_4@MOF core–shell magnetic microspheres with a designable metal–organic framework shell, J. Mater. Chem. 22 (2012) 9497–9500.

[55] S.H. Huo, X.P. Yan, Facile magnetization of metal–organic framework MIL-101 for magnetic solid-phase extraction of polycyclic aromatic hydrocarbons in environmental water samples, Analyst 137 (2012) 3445–3451.

[56] M. Saikia, D. Bhuyan, L. Saikia, Facile synthesis of Fe_3O_4 nanoparticles on metal organic framework MIL-101 (Cr): characterization and catalytic activity, N. J. Chem. 39 (2015) 64–67.

[57] S. Fan, W. Dong, X. Huang, H. Gao, J. Wang, Z. Jin, et al., In situ-induced synthesis of magnetic Cu-$CuFe_2O_4$@ HKUST-1 heterostructures with enhanced catalytic performance for selective aerobic benzylic C–H oxidation, ACS Catal. 7 (2017) 243–249.

[58] Z. Moradi, E.A. Dil, A. Asfaram, Dispersive micro-solid phase extraction based on Fe_3O_4@SiO_2@TI-MOF as a magnetic nanocomposite sorbent for the trace analysis of caffeic acid in the medical extracts of plants and water samples prior to HPLC-UV analysis, Analyst 144 (2019) 4351–4361.

[59] L. Xia, L. Liu, X. Lv, F. Qu, G. Li, J. You, Towards the determination of sulfonamides in meat samples: a magnetic and mesoporous metal-organic framework as an efficient sorbent for magnetic solid phase extraction combined with high-performance liquid chromatography, J. Chromatogr. A 1500 (2017) 24–31.

[60] S. He, T. Zeng, S. Wang, H. Niu, Y. Cai, Facile synthesis of magnetic covalent organic framework with three-dimensional bouquet-like structure for enhanced extraction of organic targets, ACS Appl. Mater. Interfaces 9 (2017) 2959–2965.

[61] Q. Kaili, T. Weijun, B. Jie, W. Liang, Z. Jing, D. Zhaoyang, et al., Application of magnetic adsorbents based on iron oxide nanoparticles for oil spill remediation: a review, J. Taiwan. Inst. Chem. E 97 (2019) 227–236.

[62] B. Socas-Rodríguez, J. Hernández-Borges, A.V. Herrera-Herrera, M.Á. Rodríguez-Delgado, Multiresidue analysis of oestrogenic compounds in cow, goat, sheep and human milk using core-shell polydopamine coated magnetic nanoparticles as extraction sorbent in micro-dispersive solid-phase extraction followed by ultra-high-performance liquid chromatography tandem mass spectrometry, Anal. Bioanal. Chem. 410 (2018) 2031–2042.

[63] A.A. Asgharinezhad, S. Karami, H. Ebrahimzadeh, N. Shekari, N. Jalilian, Polypyrrole/magnetic nanoparticles composite as an efficient sorbent for dispersive micro-solid-phase extraction of antidepressant drugs from biological fluids, Int. J. Pharm. 494 (2015) 102–112.

[64] E. Turiel, A.M. Esteban, Molecularly imprinted polymers, Solid-Phase Extraction: Handbooks in Separation Science, Elsevier, 2020, pp. 215–233.

[65] M. Peyrovi, M. Hadjmohammadi, I. Saeidi, Synthesis of magnetic nanoparticle-based molecularly imprinted polymer as a selective sorbent for efficient extraction of ezetimibe from biological samples, Biomed. Chromatogr. 33 (2019) e4404.

[66] A. Sorribes-Soriano, F.A. Esteve-Turrillas, S. Armenta, A. Montoya, J.M. Herrero-Martínez, M. de la Guardia, Magnetic molecularly imprinted polymers for the selective determination of cocaine by ion mobility spectrometry, J. Chromatogr. A 1545 (2018) 22–31.

[67] K. Hu, J. Qiao, X. Wu, H. Yang, Y. Huang, S. Zhang, Poly(calixarene ionic liquid) modified Fe_3O_4 nanoparticles as new sorbent for extraction of flavonoids in fruit juice and green tea, Microchem. J. 143 (2018) 39–46.

[68] S. Cao, J. Chen, G. Lai, C. Xi, X. Li, L. Zhang, et al., A high efficient adsorbent for plant growth regulators based on ionic liquid and β-cyclodextrin functionalized magnetic graphene oxide, Talanta 194 (2019) 14–25.

[69] S. Badragheh, M. Zeeb, M.R.T.B. Olyai, Silica-coated magnetic iron oxide functionalized with hydrophobic polymeric ionic liquid: a promising nanoscale sorbent for simultaneous extraction of antidiabetic drugs from human plasma prior to their quantitation by HPLC, RSC Adv. 8 (2018) 30550–30561.

[70] S. Hayashi, H.-o Hamaguchi, Discovery of a magnetic ionic liquid [bmim]$FeCl_4$, Chem. Lett. 33 (2004) 1590–1591.

[71] K.D. Clark, O. Nacham, H. Yu, T. Li, M.M. Yamsek, D.R. Ronning, et al., Extraction of DNA by magnetic ionic liquids: tunable solvents for rapid and selective DNA analysis, Anal. Chem. 87 (2015) 1552–1559.

[72] M. Ahmadi, H. Elmongy, T. Madrakian, M. Abdel-Rehim, Nanomaterials as sorbents for sample preparation in bioanalysis: a review, Anal. Chim. Acta 958 (2017) 1–21.

[73] M. Faraji, Recent analytical applications of magnetic nanoparticles, Nanochem. Res. 1 (2016) 264–290.

[74] B. Ebrahimpour, Y. Yamini, S. Seidi, M. Tajik, Nano polypyrrole-coated magnetic solid phase extraction followed by dispersive liquid phase microextraction for trace determination of megestrol acetate and levonorgestrel, Anal. Chim. Acta 885 (2015) 98–105.

[75] C.-G. Héden, The potential of magnetic separation of biologically active proteins, Biotechnol. Bioeng. Symp. 3 (1972) 173–174.

[76] P.J. Robinson, P. Dunnill, M.D. Lilly, The properties of magnetic supports in relation to immobilized enzyme reactors, Biotechnol. Bioeng. 15 (1973) 603–606.

[77] M. Wierucka, M. Biziuk, Application of magnetic nanoparticles for magnetic solid-phase extraction in preparing biological, environmental and food samples, TrAC-Trends Anal. Chem. 59 (2014) 50–58.
[78] M. Faraji, Y. Yamini, M. Rezaee, Magnetic nanoparticles: synthesis, stabilization, functionalization, characterization, and applications, J. Iran. Chem. Soc. 7 (2010) 1–37.
[79] L. Xie, R. Jiang, F. Zhu, H. Liu, G. Ouyang, Application of functionalized magnetic nanoparticles in sample preparation, Anal. Bioanal. Chem. 406 (2014) 377–399.
[80] D. Yuvali, I. Narin, M. Soylak, E. Yilmaz, Green synthesis of magnetic carbon nanodot/graphene oxide hybrid material (Fe_3O_4@C-nanodot@GO) for magnetic solid phase extraction of ibuprofen in human blood samples prior to HPLC-DAD determination, J. Pharm. Biomed. Anal. 179 (2020) 1–10.
[81] Z. Lotfi, H.Z. Mousavi, S.M. Sajjadi, Amino-terminated hyper-branched polyamidoamine polymer grafted magnetic graphene oxide nanosheets as an efficient sorbent for the extraction of selective serotonin reuptake inhibitors from plasma samples, Anal. Methods 9 (2017) 4504–4513.
[82] D. Li, Z. Bie, F. Wang, E. Guo, Efficient synthesis of riboflavin-imprinted magnetic nanoparticles by boronate affinity-based surface imprinting for the selective recognition of riboflavin, Analyst 143 (2018) 4936–4943.
[83] A. Barati, E. Kazemi, S. Dadfarnia, A.M. Haji Shabani, Synthesis/characterization of molecular imprinted polymer based on magnetic chitosan/graphene oxide for selective separation/pre-concentration of fluoxetine from environmental and biological samples, J. Ind. Eng. Chem. 46 (2017) 212–221.
[84] T. Madrakian, R. Haryani, M. Ahmadi, A. Afkhami, Spectrofluorometric determination of venlafaxine in biological samples after selective extraction on the superparamagnetic surface molecularly imprinted nanoparticles, Anal. Methods 7 (2015) 428–435.
[85] W. Huang, X. Hou, Y. Tong, M. Tian, Determination of sialic acid in serum samples by dispersive solid-phase extraction based on boronate-affinity magnetic hollow molecularly imprinted polymer sorbent, RSC Adv. 9 (2019) 5394–5401.
[86] J. Wang, J. Qiu, A review of carbon dots in biological applications, J. Mater. Sci. 51 (2016) 4728–4738.
[87] J. Chen, Y. Wang, Y. Huang, K. Xu, N. Li, Q. Wen, et al., Magnetic multiwall carbon nanotubes modified with dual hydroxy functional ionic liquid for the solid-phase extraction of protein, Analyst 140 (2015) 3474–3483.
[88] Q. Liu, J. Shi, M. Cheng, G. Li, D. Cao, G. Jiang, Preparation of graphene-encapsulated magnetic microspheres for protein/peptide enrichment and MALDI-TOF MS analysis, Chem. Commun. 48 (2012) 1874–1876.
[89] A. Spietelun, Ł. Marcinkowski, M. de la Guardia, J. Namieśnik, Recent developments and future trends in solid phase microextraction techniques towards green analytical chemistry, J. Chromatogr. A 1321 (2013) 1–13.
[90] Y.Da Zhang, Q.W. Huang, C. Ma, X.Y. Liu, H.X. Zhang, Magnetic fluorescent molecularly imprinted nanoparticles for detection and separation of transferrin in human serum, Talanta 188 (2018) 540–545.
[91] L. Qi, R. Xu, J. Gong, Monitoring DNA adducts in human blood samples using magnetic Fe_3O_4@graphene oxide as a nano-adsorbent and mass spectrometry, Talanta 209 (2020) 120523.
[92] H. Heidari, B. Limouei-Khosrowshahi, Magnetic solid phase extraction with carbon-coated Fe_3O_4 nanoparticles coupled to HPLC-UV for the simultaneous determination of losartan, carvedilol, and amlodipine besylate in plasma samples, J. Chromatogr. B 1114–1115 (2019) 24–30.

[93] S. Ansari, S. Masoum, A multi-walled carbon nanotube-based magnetic molecularly imprinted polymer as a highly selective sorbent for ultrasonic-assisted dispersive solid-phase microextraction of sotalol in biological fluids, Analyst 143 (2018) 2862–2875.

[94] K. Xu, Y. Wang, X. Ding, Y. Huang, N. Li, Q. Wen, Magnetic solid-phase extraction of protein with deep eutectic solvent immobilized magnetic graphene oxide nanoparticles, Talanta 148 (2016) 153–162.

[95] L. Cao, Y. Zhao, Z. Chu, X. Zhang, W. Zhang, Core-shell magnetic bimetallic MOF material for synergistic enrichment of phosphopeptides, Talanta 206 (2020) 120165.

[96] J. Zhu, Y. Li, Y. Xiang, L. Zhou, Y. Li, Magnetic solid phase extraction followed with LC-MS/MS for determination of glimepiride in beagle dog plasma and its application to bioequivalence study, J. Pharm. Biomed. Anal. 184 (2020) 113180.

[97] Q.L. Li, L.L. Wang, X. Wang, M.L. Wang, R.S. Zhao, Magnetic metal-organic nanotubes: an adsorbent for magnetic solid-phase extraction of polychlorinated biphenyls from environmental and biological samples, J. Chromatogr. A 1449 (2016) 39–47.

[98] K. Cherkashina, M. Voznesenskiy, O. Osmolovskaya, C. Vakh, A. Bulatov, Effect of surfactant coating of Fe_3O_4 nanoparticles on magnetic dispersive micro-solid phase extraction of tetracyclines from human serum, Talanta 214 (2020) 120861.

[99] D.D. Zhou, Q. Zhang, H. Zhang, Y.Z. Wang, F.Q. Yang, S.P. Wang, et al., Cupric ion functionalized polydopamine coated magnetic microspheres as solid-phase adsorbent for the extraction of purines in plasma, J. Chromatogr. B 1120 (2019) 95–103.

[100] M. Ghazaghi, H.Z. Mousavi, H. Shirkhanloo, A. Rashidi, Ultrasound assisted dispersive micro solid-phase extraction of four tyrosine kinase inhibitors from serum and cerebrospinal fluid by using magnetic nanoparticles coated with nickel-doped silica as an adsorbent, Microchim. Acta 183 (2016) 2779–2789.

[101] W. Yantasee, K. Hongsirikarn, C.L. Warner, D. Choi, T. Sangvanich, M.B. Toloczko, et al., Direct detection of Pb in urine and Cd, Pb, Cu, and Ag in natural waters using electrochemical sensors immobilized with DMSA functionalized magnetic nanoparticles, Analyst 133 (2008) 348–355.

[102] J. Sun, Q. Liang, Q. Han, X. Zhang, M. Ding, One-step synthesis of magnetic graphene oxide nanocomposite and its application in magnetic solid phase extraction of heavy metal ions from biological samples, Talanta 132 (2015) 557–563.

[103] M.M. Churley, T.P. Lyons, P.V. Robandt, M.R. Bruins, Evaluation of a solid-phase extraction method for codeine and morphine in urine using Cerex Polycrom™ Clin II columns and the Speedisk™ 48 pressure processor at a high-throughput forensic drug-testing laboratory, J. Anal. Toxicol. 27 (2003) 530–532.

[104] K.U. Alwis, B.C. Blount, A.S. Britt, D. Patel, D.L. Ashley, Simultaneous analysis of 28 urinary VOC metabolites using ultra high performance liquid chromatography coupled with electrospray ionization tandem mass spectrometry (UPLC-ESI/MSMS), Anal. Chim. Acta 750 (2012) 152–160.

[105] T. Guan, Y. Kuang, X. Li, J. Fang, W. Fang, D. Wu, The recovery of phosphorus from source-separated urine by repeatedly usable magnetic $Fe_3O_4@ZrO_2$ nanoparticles under acidic conditions, Environ. Int. 134 (2020) 105322.

[106] M. Cruz-Vera, R. Lucena, S. Cárdenas, M. Valcárcel, Sorptive microextraction for liquid-chromatographic determination of drugs in urine, TrAC-Trends Anal. Chem. 28 (2009) 1164–1173.

[107] Z. Li, L. Wei, M. Gao, H. Lei, One-pot reaction to synthesize biocompatible magnetite nanoparticles, Adv. Mater. 17 (2005) 1001–1005.

[108] A. Amiri, M. Baghayeri, S. Nori, Magnetic solid-phase extraction using poly(para-phenylenediamine) modified with magnetic nanoparticles as adsorbent for analysis of monocyclic aromatic amines in water and urine samples, J. Chromatogr. A 1415 (2015) 20–26.
[109] W. Guo, W. Hu, J. Pan, H. Zhou, W. Guan, X. Wang, et al., Selective adsorption and separation of BPA from aqueous solution using novel molecularly imprinted polymers based on kaolinite/Fe_3O_4 composites, Chem. Eng. J. 171 (2011) 603–611.
[110] I. Dakova, I. Karadjova, V. Georgieva, G. Georgiev, Ion-imprinted polymethacrylic microbeads as new sorbent for pre-concentration and speciation of mercury, Talanta 78 (2009) 523–529.
[111] N. Khoddami, F. Shemirani, A new magnetic ion-imprinted polymer as a highly selective sorbent for determination of cobalt in biological and environmental samples, Talanta 146 (2016) 244–252.
[112] P. Chutia, S. Kato, T. Kojima, S. Satokawa, Adsorption of As(V) on surfactant-modified natural zeolites, J. Hazard. Mater. 162 (2009) 204–211.
[113] P. Baile, L. Vidal, A. Canals, A modified zeolite/iron oxide composite as a sorbent for magnetic dispersive solid-phase extraction for the pre-concentration of non-steroidal anti-inflammatory drugs in water and urine samples, J. Chromatogr. A 1603 (2019) 33–43.
[114] A. Jouyban, S. Hamidi, Dispersive micro-solid-phase extraction using carbon-based adsorbents for the sensitive determination of verapamil in plasma samples coupled with capillary electrophoresis, J. Sep. Sci. 40 (2017) 3318–3326.
[115] A. Taghvimi, S. Hamidi, Y. Javadzadeh, Mixed hemimicelle magnetic dispersive solid-phase extraction using carbon-coated magnetic nanoparticles for the determination of tramadol in urine samples, J. Sep. Sci. 42 (2019) 582–590.
[116] E. Ghasemi, M. Sillanpaa, Ultrasound-assisted solid-phase extraction of parabens from environmental and biological samples using magnetic hydroxyapatite nanoparticles as an efficient and regenerable nanosorbent, Microchim. Acta 186 (2019) 622.
[117] Q. Lu, H. Guo, Y. Zhang, X. Tang, W. Lei, R. Qi, et al., Graphene oxide-Fe_3O_4 nanocomposite magnetic solid phase extraction followed by UHPLC-MS/MS for highly sensitive determination of eight psychoactive drugs in urine samples, Talanta 206 (2020) 120212.
[118] F. Qiao, M. Wang, Preparation of selective magnetic dispersive solid-phase sorbent and its application for recognition clenbuterol from bovine urine, J. Chromatogr. B 1017–1018 (2016) 18–27.
[119] T. Madrakian, F. Fazl, M. Ahmadi, A. Afkhami, Efficient solid phase extraction of codeine from human urine samples using a novel magnetic molecularly imprinted nanoadsorbent and its spectrofluorometric determination, N. J. Chem. 40 (2016) 122–129.
[120] E. Yilmaz, S. Salem, G. Sarp, S. Aydin, K. Sahin, I. Korkmaz, et al., TiO_2 nanoparticles and C-nanofibers modified magnetic Fe_3O_4 nanospheres (TiO_2@Fe_3O_4@C-NF): a multifunctional hybrid material for magnetic solid-phase extraction of ibuprofen and photocatalytic degradation of drug molecules and azo dye, Talanta 213 (2020) 120813.
[121] J. Xu, Z. Zhang, X.M. He, R.Q. Wang, D. Hussain, Y.Q. Feng, Immobilization of zirconium-glycerolate nanowires on magnetic nanoparticles for extraction of urinary ribonucleosides, Microchim. Acta 185 (2018) 1–8.
[122] Y. Li, Y. Liao, Y. Huang, Z. Ye, X. Huang, Dual functional monomers modified magnetic adsorbent for the enrichment of non-steroidal anti-inflammatory drugs in water and urine samples, Talanta 201 (2019) 496–502.

[123] K. Murtada, F. de Andrés, I. Galván, Á. Ríos, M. Zougagh, LC-MS determination of catecholamines and related metabolites in red deer urine and hair extracted using magnetic multi-walled carbon nanotube poly(styrene-co-divinylbenzene) composite, J. Chromatogr. B 1136 (2020) 121878.
[124] L. Fu, H. Zhou, E. Miao, S. Lu, S. Jing, Y. Hu, et al., Functionalization of amino terminated carbon nanotubes with isocyanates for magnetic solid phase extraction of sulfonamides from milk and their subsequent determination by liquid chromatography-high resolution mass spectrometry, Food Chem. 289 (2019) 701–707.
[125] H. Sahebi, E. Konoz, A. Ezabadi, Synthesis of DABCO-based ionic liquid functionalized magnetic nanoparticles as a novel sorbent for the determination of cephalosporins in milk samples by dispersive solid-phase extraction followed by ultra-performance liquid chromatography-tandem mass spectrometry, N. J. Chem. 43 (2019) 13554–13570.
[126] S. Di, J. Yu, P. Chen, G. Zhu, S. Zhu, Net-like mesoporous carbon nanocomposites for magnetic solid-phase extraction of sulfonamides prior to their quantitation by UPLC-HRMS, Microchim. Acta 187 (2020) 112.
[127] S. Hu, M. Arellano, P. Boontheung, J. Wang, H. Zhou, J. Jiang, et al., Salivary proteomics for oral cancer biomarker discovery, Clin. Cancer Res. 14 (2008) 6246–6252.
[128] S.A. Lim, D. Koh, M.U. Ahmed, Single wall carbon nanotube and magnetic bead based electrochemical immunosensor for sensitive detection of salivary secretory immunoglobulin A, Curr. Anal. Chem. 14 (2018) 399–405.
[129] B. Socas-Rodríguez, M. Asensio-Ramos, J. Hernández-Borges, A.V. Herrera-Herrera, M. Ángel Rodríguez-Delgado, Chromatographic analysis of natural and synthetic estrogens in milk and dairy products, TrAC-Trends Anal. Chem. 44 (2013) 58–77.
[130] S. Paterson, R. Cordero, E. Stearns, Chronic drug use confirmed by hair analysis: its role in understanding both the medical cause of death and the circumstances surrounding the death, J. Forensic Leg. Med. 16 (2009) 143–147.
[131] F. Amado, M.J.C. Lobo, P. Domingues, J.A. Duarte, R. Vitorino, Salivary peptidomics, Expert Rev. Proteomics 7 (2010) 709–721.
[132] H. Chu, X. Hu, J. Yao, G. Yan, N. Sun, C. Deng, One-pot preparation of hydrophilic citric acid-magnetic nanoparticles for identification of glycopeptides in human saliva, Talanta 206 (2020) 120178.
[133] Z. Huang, H.K. Lee, Study and comparison of polydopamine and its derived carbon decorated nanoparticles in the magnetic solid-phase extraction of estrogens, J. Chromatogr. A 1414 (2015) 41–50.
[134] P.N. Jouan, Y. Pouliot, S.F. Gauthier, J.P. Laforest, Hormones in bovine milk and milk products: a survey, Int. Dairy J. 16 (2006) 1408–1414.

Application of magnetic nanomaterials in environmental monitoring

7

Mohammad Faraji[1], Yadollah Yamini[2], Farshid Noormohammadi[3] and Mahnaz Adeli[4]

[1]Research Group of Food, Halal and Agricultural Products, Research Department of Food Technology and Agricultural Products, Standard Research Institute (SRI), Karaj, Iran, [2]Department of Chemistry, Faculty of Sciences, Tarbiat Modares University, Tehran, Iran, [3]Chemical Engineering Division, College of Environment, UoE, Karaj, Iran, [4]Knowledge Development & University Relationship Department, Iran Khodro Company, Tehran, Iran

7.1 Introduction

For many years, the pretreatment and analysis of chemical contaminants in environmental samples have been a significant concern for scientific researchers [1]. In recent years, novel, efficient, fast, simple, and especially environmentally friendly approaches for extraction and removal methods have received much attention. In this regard, magnetic nanomaterials (MNMs) have attracted significant attention in the scientific community due to their unique properties, such as a high surface-to-volume ratio and a short diffusion route, which lend themselves to high extraction efficiencies, fast extraction dynamics, and high extraction capacities in comparison with common micrometer-sized sorbents [1,2]. These types of sorbents are completely and efficiently dispersed in the sample solutions and are readily and rapidly separated from the sample by using a strong magnet, even in the case of large sample sizes. As a result, extraction techniques based on MNMs are time-saving, convenient, and economical techniques by avoiding centrifugation or filtration procedures for the separation of the sorbent.

Application of MNMs in the field of environmental analysis can be divided into two sections:

1. Application of MNMs as sorbents in the magnetic solid-phase extraction (MSPE) of trace levels of environmental pollutants ($\mu g\ L^{-1}$ or $ng\ L^{-1}$ concentration levels) such as heavy metals, pesticides, pharmaceutical, polycyclic aromatic hydrocarbons (PAHs), etc.
2. Application of MNMs as sorbents for removal of large amounts of environmental pollutants ($mg\ L^{-1}$ or higher concentration levels) such as dyes, heavy metals, oils, etc.

Magnetic Nanomaterials in Analytical Chemistry. DOI: https://doi.org/10.1016/B978-0-12-822131-0.00005-4

7.2 Magnetic nanomaterials as sorbents in MSPE

Today, a new approach to solid-phase extraction (SPE) and MSPE has attracted more attention in trace analysis. The first use of MSPE in analytical applications was proposed in 1999 by Šafaříková and Šafařík to separate copper-phthalocyanine dye using silanized magnetite [3]. Nowadays, various MNMs have been synthesized and used as efficient and favorable sorbents in MSPE techniques. In these techniques, MNMs are dispersed into the sample solution. After the extraction is completed, the sorbent is rapidly and easily collected by a magnet, and then the adsorbed analytes are eluted with a suitable solvent. Because MNMs have advantages such as high specific surface area, suitable porosity, short diffusion routes, and usually good dispersity and magnetism, the use of such sorbents results in higher extraction capacity and extraction efficiency and short equilibrium time compared with conventional SPE techniques. Also, as a result of the easy collection of the sorbents using external magnets, centrifugation and filtration process can be avoided. This advantage is most favorable for a large volume of environmental samples. In addition, in MNMs—based MSPE methods, the use of organic solvents is significantly reduced, and the range of sorbents with the appropriate variety and selectivity has increased [1,2,4—9].

7.3 Magnetic nanomaterials as sorbents in removal process

Industrialization leads to the release of high concentrations of organic pollutants (such as dyes) and inorganic pollutants (such as heavy metals) into the environment. Hence, removing these toxic chemicals has become environmentally important. Among different contaminant treatment techniques, adsorption is extensively applied to removing contaminants from environmental samples due to its advantages, such as the availability of a variety of adsorbents, simplicity of operational conditions, low cost, and the ability to use it on a large scale. A well-known and efficient sorbent in adsorption techniques is activated carbon. Despite the high efficiency of this sorbent, its distinct limitations are low reusability and high cost [6]. Recently, the application of MNMs as advanced sorbents has attracted much attention from researchers owing to the chemicals' high adsorption capacity, magnetic properties, fast removal kinetic, and low cost. Also, MNMs can be regenerated and reused for several cycles [1,2,4—9].

7.4 Emerging magnetic nanomaterials for magnetic solid phase extraction and removal purposes

To date, a variety of magnetic nanomaterials have been used as sorbents in MSPE and removal techniques [1,2,4—9]. In this regard, by adding magnetic properties to advanced materials such as graphene/graphene oxides, carbon nanotubes (CNTs), metal-organic frameworks (MOFs), layered double hydroxides (LDHs), covalent organic frameworks (COFs), and mesoporous materials, a variety of novel magnetic sorbents are prepared that are widely used in environmental analysis. In some recent studies, magnetic nanocomposites using two or more advanced materials have also been developed to take advantage of these advanced materials at the same time. In the following sections, some of the most widely used advanced materials and their important features will be reviewed.

7.4.1 Graphene/graphene oxides

Graphene (G) is a carbon-based material with a single-layer two-dimensional honeycomb lattice structure with many superior physical and chemical properties, such as large surface area, ready chemical functionalization, high electrical and thermal conductivity, and great mechanical strength [2,4,10–13]. Graphene oxide (GO) is an oxidized form of graphene. Fig. 7.1A presents a schematic illustration of the fabrication of the Fe_3O_4@SiO_2-G nanocomposite from graphite [13]. GO has more hydrophilic groups (oxygen-containing), including hydroxyl, carboxyl, epoxy, and carbonyl groups, on its surface compared to reduced GO or G. Besides, GO has a strong affinity to aromatic target analytes due to its huge aromatic domains, leading to present strong $\pi-\pi$ stacking, van der Waals, and other noncovalent interactions [10]. GO magnetizing leads to extended applications of GO-based sorbents and their easy and effective separation [11–13].

Zhang et al. prepared a magnetic graphene oxide (MGO) composite for MSPE of three phenolic compounds (phenol, 4-nitrophenol, and *m*-methyl phenol) in environmental water samples. Owing to the high surface area and excellent adsorption capacity of the prepared material, satisfactory extraction with the limit of detection in the range of 0.5 and 0.8 $\mu g\ L^{-1}$ was achieved [12].

Huang et al. investigated the efficiency of different magnetic graphene-based sorbent including magnetic annealing-reduced graphene, magnetic chemically reduced graphene, and MGO for adsorption of arsenate (As(V)), cadmium (Cd(II)), and tetracycline (TC) in order to understand the mechanism of the adsorption. Among the tested sorbents, MGO has shown the highest adsorption capacities (252 $mg\ g^{-1}$ for TC, 234 $mg\ g^{-1}$ for Cd(II), and 14 $mg\ g^{-1}$ for As(V)). The superiority of MGO has been mainly assigned to its thin nanosheets, various O-containing functional groups, as well as its high dispersibility. Their mechanism investigation showed that the $n-\pi$ electron-donor-acceptor effect is the main driving force in strong TC adsorption onto MGO along with $\pi-\pi$ interactions and H-bonding (Fig. 7.1B). In addition, they observed that As(V) adsorption remarkably decreased in the presence of TC, while As(V) hardly affected TC adsorption. They attributed these observations to the formation of type A ternary surface complexation and electrostatic attraction [14].

7.4.2 Carbon nanotubes

CNTs are usually curled by single or multiple layers of graphene to form hollow tubular structures [2,4]. Magnetic carbon nanotubes (MCNTs) are prepared by depositing MNPs on CNTs through one-step and/or two-step routes. CNTs have

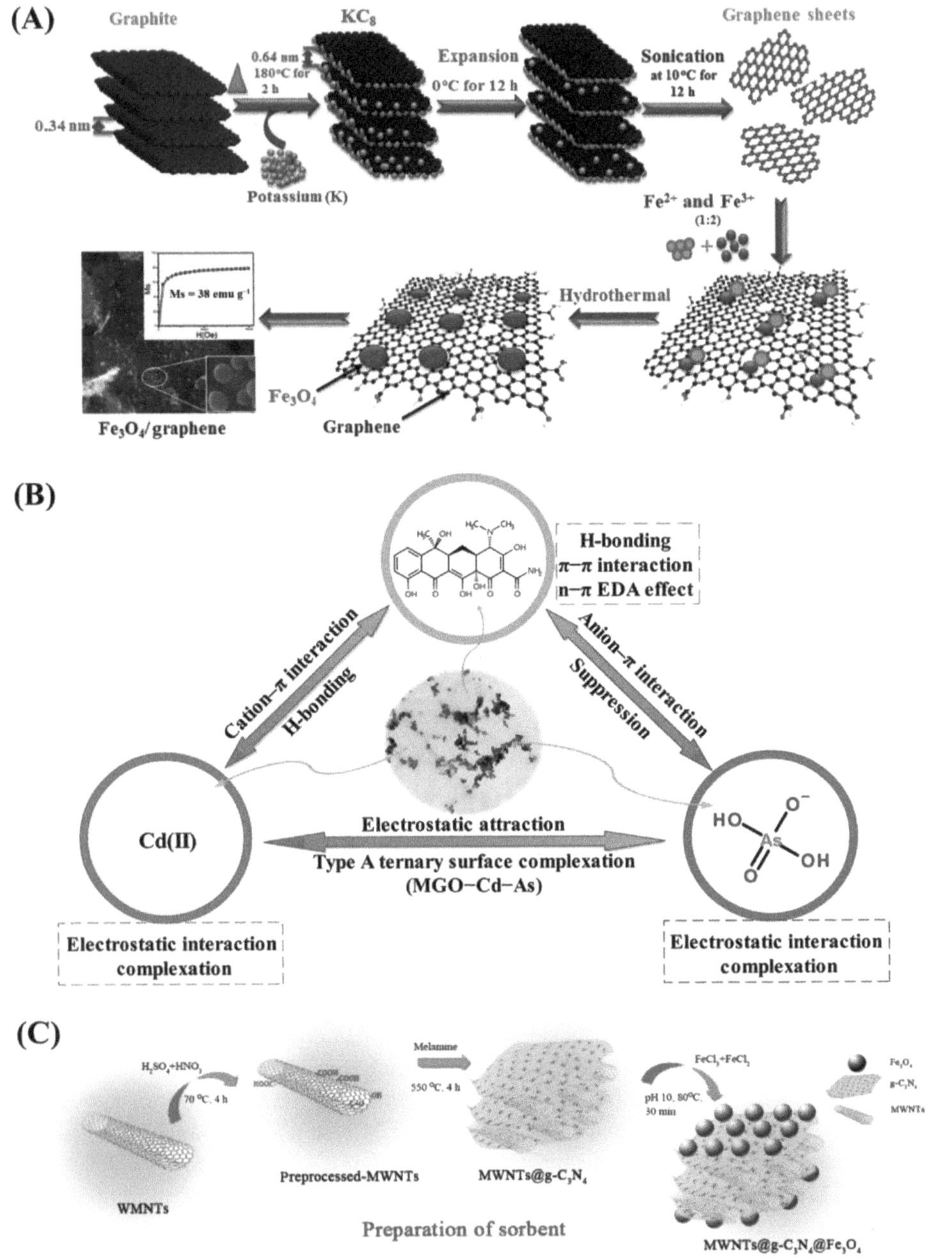

Figure 7.1 (A) Schematic illustrations for the scalable preparation of porous graphene sheets from potassium–graphite intercalation compound (KC8) and formation mechanism of Fe_3O_4/porous graphene nanocomposites [13]. (B) Schematic illustration of possible interactions in adsorption of tetracycline (TC), cadmium, and arsenate [As(V)] onto magnetic graphene oxide (MGO) [14]. (C) Scheme of the preparation of 3D MWNTs@g-C_3N_4@Fe_3O_4 nanocomposites [15].

high adsorption capacity and large specific surface area; however, their dispersion in aqueous samples is not favorable. In this regard, in the preparation of MCNTs, their dispersity and the separation of the CNTs can be improved [2,4]. However, further functionalization is usually performed with hydrophilic functional groups for enhancing the dispersibility and stability of MCNTs in aqueous solutions [15–19]. MCNTs are widely used as superior sorbents for various analytes at trace level concentrations as well as sorbents for the removal of a large concentration of pollutants from environmental samples [2,4].

Zhang et al. prepared 3D MWNTs@g-C_3N_4@Fe_3O_4 nanocomposite (Fig. 7.1C) as a novel sorbent for the extraction of 16 PAHs in industrial wastewater, rainwater, and river water samples. The nanocomposite presents a large specific surface area and excellent $\pi-\pi$ interactions and hydrogen bonds for the efficient extraction of PAHs (LOD in the range of 0.001–0.5 $\mu g\ L^{-1}$) [15].

El-Sheikh et al. investigated the effect of the ratio and dimension of CNT in prepared MCNT for MSPE of some nonsteroidal antiinflammatory drugs. Their results revealed that adsorbents consisting of long CNT have better efficiency than short CNT and that the best external diameter was 60–100 nm. Also, the ratio of 1:1 (CNT:Fe_3O_4) was the best [16].

7.4.3 Metal-organic frameworks

MOFs, a class of emerging materials, are formed as a crystalline network through coordination bonds between multidentate organic ligands and metal ions [20,21]. The combination of inorganic and organic units in their construction results in favorable properties, such as high specific surface area and high porosity, for use as selective sorbents. The porosity of MOFs with specific surface areas up to over 7000 $m^2\ g^{-1}$ may reach 90% of their volume [22]. Another feature of MOFs is their designable structure. First, various inorganic metal and organic ligands can be chosen for producing MOFs for specific applications [23]. Second, the pore size can be adjusted by varying the ligands and metals, as well as by controlling the reaction conditions [24]. However, their applications can be limited for the effective enrichment of target analytes from aqueous samples due to their poor dispersibility [25,26]. To solve such limitations, MOFs can be further functionalized to meet the requirements of specific applications. By postsynthetic modification, functional groups with different properties such as graphene oxide (GO) [25] or polyethylenimine (PEI) can be introduced [26]. These characteristics allow MOFs to be utilized as a sorbent for MSPE.

Abdi et al. synthesized a magnetic MOF nanocomposite (ZIF-8@SiO_2@$MnFe_2O_4$) as a novel magnetic microporous nanocomposite for extraction of methyl orange (MO) and malachite green (MG) (Fig. 7.2A). The MOF (ZIF8) had a high surface area (1485.1 $m^2\ g^{-1}$). The synthesized sorbent had high selectivity for MG (cationic dyes) compared to MO (anionic dyes) from the MG/MO mixture (removal efficiencies 92.5% for MG and 5.9% for MO, respectively) [27].

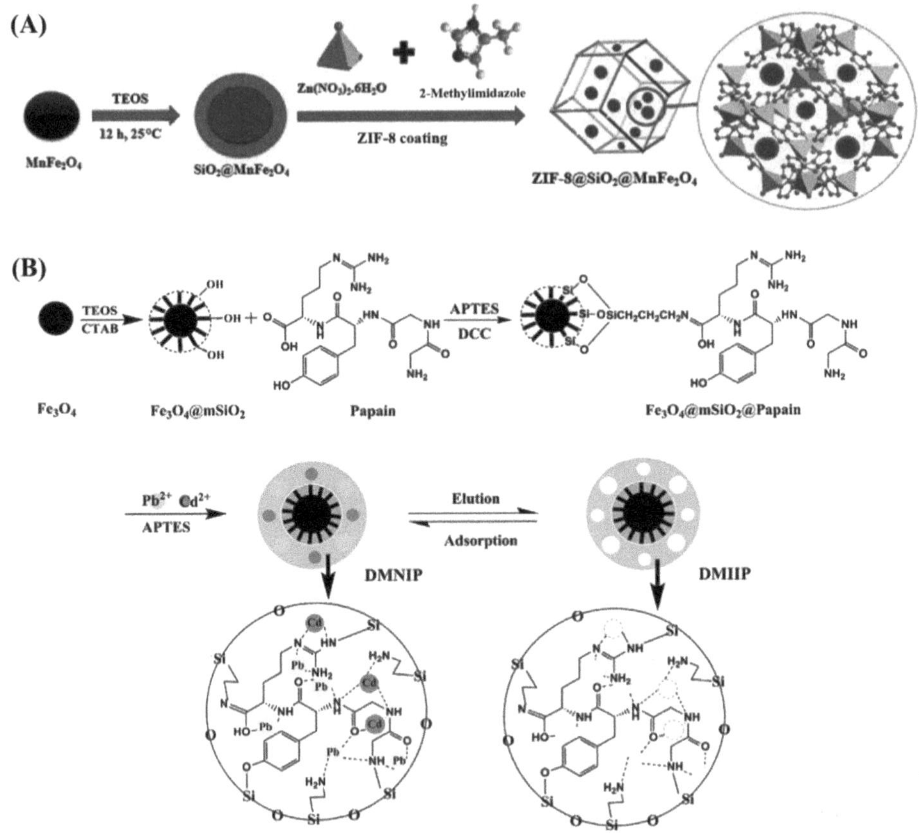

Figure 7.2 (A) Schematic diagram of the fabrication procedure for ZIF-8@SiO_2@$MnFe_2O_4$ nanocomposites [27]; (B) schematic illustration for the DMIIP preparation [28].

7.4.4 Molecular/ion-imprinted polymer

Molecular/ion-imprinted polymers (MIPs/IIPs) are emerging and novel sorbents with excellent selectivity for template molecules or ions synthesized by imprinting techniques. Under certain synthesis conditions, in the presence of template ions or molecules, monomers, cross-linking reagents, and initiators of highly cross-linked 3D network polymers are formed. Then template molecules or ions are removed from 3D network polymers, leading to the creation of a cavity in the structure of the polymer, which corresponds to the spatial structure, size, and binding sites of the template. Such MIPs/IIPs have excellent selectivity to template ions or molecules [2,4,29,30]. The introduction of magnetism to MIPs improves their separations, which would otherwise need repeated time-consuming filtration and centrifugation steps. Therefore magnetic molecular imprinted polymer (MMIPs) can be used for the selective extraction of target compounds from environmental samples along with the fast and easy collection of the sorbent [28,30–32].

Xie et al. prepared a dual-template magnetic ion-imprinted polymer (DMIIP) for removal of Cd^{2+} and Pb^{2+} by a surface imprinting technique followed by the sol–gel process using magnetic mesoporous $Fe_3O_4@mSiO_2$ as the substrate, papain as a functional monomer, and Cd^{2+} and Pb^{2+} as dual templates (Fig. 7.2B). The DMIIP showed efficient removal of the metals in a mixture solution (76.39 mg g^{-1} for Pb^{2+} and 41.69 mg g^{-1} for Cd^{2+}). Imprinting leads to obtaining higher efficiency for the sorption of Cd^{2+} and Pb^{2+}. In addition, the adsorption capacity of DMIIP decreased only about 6.0% after eight adsorption–desorption cycles [28].

Lu et al. prepared novel magnetic molecularly imprinted nanopolymers (MMIPs) for the selective adsorption and convenient collection of triclosan (TCS), a kind of chronic toxicity to aquatic organisms. The MMIPs present rapid adsorption equilibrium time (2 minutes), superior adsorption capacity ($Q_e = 53.12$ mg g^{-1}), good reusability (five cycles), and high selectivity ($k' = 6.321$) for TCS. As a result of their magnetic property, sample handling and recycling can be simplified. As a practical application, the toxicity test in microalgae confirmed that a small amount of MMIPs can eliminate the toxic effect of TCS on *Chlamydomonas reinhardtii* through effective binding to TCS [32].

7.4.5 Layered double hydroxides

LDHs, also known as anionic clays or hydrothalite-like compounds, have attracted considerable attention as extraction adsorbents in recent years [2,4,5,33]. LDHs are substances with 2D nanostructural layers, which consist of positively charged layers of metal hydroxides separated by charge-balancing anions in addition to water molecules [33,34]. LDHs have many advantageous attributes as an adsorbent, such as large interlayer spaces (high porosity), large specific surface area, good anion-exchange capacity (a huge number of exchangeable anions), water-resistant structure, and easy synthesis/preparation [2,4,5,33]. However, LDHs cannot be directly applied for enriching pollutants from sample solutions due to their poor dispersive characteristics and difficult separation from aqueous solution. Thus some modifications with hydrophilic compounds such as polyimide [34] or sodium dodecyl sulfate [35] are required to improve the extraction efficiency of LDH sorbents.

Wu et al. successfully synthesized a novel composite containing a porous polyimide (PI)-coated Fe_3O_4 magnetic core and LDH (Fe_3O_4@PI@LDO) for removal of 2,4-dichlorophenol (2,4-DCP), glyphosate (GP), and TC from aqueous solution (Fig. 7.3A). The prepared nanocomposite showed spontaneous and endothermic adsorption with maximum adsorption capacities of 176.06, 190.84, and 185.53 mg g^{-1} for 2,4-DCP, GP, and TC, respectively. In addition, the sorbent can be reused up to five cycles [34].

7.4.6 Mesoporous materials

Mesoporous materials are porous materials with pore diameters in the range of 2–50 nm. These materials have unique features such as large pore volume, large specific surface areas, adjustable mesoporous channels, and easy surface

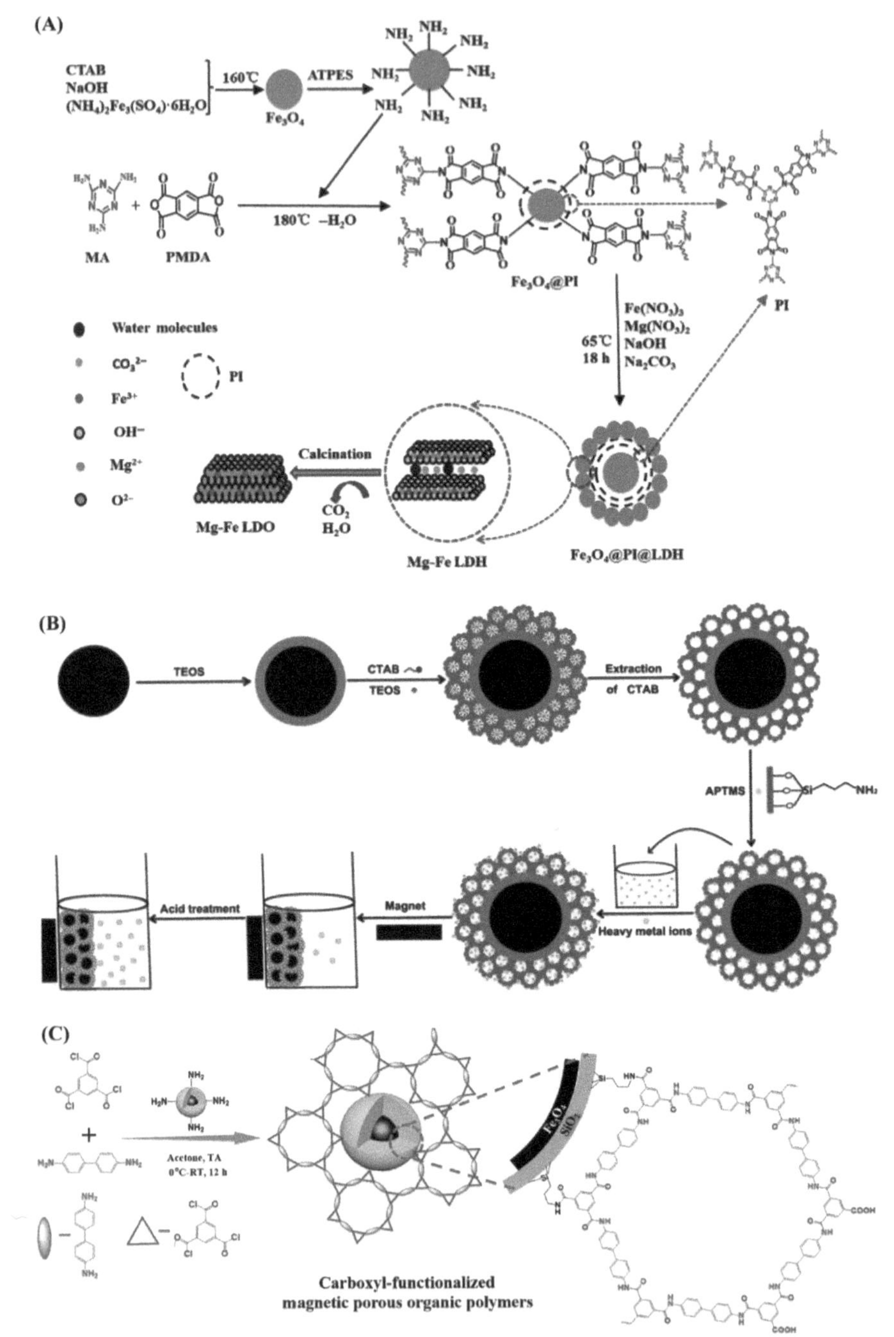

Figure 7.3 (A) Synthetic routes and schematic illustrations of magnetic PI@LDO composites [34]; (B) synthesis route of $Fe_3O_4@SiO_2@meso\text{-}SiO_2\text{-}R1\text{-}NH_2$ microsphere and its use for heavy metal ions removal [36]; (C) the synthesis procedure for CMPOPs [37].

modification [2]. Because of the low cost and simple preparation of mesoporous silica gel, it is a widely used mesoporous material. The general route for the preparation of magnetic mesoporous materials consists of the sol–gel reaction on the surface of MNPs in the presence of a template such as surfactants and then removal of the template [36,38]. Mesoporous silica MCM-41 and MCM-48 are the most commonly used silica nanoporous materials. MCM-48 has a cubic array with a 3D pore network structure, which leads to fast and favorable mass transfer kinetics compared to MCM-41 with a hexagonal array of unidirectional pore structure. Therefore, as a sorbent, MCM-48 is usually more efficient as an absorbent support than MCM-41 [38].

Surface grafting depends on the pore size of the mesoporous materials. Yuan et al. synthesized a novel magnetic mesoporous silica shell with a large pore size (about 10.3 nm) using cetyltrimethylammonium bromide (CTAB) as a template (Fig. 7.3B). Grafting amino groups to large-pore-size mesoporous silica magnetic microspheres (447 mg g^{-1}) is much more efficient than small-pore-size ones (c. 2–4 nm), which leads to more efficiency in heavy metal ion adsorption (Cd^{2+}: 492.4 mg g^{-1}, Cu^{2+}: 628.3 mg g^{-1}, Pb^{2+}: 880.6 mg g^{-1}).

7.4.7 Covalent organic frameworks

COFs are a new class of carbonaceous porous organic polymers (POPs) that have attracted great interest from analytical chemists [2,37,39–45]. COFs possess the obvious advantages of low crystal density, large specific surface area, highly homogeneous and rigid pore structure, tunable pore size, relatively low cost, and excellent thermal stability [39,40]. These superior features make COFs promising sorbents [37,39–45]. A typical COF in recent years is based on benzidine (BD), 1,3,5-triformylphloroglucinol (Tp) known as TpBD [39,40]. Most reported COFs, such as Fe_3O_4@COF (TpBD), have limited applications in MSPE due to their hydrophobic networks that lead to poor dispersion in aqueous samples. Application of hydrophilic building blocks such as covalent triazine–based frameworks [43] or further functionalization with different polar functional groups such as –COOH [37], $-NH_2$ [41,44], $-SO_3H$, and –SH [42] are used to improve the dispersibility as well as the extraction capacity of the COFs. Hydrophobic, π–π, or hydrogen bonding interactions are the main driving forces in extractions of analytes using COFs as sorbents.

Romero et al. prepared a magnetic POP with abundant carboxyl groups (CMPOPs) for removal of MG from aquaculture wastewater. The results showed strong electrostatic interaction between the amine groups of MG and the carboxyl groups of sorbent that leads to fast (2 minutes for an initial concentration of 100.0 mg L^{-1}) and efficient removal of MG with excellent q_{max} in the range of 569.4–779.1 mg g^{-1} [37].

Huang et al. prepared thiol-functionalized magnetic COFs with abundant accessible chelating sites for the selective and efficient adsorption of Hg^{2+} from environmental samples. The nanocomposite showed excellent q_{max} of 383 mg g^{-1} with rapid kinetic (adsorption equilibrium within 10 minutes) and reusability (up to five times). [45].

7.5 Magnetic nanomaterials as sorbents for preconcentration purposes

7.5.1 Analysis of pesticides

The use of pesticides in agriculture, horticulture, public health, households, and veterinary medicine has increased dramatically and undeniably during the last few decades. Nowadays, pesticide pollution is a serious concern, and severe restrictions have been set for residue levels. Some types, such as organochlorine pesticides and triazole pesticides, are highly stable, have long chemical and photochemical half-lives, can be easily transported, and accumulate in environmental soil and water [25,46]. Pesticide concentrations, which are commonly found in environmental samples, are at trace levels ($\mu g\,L^{-1}$) [17,25,46–58]. Therefore it is important to develop a rapid, highly sensitive, and convenient analytical method for trace pesticide residues analysis to assess pesticide contamination in environmental samples. Toward this end, the application of magnetic nanomaterials such as novel sorbents in MSPE techniques has been extensively investigated in the field of pesticide residues analysis. In this sense, the most recent and significant examples are shown in Table 7.1. The different types of pesticides such as organochlorine pesticides [46,50,55,58], sulfonylurea herbicides [53,57], organophosphorus pesticides [52,56], benzoylurea pesticides [48], heterocyclic pesticides [49], pyrethroid pesticides [51], and triazole pesticides [25] have been extracted via MSPE in environmental samples.

7.5.2 Analysis of heavy metal ions

With the development of industry and technology, the release of heavy metals worldwide has increased. Since trace level determination of heavy metals in environmental samples has received significant attention due to the extreme toxicity and nonbiodegradability of these metals. The existence of heavy metals such as cadmium (Cd), chromium (Cr), copper (Cu), mercury (Hg), lead (Pb), nickel (Ni), and zinc (Zn), even at trace levels in environmental samples, raises serious environmental concerns related to human health and ecosystem quality. These metals easily accumulate in living organisms, and most of them are known to be highly toxic or carcinogenic [1,2]. Therefore the development of reliable analytical methods is required for the determination of heavy metals in environmental samples. In this sense, the most recent and significant examples are shown in Table 7.1 [59–71]. Different heavy metals ions such as Cu [59,64,70], Cd [19,60,63,64,70], Hg [59–63,65], Pb [19,63,64,69,70], Zn [64], Cr [64,66,70], Ni [70], Co [59,70], Ag, Au, Ir, Os, Pd, Pt, Sb [64], Se [68], Te [67,68], and Tl [69] have been extracted via MSPE in environmental samples.

7.5.3 Analysis of polycyclic aromatic hydrocarbons

PAHs are persistent organic pollutants in the environment that accumulate in biological systems due to their hydrophobic nature [99]. PAHs, known as mutagenic,

Table 7.1 Some of the most recent and significant applications of MNMs as sorbents for preconcentration purposes.

Analyte	Sorbent	Sample	Detection system	LOD ($\mu g\ L^{-1}$)	Eluent	V_{sample} (mL)	Ref.
Pesticides							
96 Pesticides	Hydrophilic-lipophilic-balanced magnetic particles	Seawater	LC-MS/MS	$(0.13-0.42) \times 10^{-3}$	3 mL acetonitrile (ACN) 0.1% formic acid	500	[47]
Benzoylurea pesticides	Polyethylene glycol–modified magnetic covalent organic framework material (PEG/Fe_3O_4@SNW-1)	Water	HPLC	<1.0	10 μL water/ACN	N/R	[48]
Heterocyclic pesticides	Magnetic metal-organic framework (MOF-5)	River water	HPLC	0.04–0.11	3 mL ethyl acetate	60	[49]
Organochlorine pesticides	Bimetallic metal-organic framework	Water	GC-MS/MS	$(0.39-0.70) \times 10^{-3}$	2 mL dichloromethane (DCM)	10	[50]
Pyrethroid pesticides	Mesoporous KIT-6-magnetite composite (Fe_3O_4@SiO_2@KIT-6)	River, lake, and tap water	HPLC	$(0.006-0.01) \times 10^{-3}$	2 mL ACN	100	[51]
Triazole pesticides	Magnetic graphene oxide–functionalized copper-based metal-organic framework (MOF-199)	Tap and well water	HPLC MS/MS	0.05–0.1	3 mL acetone	4	[25]
Organophosphorus pesticide	Magnetic graphene nanocomposite (graphene@SiO_2@Fe_3O_4)	Water	GC-FPD	16–33	3 mL acetone	10	[52]
18 Chiral pesticides	Magnetic amino-modified multiwalled carbon nanotubes (m-MWCNT-NH_2)	Water, soil, and river sediment	LC-MS/MS	0.10–0.50	N/R	200	[17]
Organochlorine pesticides	Amino-functionalized magnetic framework composite (Fe_3O_4-NH_2@MIL 101(Cr))	Soil samples	GC-ECD	0.15–0.28 (ng g^{-1})	800 μL acetic acid (AA)-acetone	N/R	[46]

(Continued)

Table 7.1 (Continued)

Analyte	Sorbent	Sample	Detection system	LOD (μg L^{-1})	Eluent	V_{sample} (mL)	Ref.
Sulfonylureas pesticides	Hybrid magnetic carbon nanotubes—a C18-modified nano-SiO_2	Well, river, tap water	HPLC-UV	70–600	1.5 mL ACN	10	[53]
Organochlorine pesticides	Magnetic metal-organic-framework (MMOF) (Fe_3O_4@MIL-100 (Fe))	Tea leaves	GC-MS	(0.62–3.92) × 10^{-3}	1 mL ethyl acetate	15	[54]
Acidic and basic pesticides	Triazine-based polymeric network graphene oxide (TPN/Fe_3O_4/GO)	Water samples	HPLC-UV	0.17 and 1.7	ACN, water 0.5% AA	50	[55]
Diazinon and malathion	Magnetic mesoporous silica composite	Well, tap water	HPLC	0.09 and 0.14	183 μL methanol	24	[56]
Sulfonylurea herbicides	Magnetic polydopamine–modified with deep eutectic solvent (Fe_3O_4@PDA-DES)	Water	UPLC	0.0098–0.0110	0.5 mL ACN	100	[57]
Organochlorine pesticide	Porous $ZnFe_2O_4$/carbon, derived from Zn-Fe zeolitic imidazolate framework (Zn-Fe-ZIF)	Water	GC-ECD	(0.005–0.3) × 10^{-3}	1 mL ethyl acetate	10	[58]
Heavy metals							
Cu^{2+}, Co^{2+}, Hg^{+}	Monolithic capillary microextraction column embedded modified Fe_3O_4 magnetic nanoparticles (MCEN)	Water	HPLC-DAD	0.050–150.0	0.08 mL methanol	4.0	[59]
$Cd^{2+}$and Hg^{2+}	Magnetic polyamidoamine (PAMAM) dendrimers	Natural water samples	HPLC	0.016 and 0.040	6 mL ACN	100	[60]
Hg^{2+}, $MeHg^{+}$, $PhHg^{+}$	Fe_3O_4@SiO_2@γ-mercaptopropyl trimethoxysilane (γ-MPTS) magnetic nanoparticles	Water samples	HPLC-ICP-MS	(0.49–0.74) × 10^{-3}	500 μL HNO_3 4.0% (m/v) thiourea	100	[61]
Hg^{2+}	Fe_3O_4@SiO_2	Water samples	ICP-MS	0.52×10^{-3}	0.5 mL Na_2S	50	[62]
Cd^{2+}, Pb^{2+}, Hg^{2+}	Fe@Ag@dimercaptobenzene (DMB)	Environmental water samples	HPLC	0.011–0.031	2 mL sodium diethyl dithiocarbamate (DDTC-Na)	100	[63]

Cu^{2+}, Pb^{2+}, Zn^{2+}, Cr^{3+}, Cd^{2+}	SiO_2-coated magnetic graphene oxide modified with polypyrrole–polythiophene composite	River, well water	FAAS	0.15–0.65	0.5 mL 1 mol L^{-1} HNO_3	50	[64]
Ag, Au, Ir, Os, Pd, Pt, Sb, Hg	Polythiophene magnetic nanoparticles (PTH-MNP)	Lake, sea, tap water	ICP OES	0.05–2.6	0.5 mL 1 mol L^{-1} HNO_3	20	[65]
Cr	Octyl-coated cobalt-ferrite/silica core-shell nanoparticles	Tap, river, and wastewater	FAAS	0.13	0.5 mL methanol	100	[66]
Cd, Pb	Organic acid-modified magnetic carbon nanotube	Tap, well water	FAAS	1.8	10 mL 0.3 mol L^{-1} HNO_3	100	[19]
$Te^{4+}$and Te^{6+}	Magnetic polymeric ionic liquid nanocomposite	Waters, soils	AFS	$1.9–3.7 \times 10^{-3}$	500 μL HCl 5 mol L^{-1}	50	[67]
$Se^{4+}$and Te^{4+}	Polyaniline-functionalized magnetic nanoparticles (Fe_3O_4@SiO_2@PANI)	Environmental water samples	ICP-MS	$(5.3, 1.2) \times 10^{-3}$	1.0 mL NaOH	50	[68]
Pb, Tl	Magnetic multiwalled carbon nanotubes@zeolite (MWCNT-Fe_3O_4@Zeo) nanocomposite	Wastewater	ICP-OES	0.28–0.23	5 mL 1 mol L^{-1} HNO_3	100	[69]
Cr, Co, Ni, Cu, Cd, Pb, Ag	Fe_3O_4-GO@SiO_2	Water	ICP-MS	$(2.023–13.81) \times 10^{-3}$	5 mL 1 mol L^{-1} HNO_3	50	[70]
Rare earth elements Rres: Sc, Y, La, Ce, Pr, Nd, Sm, Eu, Gd, Tb, Dy, Ho, Er, Tm, Yb, and Lu	Diglycolamide-grafted Fe_3O_4/polydopamine nanomaterial (Fe_3O_4@PDA@DGA)	Water	ICP-OES	0.039	10 mL 0.1 mol L^{-1} ethylene diamine tetraacetic acid (EDTA)	10	[71]
Polycyclic aromatic hydrocarbons (PAHs)							
PAHs	Magnetic β-cyclodextrin-carbon nano-tube Fe_3O_4@CNT-CD (M-CNT-CD)	Rainwater, well	GC-FID	$(0.6–3.0) \times 10^{-3}$	200 μL toluene	10	[18]
PAHs	Polydopamine β-cyclodextrin/maltodextrinFe_3O_4@SiO_2@PDA-CD/MD	Environmental water	GC-FID	0.04–0.57	100 μL toluene	10	[72]
PAHs	Hybrid cobalt ferrite/natural organic matter(NOM)	Produced water	HPLC	0.500–100	10 mL ethyl acetate	10.0	[73]

(Continued)

Table 7.1 (Continued)

Analyte	Sorbent	Sample	Detection system	LOD ($\mu g\ L^{-1}$)	Eluent	V_{sample} (mL)	Ref.
PAHs	2-Mercapto-4-methyl-5-thiazoleAA bonding Au (Fe_3O_4@SiO_2@MMTA-Au)	Rainfall water, river	HPLC-FLD	$(0.25-37.5) \times 10^{-3}$	ACN	100	[74]
PAHs	Magnetic molecularly imprinted polymers (MMIPs)	River and produced water	APGC-MS/MS	$(1-100) \times 10^{-3}$	0.5 mL hexane	20	[75]
PAHs	MMIP	Seawater	GC-MS	$(30-375) \times 10^{-3}$	5 mL acetone	20	[74]
PAHs	Three-dimensional multiwalled carbon nanotubes@g-C_3N_4@Fe_3O_4nanocomposites	Rain water, river, wastewater	GC-FID	0.001–0.5	0.5 mL ACN	20.0	[15]
PAHs	Metal-organic framework MIL-101 (Cr) modified zero-valent iron nanoparticles	River, tombs, reservoir	HPLC	0.064	5 mL ACN	100	[76]
PAHs	Magnetic polyaniline–coated newly synthesized dicationic ionic liquid (MNP/PANI/DICAT)	River, lake, well, tap water, soil	GC-MS	$(0.8-208.6) \times 10^{-3}$	1 mL ACN	30	[77]
PAHs	Phenyl-modified magnetic mesoporous silica synthesized with *p*-toluenesulfonic acid (Fe_3O_4@mSiO_2-Ph-PTSA)	Soils	GC–MS	0.07–0.41 ng g^{-1}	7 mL acetone/DCM 1:1	N/R	[78]
PAHs	Graphene oxide silica-coated magnetite phenylethylamine (Fe_3O_4@SiO_2@GO–PEA)	Wastewater/river water	GC-FID	0.005–0.1	2 mL ACN/C_2C_{14} (4:1)	50	[79]
Other analytes							
Phthalates	Fe_3O_4@r-gom-NPs	Pond, wastewater	UPLC-MS	N/A	4 mL DCM	25	[80]
Phthalate products	Graphene oxide	Water	HPLC	4–13	5 mL ethyl acetate	50	[81]

Phthalate esters	Magnetic microsphere-confined graphene (Fe_3O_4@SiO_2-G)	Environmental water	HPLC	$(70-100) \times 10^{-3}$	0.6 mL acetone	100	[82]
Anionic azo dyes	Magnetic halloysite nanotubes (MHNTs)	Lake, tap water	HPLC	0.042, 0.050	2 mL methanol 1% AA	200	[83]
Ibuprofen, antiinflammatory drugs, and azo dyes	Fe_3O_4@TiO_2@C–N	Lake, sea, surface water	HPLC	0.95	2 mL acetone	100	[84]
Bisphenol A, phenol, and hydroquinone	Fe@SiO_2@poly *N*-isopropylacrymide-*co*-methacrylic acid (Fe@SiO_2@PNIPAM-co-MAA)	Water	HPLC	0.019–0.031	3 mL methanol	40	[85]
Estrogens and bisphenol A	Fe_3O_4@nSiO_2@mSiO_2-chomnps	Water samples	UHPLC-HRMS	$(0.2-3.0) \times 10^{-3}$	3 mL ACN	100	[86]
Tetrabromobisphenol A 4-Nonylphenol	Decorated Fe_3O_4 nanoparticle	Environmental water samples	HPLC	0.011–0.017	3 mL acetone	60	[87]
Chlorophenols	Planar graphene oxide-based magnetic ionic liquid nanomaterial (PGO-MILN)	Tap, river, well	LC-MS	$(0.2-2.6) \times 10^{-3}$	1 mL NaOH 10 mM in ACN/methanol (1/1, v/v)	5	[88]
Chlorophenols	Magnetic polyamidoamine (PAMAM)	Tap, sea, well	GC-MS	$(0.6-10) \times 10^{-3}$	N/R	12	[89]
Perfluorinated compounds	Magnetic covalent triazine-based frameworks	River, pond, snow	LC–MS	$(0.62-1.39) \times 10^{-3}$	3 mL acetone	25	[43]
Polychlorinated biphenyls	Fe_3O_4@MOF core–shell magnetic microspheres	Environmental water	GC-MS/MS	$(1.07-1.57) \times 10^{-3}$	5 mL hexane	10	[90]
Fluoroquinolone antibiotics	Molybdenum disulfide-graphene oxide supported magnetic nanoparticles (Fe_3O_4/GO/MoS_2)	Water	HPLC	$(0.25-0.50) \times 10^{-3}$	2 mL ACN	10	[91]
Brominated flame retardants	Taiji-principle-designed magnetic porous C-doped graphitic carbon nitride	Water	HPLC	0.1–0.2	2 mL methanol	100	[92]

(Continued)

Table 7.1 (Continued)

Analyte	Sorbent	Sample	Detection system	LOD (μg L^{-1})	Eluent	V_{sample} (mL)	Ref.
Organic micropollutants	Fe_3O_4	River	GC-MS	$<4 \times 10^{-3}$	12 mL DCM and ethyl acetate (1:1)	1000	[93]
Phthalic acid esters	Core–shell poly(dopamine) magnetic nanoparticles (Fe_3O_4@pDANPs)	Water	GC-MS	0.36	6 mL DCM	25	[94]
Estrogens	Reduced graphene oxide/$ZnFe_2O_4$ (rGO/$ZnFe_2O_4$)	Water, soil	HPLC	$(0.01-0.02) \times 10^{-3}$	3 mL ACN	30	[95]
Polychlorinated biphenyls (PCBs)	Fe_3O_4 @ cobalt(Ⅱ)-based metal-organic nanotube (Fe_3O_4 @ Co-MONT)	Water	GC-MS	$(0.31-0.49) \times 10^{-3}$	5 mL hexane	10	[96]
Polychlorinated biphenyl 28	Fe_3O_4 nanoparticle grafted graphene oxide (Fe_3O_4@GO)	Water	GC-MS	$(0.027-0.059) \times 10^{-3}$	14 mL *n*-hexane/DCM (70:30)	200	[97]
UV filters	Magnetism-enhanced monolith-based in-tube	Water	HPLC	0.04–0.26	0.10 mL methanol	1.50	[98]

N/R, Not reported.

teratogenic, and carcinogenic compounds, threaten the ecosystem and human health [100]. In general, they occur in the environment by natural and anthropogenic processes [99,100]. In this regard, 16 PAHs have been included in the list of priority organic pollutants by the U.S. Environmental Protection Agency [101]. Given the trace level of PAHs and the complexity of environmental samples, the development of accurate and reliable extraction methods is required to detect PAHs in environmental samples. In this sense, the most recent and significant examples are shown in Table 7.1 [15,18,72–79]. Various types of magnetic nanomaterials as sorbents have been used in the MSPE of PAHs such as three-dimensional multiwalled carbon nanotubes@g-C_3N_4@Fe_3O_4 nanocomposites [15] magnetic β-cyclodextrin-carbon nanotube [18], hybrid cobalt ferrite/natural organic matter (NOM) [73], magnetic molecularly imprinted polymers [74,75], magnetic metal-organic framework MIL-101 (Cr) [76], magnetic polyaniline–coated dicationic ionic liquid [77], phenyl-modified magnetic mesoporous silica [78], graphene oxide silica-coated magnetite phenylethylamine [79].

7.5.4 Analysis of other contaminants

Apart from the major environmental applications mentioned previously (pesticides, heavy metals, PAHs), magnetic nanomaterials as sorbents in MSPE are also used to determine other families of environmental contaminants [43,80–98] such as phthalates [80–82], anionic azo dyes [83], ibuprofen [84], bisphenol A (BPA), phenol, and hydroquinone [85], estrogens and BPA [86], tetrabromobisphenol A, 4-nonylphenol [87], chlorophenols [88,89], perfluorinated compounds [43], polychlorinated biphenyls [90], fluoroquinolone antibiotics [91], and brominated flame retardants [92]. The most recent and significant examples are shown in Table 7.1.

Mei and Huang developed an online setup for the in-tube solid-phase microextraction of five organic UV filters [98]. Their setup and synthesis route are schematically shown in Fig. 7.4. The sorbent has been prepared by doping magnetic nanoparticles into an ionic liquid-based monolithic capillary column. The use of such setups leads to the automation of MSPE or magnetic solid-phase microextraction of the target compounds before instrumental analysis.

7.6 Magnetic nanomaterials as sorbents for removal purposes

One of the most serious global issues is water pollution. This challenge mounts every year, as the amount of wastewater discharged into environmental waters increases every year. Since the last century and with continuous industrialization, large amounts of pollutants such as heavy metals, dyes, pharmaceuticals, and endocrine-disrupting compounds (EDCs) are released into the environment.

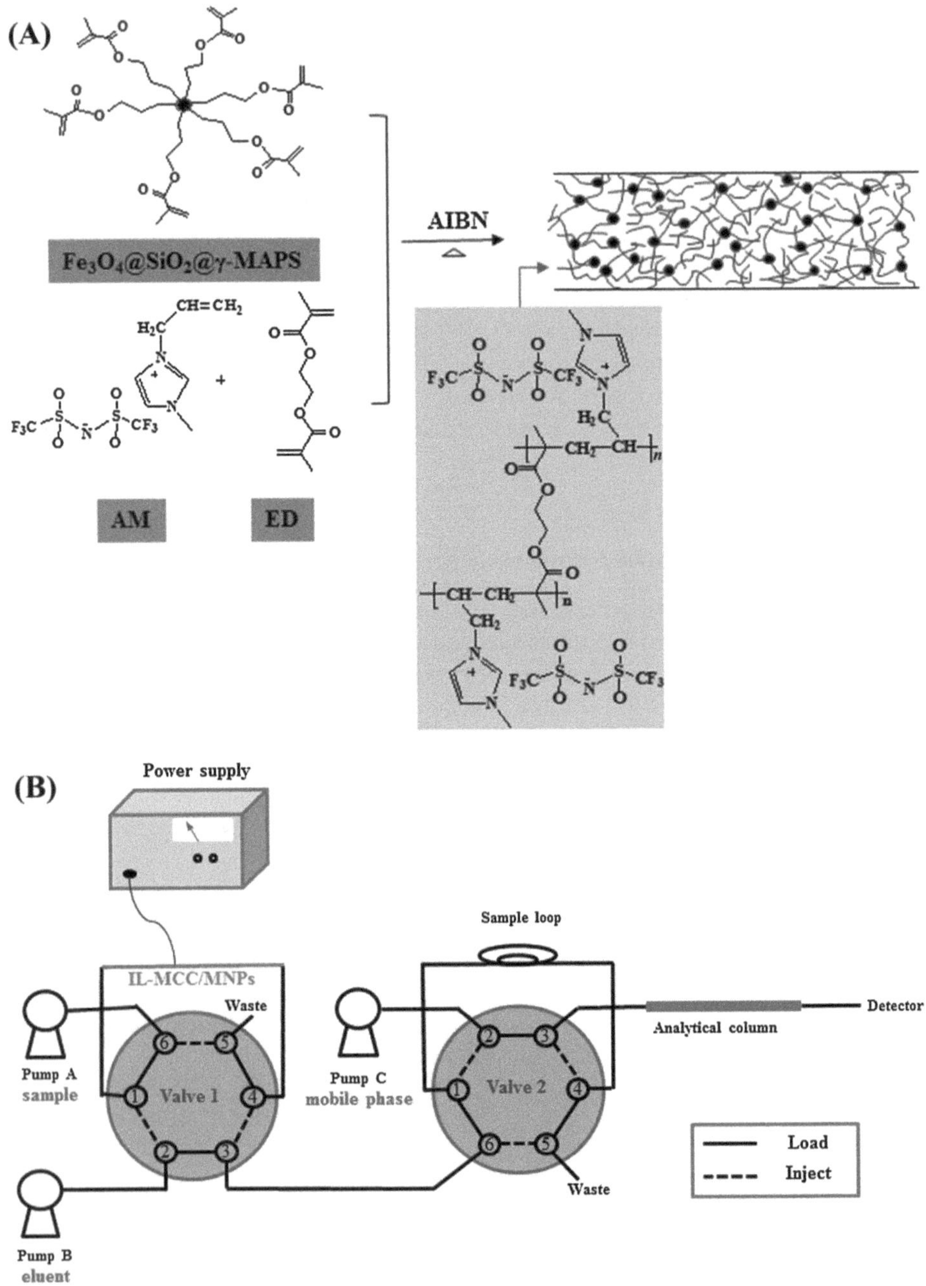

Figure 7.4 (A) Synthesis sketch of PIL-based MCC doped with MNPs; (B) schematic of the online ME-MB/IT-SPME-HPLC system [98].

7.6.1 Removal of heavy metals

The effluent runoffs of various industries, such as alloying, metal fabrication, batteries, electroplating leather, mining, fertilizers, pigments, and paints, results in the discharge of a large amount of heavy metals into the environment [6,102]. As a result, nonbiodegradable heavy metal ions, at every level, can be accumulated in the environment through biomagnification, leading to the heavy metal poisoning of living organisms.

In recent years, magnetic nanomaterials have attracted a great deal of attention in the decontamination of wastewaters compared to conventional adsorbents, such as activated carbon, clay, and metal oxides, due to their large specific surface area, a large number of binding sites, and active groups, as well as their facile magnetic separation [6]. Besides, unlike activated carbon, magnetic nanomaterials can be easily recycled and reused several times, making them more attractive and less expensive. The most recent and significant examples of the application of magnetic nanomaterials in the removal of heavy metals are shown in Table 7.2 [13,14,103–116].

7.6.2 Removal of dyes

Over the last century, large amounts of dyes are discharged to the environment as a result of the continuous development of dyeing and the printing industrialization process. Dyes are widely used for the coloration of many products such as printing, leather, textile, paper, rubber, and plastics [144]. Most of the synthetic dyes have an aromatic molecular structure, so they can lead to mutagenicity and carcinogenicity. In addition, by light penetration in water, they can harm the aquatic life [145,146]. Therefore these effluents need to be treated to remove the dyes before discharge into the environment.

The application of magnetic nanomaterials as efficient, low-cost, environmentally friendly, easily recoverable, and reusable sorbents has garnered increased interest in dye removal. Magnetic nanomaterials have a large surface area and short diffusion routes that are favorable for efficient and fast adsorption as well as for their magnetic properties that allow the sorbents to be easily separated from treated water. The most recent and significant examples of the application of magnetic nanomaterials in the removal of the dyes are shown in Table 7.2 [117–124].

7.6.3 Removal of pharmaceutical products

Recently, the presence of drug residues in the environment has attracted a lot of scientific attention because most pharmaceutical wastes are seriously dangerous because of their greater stability against biological degradation and negative effects such as resistance to pathogens and allergic reactions [125]. Drug residues can be released into the environment during the different steps of production, delivery, and application [147]. However, the main sources of pharmaceuticals in the environmental water samples include human activities, pharmaceutical companies' waste, veterinary treatment, and hospital waste

Table 7.2 Some of the most recent and significant applications of MNMs as sorbents for removal purposes.

Analyte	Sorbent	Sample	Sorbent capacity ($mg\ g^{-1}$)	V_{sample} (mL)	Ref.
Heavy metal ions					
Pb^{2+}, Cu^{2+}, and Cd^{2+}	(Fe_3O_4)/porous graphene	Wastewater	460	50	[13]
Tetracycline (TC), Cd^{2+}, and As^{5+}	Magnetic graphene nanomaterials	Water	14–252	N/R	[14]
Heavy metals	Ethylenediaminetetraacetate silica-coated magnetic nanoparticles (Fe_3O_4/SiO_2/thermosensitive/EDTA)	Water	N/R	N/R	[103]
Cd^{2+} and As^{5+}	Magnetic graphene nanomaterials (MGNs)	Water	34.37	40	[104]
U^{6+}	Hierarchical core–shell and yolk–shell magnetic titanate nanosheets (Fe_3O_4@TNS)	Wastewater	283.29	N/R	[105]
As^{5+} and As^{3+}	Mesoporous $CoFe_2O_4$@MIL-100(Fe) hybrid MNPs	Well, river	143.6	500	[106]
Co^{2+}	Magnetic graphene oxide N-doted (NMGO)	Wastewater	14.6–43.6	N/R	[107]
Cr^{6+}	*Lagerstroemia speciosa* embedded MNPs	Water	434.78	50	[108]
Pb^{2+}, Hg^{2+}	Flower-like magnetic MoS_2nanohybrid	Soil and wastewater	Pb: 263.6 and Hg: 428.9	N/R	[109]
Cu^{2+}	Polyethylenimine-derived quaternary ammonium-coated magnetic silica	Water	73.0	20	[110]
Ag nanoparticles	Polyethylenimine magnetic mesoporous silica composites(Fe_3O_4@SiO_2-PEI)	Water	909.1	100	[111]
Pb^{2+}	$SrHPO_4/Fe_3O_4$	Water	1215.5	30	[112]
Pb, Ni, Co, and Cu	Quasi-superparamagnetic Fe_3O_4	Wastewater	476.2	200	[113]
Pb^{2+}, Cd^{2+}	Fe_3O_4-SO_3H magnetic nanoparticles	Water	80.9–108.93	50	[114]
Hg^{2+}	NH_2-Fe_3O_4	Water	246	50	[115]
Cr^{6+}	Porous magnetite nanoparticles	Wastewater	14.49	50	[116]

Dyes					
Methylene blue (MB)	$FeFe_2O_4$ magnetic nanomaterial	Water	42.351	50	[117]
MB dye	Mesoporous magnetite nanoparticles	Water	N/R	20	[118]
Methylene blue	Magnetic carrageenan/silica hybrid nanoadsorbents	Water	530	20	[119]
Methylene blue	Ionic liquid modified molybdenum disulfide and reduced graphene oxide magnetic nanocomposite	Water	143.9	N/R	[120]
Methyl orange	Magnetic zeolitic imidazolate framework-67 composites	Wastewater	738	10	[121]
Crystal violet, Congo red	GO/Fe_3O_4	Water	101.74	50	[122]
Methyl violet	Magnetic nanocomposite of multiwalled carbon nanotube	Water	666.7	50	[123]
Rhodamine B	L-Serine (L-ser) capped magnetite nanoparticles	Water	82	50	[124]
Pharmaceuticals					
TC	Magnetic graphene sponge	Water	473	8	[125]
TC	Fe_3O_4@SiO_2-chitosan/graphene oxide nanocomposite	Water	183.47 mmol kg^{-1}	50	[126]
Antibiotics	Magnetic Fe_3O_4/red mud nanoparticles	Water	110.15	50	[127]
Chlortetracycline, TC, oxytetracycline	Magnetic graphene oxide	Water	106.60–162.42	30	[128]
TC	Fe_3O_4@graphene oxide	Wastewater	1272	N/R	[129]
TC	Nitrilotri AA-functionalized magnetic graphene oxide (NDMGO)	Water	212	50	[130]
Ciprofloxacin (CIP)	NDMGO	Water	334	N/R	[131]
Antibiotic amoxicillin (AA)	Magnetically modified graphene nanoplatelets	Water	14.10	10	[132]

(*Continued*)

Table 7.2 (Continued)

Analyte	Sorbent	Sample	Sorbent capacity ($mg\ g^{-1}$)	V_{sample} (mL)	Ref.
TC	Thiourea-dioxide–reduced magnetic graphene oxide	Water	1233	50	[133]
TC, CIP	Diethylenetriaminepenta AA-functionalized magnetic graphene oxide	Water	294.1 (TC), 111.7 (CIP)	N/R	[134]
Fluoroquinolone	Polydopamine-coated graphene oxide/ Fe_3O_4 (PDA@GO/Fe_3O_4)	Water	70.9	8	[135]
Endocrine-disrupting chemicals					
Paraquat	Magnetic hybrid nanobiosorbents	Water	257	5	[136]
PAHs	Fe_3O_4@polyaniline	Water	15.8479	25	[137]
Bisphenol A (BPA) and bisphenol AF (BPAF)	Core–shell magnetic covalent-organic framework nanospheres (Fe_3O_4@TpBD)	Water	BPA 160.6 BPAF 236.7	N/R	[138]
Chlorophenols	Mixed hemi/ad-micelle ionic liquid-coated magnetic graphene oxide	Water	120, 141	150	[139]
Polychlorinated biphenyls (PCBs)	Poly(ethylene glycol) magnetic core–shell nanoparticles	Water	2	N/R	[140]
PCBs	Fe_3O_4@β-CD	Water	30.32, 40.01 $mmol\ kg^{-1}$	5	[141]
Phenolic compounds	Poly(ethylene terephthalate) carbon-encapsulated iron nanoparticles	Water	230	50	[142]
Alkylphenols	Molecularly imprinted polymer mesoporous silica-coated magnetic graphene oxide (MGO@mSiO_2)	Water	16.81–61.73	50	[143]

N/R, Not reported.

[148]. As a result, drug residues have been detected and reported in rivers, wastewater, sewage, groundwater, and lakes [149].

Because of their unique properties such as excellent efficiency, easy separation, and reusability as sorbent, magnetic nanomaterials are great alternatives to conventional adsorbents (such as activated carbon) in the removal of pharmaceuticals from environmental samples. The most recent and significant examples of applications of magnetic nanomaterials in the removal of pharmaceuticals are shown in Table 7.2 [126–135].

7.6.4 Removal of endocrine-disrupting chemicals

EDCs are a class of emerging pollutants with a variety of molecular structures that have been identified in environmental samples such as wastewater, runoff, surface water, landfill leachates, and groundwater [150,151]. The function or operation of the endocrine system can be affected in a variety of ways, such as a block or mimic of a natural hormone, leading to underproduction or overproduction of hormones [151,152]. Industrially relevant chemicals (alkylphenols and BPA), environmental pollutants (DDT, polychlorinated biphenyls, dioxins, and PAHs), pharmaceuticals (diethylstilbestrol and ethynylestradiol), and also natural compounds (plant estrogens, e.g., genistein and coumestrol) are different classes of EDCs [153].

The most recent and significant examples of the application of magnetic nanomaterials in the removal of EDCs are shown in Table 7.2 [136–143]. Different magnetic nanomaterials have been used for various EDCs, such as BPA and bisphenol AF (BPAF) [138], chlorophenols [139], polychlorinated biphenyls [140,141], phenolic compounds [142], and alkylphenols [143] in environmental samples.

7.6.5 Removal of oil

Efficient oil adsorption and recovery are of great global importance for environmental protection and future energy demand. Recently, environmental issues arising from oil pollutions [154] are becoming complicated because they not only threaten human health but also deteriorate the ecological environment. To overcome the potential consequences of oil spills, a wide variety of ecological activities and cleanup measures, including the removal of oil slicks with various sorbents, are used. An ideal oil spill cleanup sorbent should include such properties as superhydrophobicity; high uptake capacity and rate; resistance, retaining ability, and durability in aqueous media; ecofriendly effect; biodegradability; and recoverability of the absorbed oil [155]. Recently, the application of magnetic nanomaterials as promising sorbents for oil spills have attracted increasing attention. Qiao et al. comprehensively reviewed the application of magnetic adsorbents for oil spill remediation [156].

Neelamegan et al. prepared hydrophobic magnetite nanoparticles–anchored cellulose and lignin-based CNT composites for rapid oil spill cleanup [156]. The

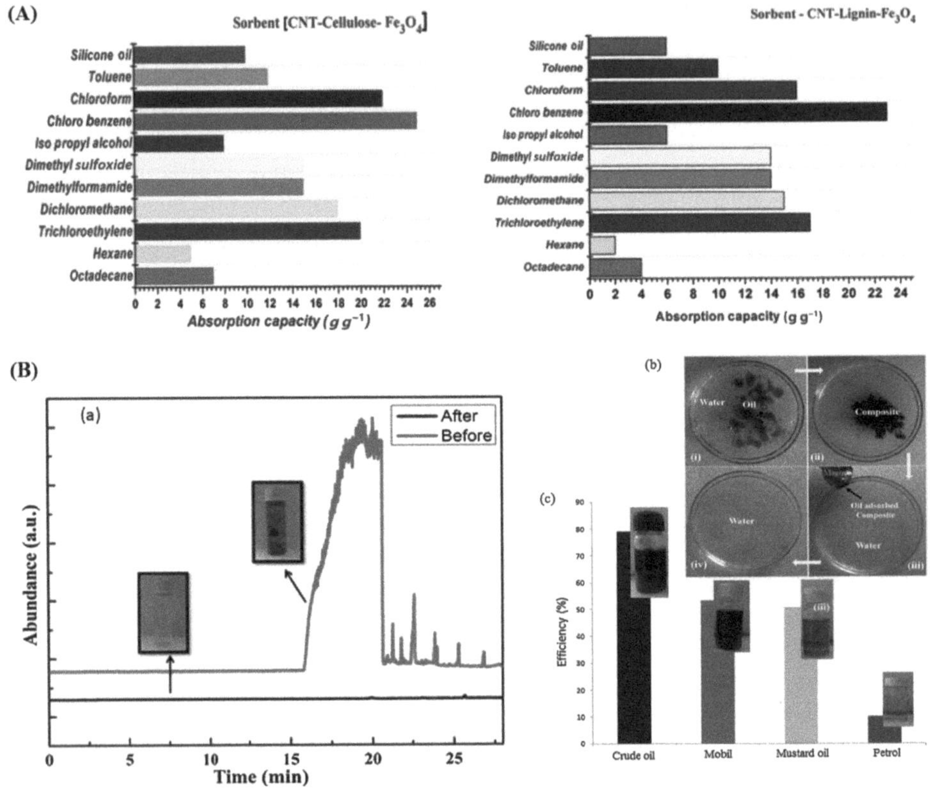

Figure 7.5 (A) Oil absorption measurements for CNT-g-cellulose/Fe_3O_4 (LHS) and CNT-g-lignin/Fe_3O_4 (RHS) in various oils/organic solvents [155]; (B) (a) GC-MS plot of the surface-modified composite before and after oil adsorptions; (b) removal process of crude oil ((i)–(iv)) from the water surface by nanocomposite under the magnetic field; (c) selectivity of surface-modified g-C_3N_4@$FeNi_3$ nanocomposite for crude oil [157].

sorption capacities of CNT-g-Cellulose/Fe_3O_4 and CNT-g-Lignin/Fe_3O_4 are calculated as 15–25 and 10–23 $g\,g^{-1}$, respectively, for various oil/organic solvents depending on the density of the liquid (Fig. 7.5A). In another work, Talukdar et al. prepared a novel nanocomposite for the efficient and rapid sorption of crude oil by combining porous and nanosheets structure (from graphitic C_3N_4), along with magnetic property (from $FeNi_3$) and further surface functionalization by fatty acid [157].

Their results showed that the graphitic sheets can selectively adsorb crude oil with the enhancement of thickness up to 9 folds and mass 4.5 times that of the pristine nanocomposite system. Fig. 7.5B shows GC-MS chromatograms of the oily sample before and after oil adsorption and the selectivity of sorbent in crude oil by the prepared composite. Also, the nanocomposite can be recovered and reused in an ecofriendly manner.

7.7 Conclusions and future trends

Perhaps the area in which MNMs have had the greatest impact is environmental applications. Nowadays, MNMs are extensively used as efficient, novel, and low-cost sorbents for the preconcentration of trace levels of pollutants and/or removal of the high concentration of them in environmental samples. MNMs universally exhibit several advantages such as large surface-area-to-volume ratio, selective adsorption capacity, fast extraction kinetic, low cost, and easy collection and reusability of the sorbent, making them very attractive and promising materials in environmental applications. Various MNMs based on advanced materials such as MOFs, GO, CNTs, COFs, and LDHs are widely used to design and prepare novel magnetic nanocomposites. However, the synthesis of new hybrid sorbents is a continuously growing field, with more efficient sorbents and different extraction mechanisms being introduced. With the increasing need to analyze more complex environmental samples, the design and preparation of novel MNMs–based sorbents will continue to develop, targeting specific organic pollutants and sample matrices. In addition, such sorbents present promising removal efficiency and sorbent capacity for the removal of large contents of environmental pollutants such as dyes, heavy metals, pharmaceutical, and more recently oil spill.

References

[1] M. Faraji, Y. Yamini, M. Rezaee, Magnetic nanoparticles: synthesis, stabilization, functionalization, characterization, and applications, J. Iran. Chem. Soc. 7 (2010) 1–37.
[2] B. Hu, M. He, B. Chen, Solid-Phase Extraction, first ed., Elsevier, 2020 (Chapter 9).
[3] M. Šafaříková, I. Šafařík, Magnetic solid-phase extraction, J. Magn. Magn. Mater. 194 (1999) 108–112.
[4] M. Faraji, Recent analytical applications of magnetic nanoparticles, Nanochem. Res. 1 (2016) 264–290.
[5] M. Faraji, Y. Yamini, M. Gholami, Recent advances and trends in applications of solid-phase extraction techniques in food and environmental analysis, Chromatographia 82 (2019) 1207–1249.
[6] S. Wadhawan, A. Jain, J. Nayyar, S.K. Meht, Role of nanomaterials as adsorbents in heavy metal ion removal from waste water: a review, J. Water Process. Eng. 33 (2020) 101038.
[7] A. Azzouza, S.K.R. Kailasab, S.S. Lee, A.J. Rascona, E. Ballesteros, M. Zhang, et al., Review of nanomaterials as sorbents in solid-phase extraction for environmental samples, Trends Anal. Chem. 108 (2018) 347–369.
[8] S. Di, T. Ning, J. Yu, P. Chen, H. Yu, J. Wang, et al., Recent advances and applications of magnetic nanomaterials in environmental sample analysis, Trends Anal. Chem. 126 (2020) 115864.
[9] K. Zhou, X. Zhou, J. Liu, Z. Huang, Application of magnetic nanoparticles in petroleum industry: a review, J. Pet. Sci. Eng. 188 (2020) 106943.

[10] N. Li, H. Jiang, X. Wang, X. Wang, G. Xu, B. Zhang, et al., Recent advances in graphene-based magnetic composites for magnetic solid-phase extraction, Trends Anal. Chem. 102 (2018) 60–74.
[11] Q. Han, Z. Wang, J. Xia, S. Chen, X. Zhang, M. Ding, Facile and tunable fabrication of Fe_3O_4/graphene oxide nanocomposites and their application in the magnetic solid-phase extraction of polycyclic aromatic hydrocarbons from environmental water samples, Talanta 101 (2012) 388–395.
[12] R.Z. Zhang, P. Su, Y. Yang, Microwave-assisted preparation of magnetic nanoparticles modified with graphene oxide for the extraction and analysis of phenolic compounds, J. Sep. Sci. 37 (2014) 3339–3346.
[13] G. Bharath, E. Alhseinat, N. Ponpandian, M.A. Khan, M.R. Siddiqui, F. Ahmed, et al., Development of adsorption and electrosorption techniques for removal of organic and inorganic pollutants from wastewater using novel magnetite/porous graphene-based nanocomposites, Sep. Purif. Technol. 188 (2017) 206–218.
[14] D. Huang, J. Wu, L. Wang, X. Liu, J. Meng, X. Tang, et al., Novel insight into adsorption and co-adsorption of heavy metal ions and an organic pollutant by magnetic grapheme nanomaterials in water, Chem. Eng. J. 358 (2019) 1399–1409.
[15] M. Zhang, G. Huang, J. Huang, W. Chen, Three-dimensional multi-walled carbon nanotubes@g-C_3N_4@Fe_3O_4 nanocomposites-based magnetic solid phase extraction for the determination of polycyclic aromatic hydrocarbons in water samples, Microchem. J. 142 (2018) 385–393.
[16] A.H. El-Sheikh, R.F. Qawariq, J.I. Abdelghani, Adsorption and magnetic solid-phase extraction of NSAIDs from pharmaceutical wastewater using magnetic carbon nanotubes: effect of sorbent dimensions, magnetite loading and competitive adsorption study, Environ. Technol. Innov. 16 (2019) 100496.
[17] P. Zhao, Z. Wang, K. Li, X. Guo, L. Zhao, Multi-residue enantiomeric analysis of 18 chiral pesticides in water, soil and river sediment using magnetic solid-phase extraction based on amino modified multiwalled carbon nanotubes and chiral liquid chromatography coupled with tandem mass spectrometry, J. Chromatogr. A 1568 (2018) 8–21.
[18] M. Yazdanpanah, S. Nojavan, Micro-solid phase extraction of some polycyclic aromatic hydrocarbons from environmental water samples using magnetic β-cyclodextrin carbon nano-tube composite as a sorbent, J. Chromatogr. A 1585 (2019) 34–45.
[19] A.H. El-Sheikh, E.M. Elayyan, B.A. Salameh, Magnetic solid phase extraction of cadmium and lead from water prior to FAAS determination: which is better? Sorbent modification with organic acids or adding organic acids to the extraction medium, J. Environ. Chem. Eng. 7 (2019) 103377.
[20] G. Li, L. Xia, J. Dong, Y. Chen, Y. Li, Solid-Phase Extraction, first ed., Elsevier, 2020 (Chapter 10).
[21] J.R. Li, J. Sculley, H.C. Zhou, Metal-organic frameworks for separations, Chem. Rev. 112 (2012) 869–932.
[22] Y. Wang, M. Rui, G. Lu, Recent applications of metal-organic frameworks in sample pretreatment, J. Sep. Sci. 41 (2018) 180–194.
[23] O.M. Yaghi, M. O'Keeffe, N.W. Ockwig, H.K. Chae, M. Eddaoudi, J. Kim, Reticular synthesis and the design of new materials, Nature 423 (2003) 705–714.
[24] S. Kitagawa, R. Matsuda, Chemistry of coordination space of porous coordination polymers, Coord. Chem. Rev. 251 (2007) 2490–2509.
[25] G. Liu, L. Li, X. Huang, S. Zheng, D. Xu, X. Xu, et al., Determination of triazole pesticides in aqueous solution based on magnetic graphene oxide functionalized MOF-199 as solid phase extraction sorbents, Microporous Mesoporous Mater. 270 (2018) 258–264.

[26] Z. Zhou, Y. Fu, Q. Qin, X. Lu, X. Shi, C. Zhao, et al., Synthesis of magnetic mesoporous metal-organic framework-5 for the effective enrichment of malachite green and crystal violet in fish samples, J. Chromatogr. A 1560 (2018) 19–25.
[27] J. Abdi, N.M. Mahmoodi, M. Vossoughi, I. Alemzadeh, Synthesis of magnetic metal-organic framework nanocomposite (ZIF-8@SiO_2@$MnFe_2O_4$) as a novel adsorbent for selective dye removal from multicomponent systems, Microporous Mesoporous Mater. 273 (2019) 177–188.
[28] C. Xie, X. Huang, S. Wei, C. Xiao, J. Cao, Z. Wang, Novel dual-template magnetic ion imprinted polymer for separation and analysis of Cd^{2+} and Pb^{2+} in soil and food, J. Clean. Prod. 262 (2020) 121387.
[29] A. Azizi, C.S. Bottaro, A critical review of molecularly imprinted polymers for the analysis of organic pollutants in environmental water samples, J. Chromatogr. A 1614 (2020) 460603.
[30] S. Ansari, Application of magnetic molecularly imprinted polymer as a versatile and highly selective tool in food and environmental analysis: recent developments and trends, Trends Anal. Chem. 90 (2017) 89–106.
[31] B. Benedetti, M.D. Carro, E. Magi, Multivariate optimization of an extraction procedure based on magnetic molecular imprinted polymer for the determination of polycyclic aromatic hydrocarbons in sea water, Microchem. J. 145 (2019) 1199–1206.
[32] Y.C. Lu, J.H. Mao, W. Zhang, C. Wang, M. Cao, X.D. Wang, et al., A novel strategy for selective removal and rapid collection of triclosan from aquatic environment using magnetic molecularly imprinted nanopolymers, Chemosphere 238 (2020) 124640.
[33] M. Sajid, C. Basheer, Layered double hydroxides: emerging sorbent materials for analytical extractions, Trends Anal. Chem. 75 (2016) 174–182.
[34] H. Wu, H. Zhang, W. Zhang, X. Yang, Z. Pan, D. Wang, Preparation of magnetic polyimide@ Mg-Fe layered double hydroxides core-shell composite for effective removal of various organic contaminants from aqueous solution, Chemosphere 219 (2019) 66–75.
[35] P. Zhang, S. Ouyang, P. Li, Y. Huang, R.L. Frost, Enhanced removal of ionic dyes by hierarchical organic three-dimensional layered double hydroxide prepared via soft-template synthesis with mechanism study, Chem. Eng. J. 360 (2019) 1137–1149.
[36] Q. Yuan, N. Li, Y. Chi, W. Geng, W. Yan, Y. Zhao, et al., Effect of large pore size of multifunctional mesoporous microsphere on removal of heavy metal ions, J. Hazard. Mater. 254–255 (2013) 157–165.
[37] L. Huang, J. Miao, Q. Shuai, Carboxyl-functionalized magnetic porous organic polymers as efficient adsorbent for wastewater remediation, J. Taiwan. Inst. Chem. Eng. (2020) 1–6. Available from: https://doi.org/10.1016/j.jtice.2020.02.008.
[38] P.N.E. Diagboya, E.D. Dikio, Silica-based mesoporous materials; emerging designer adsorbents for aqueous pollutants removal and water treatment, Microporous Mesoporous Mater. 266 (2018) 252–267.
[39] N. Li, J. Du, D. Wu, J. Liu, N. Li, Z. Sun, et al., Recent advances in facile synthesis and applications of covalent organic framework materials as superior adsorbents in sample pretreatment, Trends Anal. Chem. 108 (2018) 154–166.
[40] C.-H. Yang, J.-S. Chang, D.-J. Lee, Chemically stable covalent organic framework as adsorbent from aqueous solution: a mini-review, J. Taiwan. Inst. Chem. Eng. (2020) 1–13. Available from: https://doi.org/10.1016/j.jtice.2020.02.008.
[41] W.H. Ji, Y.S. Guo, X. Wang, X.F. Lu, D.S. Guo, Amino-modified covalent organic framework as solid phase extraction absorbent for determination of carboxylic acid pesticides in environmental water samples, J. Chromatogr. A 1595 (2019) 11–18.

[42] A. Wen, G. Li, D. Wu, Y. Yu, Y. Yang, N. Hu, et al., Sulphonate functionalized covalent organic framework-based magnetic sorbent for effective solid phase extraction and determination of fluoroquinolones, J. Chromatogr. A 1612 (2020) 460651.
[43] J.-Y. Ren, X.-L. Wang, X.-L. Li, M.-L. Wang, R.-S. Zhao, J.-M. Lin, Magnetic covalent triazine-based frameworks as magnetic solid-extraction adsorbents for sensitive determination of perfluorinat compounds in environmental water samples, Anal. Bioanal. Chem. 410 (2018) 1657–1665.
[44] V. Romero, S.P.S. Fernandes, L.R. Lorenzo, Y.V. Koleńko, B. Espiña, L.M. Salonen, Recyclable magnetic covalent organic framework for the extraction of marine biotoxins, Nanoscale 11 (2019) 6072–6079.
[45] L. Huang, R. Shen, R. Liu, Q. Shuai, Thiol-functionalized magnetic covalent organic frameworks by a cutting strategy for efficient removal of Hg, J. Hazard. Mater. 392 (2020) 122320.
[46] X. He, Y. Zhou, W. Yang, S. Li, T. Liu, T. Wang, et al., Microwave assisted magnetic solid phase extraction using a novel amino-functionalized magnetic framework composite of type Fe_3O_4-NH_2@MIL-101(Cr) for the determination of organochlorine pesticides in soil samples, Talanta 196 (2019) 572–578.
[47] Z. Liu, P. Qi, J. Wang, Z. Wang, S. Di, H. Xu, et al., Development, validation, comparison, and implementation of a highly efficient and effective method using magnetic solid-phase extraction with hydrophilic-lipophilic-balanced materials for LC-MS/MS analysis of pesticides in seawater, Sci. Total Environ. 708 (2020) 135221.
[48] J. Fan, Z. Liu, J. Li, W. Zhou, H. Gao, S. Zhang, et al., PEG-modified magnetic Schiff base network-1 materials for the magnetic solid phase extraction of benzoylurea pesticides from environmental water samples, J. Chromatogr. A 1619 (2020) 460950.
[49] J. Ma, G. Wu, S. Li, W. Tan, X. Wang, J. Li, et al., Magnetic solid-phase extraction of heterocyclic pesticides in environmental water samples using metal-organic frameworks coupled to high performance liquid chromatography determination, J. Chromatogr. A 1553 (2018) 57–66.
[50] Y. Liu, Z. Gao, R. Wu, Z. Wang, X. Chen, T.W.D. Chan, Magnetic porous carbon derived from a bimetallic metal–organic framework for magnetic solid-phase extraction of organochlorine pesticides from drinking and environmental water samples, J. Chromatogr. A 1479 (2017) 55–61.
[51] M. Zhang, J. Yang, X. Geng, Y. Li, Z. Zha, S. Cui, et al., Magnetic adsorbent based on mesoporous silica nanoparticles for magnetic solid phase extraction of pyrethroid pesticides in water samples, J. Chromatogr. A 1598 (2019) 20–29.
[52] P. Wang, M. Luo, D. Liu, J. Zhan, X. Liu, F. Wang, et al., Application of a magnetic graphene nanocomposite for organophosphorus pesticide extraction in environmental water samples, J. Chromatogr. A 1535 (2018) 9–16.
[53] V. Moreno, E.J. Llorent-Martínez, M. Zougagh, A. Ríos, Synthesis of hybrid magnetic carbon nanotubes–C18-modified nano SiO_2 under supercritical carbon dioxide media and their analytical potential for solid-phase extraction of pesticides, J. Supercrit. Fluids 137 (2018) 66–73.
[54] Y. Zhou, J. Zhu, J. Yang, Y. Lv, Y. Zhu, W. Bi, et al., Magnetic nanoparticles speed up mechanochemical solid phase extraction with enhanced enrichment capability for organochlorines in plants, Anal. Chim. Acta 1066 (2019) 49–57.
[55] S.M. Shahrebabak, M. Saber-Tehrani, M. Faraji, M. Shabanian, P. Aberoomand-Azar, Simultaneous magnetic solid phase extraction of acidic and basic pesticides using triazine-based polymeric network modified magnetic nanoparticles/graphene oxide nanocomposite in water and food samples, Microchem. J. 146 (2019) 630–639.

[56] A. Bazmandegan-Shamili, A.M. Haji Shabani, S. Dadfarnia, M. Rohani Moghadam, M. Saeidi, Preparation of magnetic mesoporous silica composite for the solid-phase microextraction of diazinon and malathion before their determination by high-performance liquid chromatography, J. Sep. Sci. 40 (2017) 1631–1854.

[57] D.-D. Wang, Y. Zhao, M.-N.O. yang, H.-M. Guo, Z.-H. Yang, Magnetic polydopamine modified with deep eutectic solvent for the magnetic solid-phase extraction of sulfonylurea herbicides in water samples, J. Chromatogr. A 1601 (2019) 53–59.

[58] S. Xia, Z. Cai, J. Donga, S. Wanga, Y. Wanga, H. Kang, et al., Preparation of porous zinc ferrite/carbon as a magnetic-assisted dispersive miniaturized solid phase extraction sorbent and its application, J. Chromatogr. A 1567 (2018) 73–80.

[59] M. Mei, J. Pang, X. Huang, Q. Luo, Magnetism-reinforced in-tube solid phase microextraction for the online determination of trace heavy metal ions in complex samples, Anal. Chim. Acta 1090 (2019) 82–90.

[60] Y. Yuan, Y. Wu, H. Wang, Y. Tong, X. Sheng, Y. Sun, et al., Simultaneous enrichment and determination of cadmium and mercury ions using magnetic PAMAM dendrimers as the adsorbents for magnetic solid phase extraction coupled with high performance liquid chromatography, J. Hazard. Mater. 386 (2020) 121658.

[61] S. Zhu, B. Chen, M. He, T. Huang, B. Hu, Speciation of mercury in water and fish samples by HPLC-ICP-MS after magnetic solid phase extraction, Talanta 17 (2017) 213–219.

[62] C. Peng, M. He, B. Chen, L. Huang, B. Hu, Magnetic sulfur-doped porous carbon for preconcentration of trace mercury in environmental water prior to ICP-MS detection, Analyst 142 (2017) 4570–4579.

[63] Q. Zhou, M. Lei, Y. Liu, Y. Wu, Y. Yuan, Simultaneous determination of cadmium, lead and mercury ions at trace level by magnetic solid phase extraction with Fe@Ag@Dimercaptobenzene coupled to high performance liquid chromatography, Talanta 175 (2017) 194–199.

[64] K. Molaei, H. Bagheri, A.A. Asgharinezhad, H. Ebrahimzadeh, M. Shamsipurd, SiO_2-coated magnetic graphene oxide modified with polypyrrole–polythiophene: a novel and efficient nanocomposite for solid phase extraction of trace amounts of heavy metals, Talanta 167 (2017) 607–616.

[65] J.C. García-Mesa, P. Montoro Leal, M.M. López Guerrero, E.I. Vereda Alonso, Simultaneous determination of noble metals, Sb and Hg by magnetic solid phase extraction on line ICP OES based on a new functionalized magnetic graphene oxide, Microchem. J. 150 (2019) 104141.

[66] A. Ghiasi, A. Malekpour, Octyl coated cobalt-ferrite/silica core-shell nanoparticles for ultrasonic assisted-magnetic solid-phase extraction and speciation of trace amount of chromium in water samples, Microchem. J. 154 (2020) 104530.

[67] M. Llaver, F.A. Casado-Carmona, R. Lucena, S. Cárdenas, R.G. Wuilloud, Ultra-trace tellurium preconcentration and speciation analysis in environmental samples with a novel magnetic polymeric ionic liquid nanocomposite and magnetic dispersive micro-solid phase extraction with flow-injection hydride generation atomic fluorescence spectrometry detection, Spectrochim. Acta Part B: Atom. Spectrosc. 162 (2019) 105705.

[68] M. He, S. Su, B. Chen, B. Hu, Simultaneous speciation of inorganic selenium and tellurium in environmental water samples by polyaniline functionalized magnetic solid phase extraction coupled with ICP-MS detection, Talanta 207 (2020) 120314.

[69] A.S. Gugushe, A. Mpupa, P.N. Nomngongo, Ultrasound-assisted magnetic solid phase extraction of lead and thallium in complex environmental samples using magnetic multi-walled carbon nanotubes/zeolite nanocomposite, Microchem. J. 149 (2019) 103960.

[70] L. Suo, X. Dong, X. Gao, J. Xu, Z. Huang, J. Ye, et al., Silica-coated magnetic graphene oxide nanocomposite based magnetic solid phase extraction of trace amounts of heavy metals in water samples prior to determination by inductively coupled plasma mass spectrometry, Microchem. J. 149 (2019) 104039.
[71] F. Li, A. Gong, L. Qiu, W. Zhang, J. Li, Z. Liu, Diglycolamide-grafted Fe_3O_4/polydopamine nanomaterial as a novel magnetic adsorbent for preconcentration of rare earth elements in water samples prior to inductively coupled plasma optical emission spectrometry determination, Chem. Eng. J. 361 (2019) 1098–1109.
[72] M. Yazdanpanah, S. Nojavan, Polydopamine-assisted attachment of β-cyclodextrin onto iron oxide/silica core-shell nanoparticles for magnetic dispersive solid phase extraction of aromatic molecules from environmental water samples, J. Chromatogr. A 1601 (2019) 9–20.
[73] V.C.L. de, B. Caetano, G. da, C. Cunha, R.V.M. Oliveira, M. da, et al., Magnetic hybrid support for ultrasound-assisted magnetic solid phase extraction of polycyclic aromatic hydrocarbons from produced water, Microchem. J. 146 (2019) 1195–1203.
[74] X. Yang, Y. Yin, Y. Zong, T. Wan, X. Liao, Magnetic nanocomposite as sorbent for magnetic solid phase extraction coupled with high performance liquid chromatography for determination of polycyclic aromatic hydrocarbons, Microchem. J. 145 (2019) 26–34.
[75] A. Azizi, F. Shahhoseini, C.S. Bottaro, Magnetic molecularly imprinted polymers prepared by reversible addition fragmentation chain transfer polymerization for dispersive solid phase extraction of polycyclic aromatic hydrocarbons in water, J. Chromatogr. A 1610 (2020) 460534.
[76] Q. Zhou, M. Lei, Y. Wu, Y. Yuan, Magnetic solid phase extraction of typical polycyclic aromatic hydrocarbons from environmental water samples with metal organic framework MIL-101 (Cr) modified zero valent iron nano-particles, J. Chromatogr. A 1487 (2017) 22–29.
[77] M.S. Shahriman, M.R. Ramachandran, N.N. Mohamad Zain, S. Mohamad, N.S. Abdul Manan, S.M. Yaman, Polyaniline-dicationic ionic liquid coated with magnetic nanoparticles composite for magnetic solid phase extraction of polycyclic aromatic hydrocarbons in environmental samples, Talanta 178 (2018) 211–221.
[78] S. Qin, Y. Fan, X. Mou, X. Li, S. Qi, Preparation of phenyl-modified magnetic silica as a selective magnetic solid-phase extraction adsorbent for polycyclic aromatic hydrocarbons in soils, J. Chromatogr. A 1568 (2018) 29–37.
[79] S. Mahpishanian, H. Sereshti, M. Ahmadvand, A nanocomposite consisting of silica-coated magnetite and phenyl-functionalized graphene oxide for extraction of polycyclic aromatic hydrocarbon from aqueous matrices, J. Environ. Sci. 55 (2017) 164–173.
[80] Á. Santana-Mayor, B. Socas-Rodríguez, M.M. Afonso, J.A. Palenzuela-López, M.Á. Rodríguez-Delgado, Reduced graphene oxide-coated magnetic-nanoparticles as sorbent for the determination of phthalates in environmental samples by micro-dispersive solid-phase extraction followed by ultra-high-performance liquid chromatography tandem mass spectrometry, J. Chromatogr. A 1565 (2018) 36–47.
[81] J.I. Abdelghani, R.S. Freihat, A.H. El-Sheikh, Magnetic solid phase extraction of phthalate products from bottled, injectable and tap waters using graphene oxide: effect of oxidation method of graphene, J. Environ. Chem. Eng. 7 (2019) 103527.
[82] W. Wang, R. Ma, Q. Wu, C. Wang, Z. Wang, Fabrication of magnetic microsphere-confined grapheme for the preconcentration of some phthalate esters from environmental water and soybean milk samples followed by their determination by HPLC, Talanta 109 (2013) 133–140.

[83] W. Liu, M. Fizir, F. Hua, A. Li, X. Huia, J. Zha, et al., Mixed hemimicelle solid-phase extraction based on magnetic halloysite nanotubes and ionic liquids for the determination and extraction of azo dyes in environmental water samples, J. Chromatogr. A 1551 (2018) 10−20.
[84] E. Yilmaz, S. Salem, G. Sarp, S. Aydin, K. Sahin, I. Korkmaz, et al., TiO_2 nanoparticles and C-nanofibers modified magnetic Fe_3O_4 nanospheres (TiO_2@Fe_3O_4@C−NF): a multifunctional hybrid material for magnetic solid phase extraction of ibuprofen and photocatalytic degradation of drug molecules and azo dye, Talanta 213 (2020) 120813.
[85] Q. Zhou, M. Lei, Y. Wu, X. Zhou, H. Wang, Y. Sun, et al., Magnetic solid phase extraction of bisphenol A, phenol and hydroquinone from water samples by magnetic and thermo dual-responsive core-shell nanomaterial, Chemosphere 238 (2020) 124621.
[86] Y. Zhao, X.-L. Bai, T. Song, G.-L. Zhan, Y.-C. Yuan, Y.-M. Liu, et al., Determination of environmental estrogens and bisphenol A in water samples by ultra-high performance liquid chromatography coupled to Q-Exactive high resolution mass spectrometry after magnetic solid-phase extraction, Microchem. J. 151 (2019) 104212.
[87] Y. Wu, C. Chen, Q. Zhou, Q.X. Li, Y. Yuan, Y. Tong, et al., Polyamidoamine dendrimer decorated nanoparticles as an adsorbent for magnetic solid-phase extraction of tetrabromobisphenol A and 4-nonylphenol from environmental water samples, J. Colloid Interface Sci. 539 (2019) 361−369.
[88] M.-Q. Cai, J. Su, J.-Q. Hu, Q. Wang, C.-Y. Dong, S.-D. Pan, et al., Planar graphene oxide-based magnetic ionic liquid nanomaterial for extraction of chlorophenols from environmental water samples coupled with liquid chromatography−tandem mass spectrometry, J. Chromatogr. A 1459 (2016) 38−46.
[89] H. Bagheri, M. Manouchehri, M. Allahdadlalouni, A magnetic multifunctional dendrimeric coating on a steel fiber for solid phase microextraction of chlorophenols, Microchim. Acta 184 (2017) 2201−2209.
[90] X. Chen, N. Ding, H. Zang, H. Yeung, R.-S. Zhao, C. Cheng, et al., Fe3O4@MOF core-shell magnetic microspheres for magnetic solid-phase extraction of polychlorinated biphenyls from environmental water samples, J. Chromatogr. A 1304 (2013) 241−245.
[91] R. Xiao, S. Wang, M.H. Ibrahim, H.I. Abdu, D. Shan, J. Chen, et al., Three dimensional hierarchical frameworks based on molybdenum disulfide graphene oxide-supported magnetic nanoparticles for enrichment fluoroquinolone antibiotics in water, J. Chromatogr. A 1593 (2019) 1−8.
[92] M. Wang, H. Yuan, W. Deng, W. Bi, X. Yang, A Taiji-principle-designed magnetic porous C-doped graphitic carbon nitride for environment-friendly solid phase extraction of pollutants from water samples, J. Chromatogr. A 1412 (2015) 12−21.
[93] Y. Liang, Z. Li, P. Shi, C. Ling, X. Chen, Q. Zhou, et al., Performance of a novel magnetic solid-phase-extraction microsphere and its application in the detection of organic micropollutants in the Huai River China, Environ. Pollut. 252 (2019) 196−204.
[94] J. González-Sálamo, B. Socas-Rodríguez, J. Hernández-Borges, M.Á. Rodríguez-Delgado, Determination of phthalic acid esters in water samples using core-shell poly (dopamine) magnetic nanoparticles and gas chromatography tandem mass spectrometry, J. Chromatogr. A 1530 (2017) 35−44.
[95] W. Li, J. Zhang, W. Zhu, P. Qin, Q. Zhou, M. Lu, et al., Facile preparation of reduced graphene oxide/$ZnFe_2O_4$ nanocomposite as magnetic sorbents for enrichment of estrogens, Talanta 208 (2020) 120440.
[96] Q.-L. Li, L.-L. Wang, X. Wanga, M.-L. Wang, R.-S. Zhao, Magnetic metal-organic nanotubes: an adsorbent for magnetic solid-phase extraction of polychlorinated biphenyls from environmental and biological samples, J. Chromatogr. A 1449 (2016) 39−47.

[97] S. Zeng, N. Gan, R. Weideman-Mera, Y. Cao, T. Li, W. Sang, Enrichment of polychlorinated biphenyl 28 from aqueous solutions using Fe_3O_4 grafted graphene oxide, Chem. Eng. J. 218 (2013) 108–115.
[98] M. Mei, X. Huang, Online analysis of five organic ultraviolet filters in environmental water samples using magnetism-enhanced monolith-based in-tube solid phase microextraction coupled with high-performance liquid chromatography, J. Chromatogr. A 1525 (2017) 1–9.
[99] R.G. Harvey, 978-0-471-18608-3 Polycyclic Aromatic Hydrocarbons, Wiley, New York, 1996.
[100] E. Manoli, C. Samara, Polycyclic aromatic hydrocarbons in natural waters: sources, occurrence and analysis, Trends Anal. Chem. 18 (1999) 417–428.
[101] U.S. Environmental Protection Agency, http://www.epa.gov/ttnamti1/files/ambient/airtox/tocomp99.pdf (accessed 21.12.13).
[102] P. Xu, G. Ming Zeng, D.L. Huang, C.L.G. Feng, S. Hu, M.H. Zhao, et al., Use of iron oxide nanomaterials in wastewater treatment: a review, Sci. Total Environ. 424 (2012) 1–10.
[103] I. Ahmad, W.A. Siddiqui, T. Ahmad, Synthesis and characterization of molecularly imprinted magnetite nanomaterials as a novel adsorbent for the removal of heavy metals from aqueous solution, J. Mater. Res. Technol. 8 (2019) 4239–4252.
[104] D. Huang, B. Xu, J. Wu, P.C. Brookes, J. Xu, Adsorption and desorption of phenanthrene by magnetic graphene nanomaterials from water: roles of pH, heavy metal ions and natural organic matter, Chem. Eng. J. 368 (2019) 390–399.
[105] L. Yin, S. Song, X. Wang, F. Niu, R. Ma, S. Yu, et al., Rationally designed core-shell and yolk-shell magnetic titanate nanosheets for efficient U(VI) adsorption performance, Environ. Pollut. 238 (2018) 725–738.
[106] J.-C. Yang, X.-B. Yin, $CoFe_2O_4$@MIL-100(Fe) hybrid magnetic nanoparticles exhibit fast and selective adsorption of arsenic with high adsorption capacity, Sci. Rep. 7 (2017) 40955.
[107] X. Wang, Y. Liu, H. Pang, S. Yu, X. Wang, Y. Ai, et al., Effect of graphene oxide surface modification on the elimination of Co(II) from aqueous solutions, Chem. Eng. J. 34415 (2018) 380–390.
[108] S. Srivastava, S. Bhushan, A. Monoj, K. Mondal, Synthesis, characterization and application of *Lagerstroemia speciosa* embedded magnetic nanoparticle for Cr(VI) adsorption from aqueous solution, J. Environ. Sci. 55 (2017) 283–293.
[109] Z. Wang, J. Zhang, T. Wenc, X. Liu, Y. Wangd, H. Yang, et al., Highly effective remediation of Pb(II) and Hg(II) contaminated wastewater and soil by flower-like magnetic MoS_2 nanohybrid, Sci. Total Environ. 699 (2020) 134341.
[110] Y. Wu, L. Chen, X. Long, X. Zhang, B. Pan, J. Qian, Multi-functional magnetic water purifier for disinfection and removal of dyes and metal ions with superior reusability, J. Hazard. Mater. 347 (2018) 160–167.
[111] X. Zhang, Y. Zhang, X. Zhang, S. Li, Y. Huang, Nitrogen rich core–shell magnetic mesoporous silica as an effective adsorbent for removal of silver nanoparticles from water, J. Hazard. Mater. 337 (2017) 1–9.
[112] Z. Lu, R. Tan, W. Chu, S. Tang, W. Xu, W. Song, Synthesis of $SrHPO_4/Fe_3O_4$ magnetic nanocomposite and its application on Pb^{2+} removal from aqueous solutions, Microchem. J. 142 (2018) 152–158.
[113] X. Wei, P.J. Sugumaran, E. Peng, X.L. Liu, J. Ding, Low-field dynamic magnetic separation by self-fabricated magnetic meshes for efficient heavy metal removal, ACS Appl. Mater. Interfaces 9 (2017) 36772–36782.

[114] K. Chen, J.Y. He, Y.L. Li, X.G. Cai, K.S. Zhang, T. Liu, et al., Removal of cadmium and lead ions from water by sulfonated magnetic nanoparticle adsorbents, J. Colloid Interface Sci. 494 (2017) 307–316.
[115] C. Hou, D. Zhao, S. Zhang, Y. Wang, Highly selective adsorption of Hg(II) by the monodisperse magnetic functional chitosan nano-biosorbent, Colloid Polym. Sci. 296 (2018) 547–555.
[116] G. Ren, X. Wang, P. Huang, Chromium (VI) adsorption from wastewater using porous magnetite nanoparticles prepared from titanium residue by a novel solid-phase reduction method, Sci. Total Environ. 607–608 (2017) 900–910.
[117] V.-P. Dinh, N.Q. Tran, N.-Q.-T. Le, Q.-H. Tran, T.D. Nguyen, V.T. Le, Facile synthesis of $FeFe_2O_4$ magnetic nanomaterial for removing methylene blue from aqueous solution, Prog. Nat. Sci. Mater. Int. 29 (2019) 648–654.
[118] A. Dash, M.T. Ahmed, R. Selvaraj, Mesoporous magnetite nanoparticles synthesis using the Peltophorum pterocarpumpod extract, their antibacterial efficacy against pathogens and ability to remove a pollutant dye, J. Mol. Struct. 1178 (2019) 268–273.
[119] S.F. Soares, T.R. Simões, T. Trindade, A.L. Daniel-da-Silva, Highly efficient removal of dye from water using magnetic carrageenan/silica hybrid nano-adsorbents, Water Air Soil Pollut. 228 (2017) 87.
[120] R. Ni, Y. Wang, X. Wei, J. Chen, P. Xu, W. Xu, et al., Ionic liquid modified molybdenum disulfide and reduced graphene oxide magnetic nanocomposite for the magnetic separation of dye from aqueous solution, Anal. Chim. Acta 1054 (2019) 47–58.
[121] Q. Yang, S.S. Ren, Q. Zhao, R. Lu, H. Cheng, Z. Chen, et al., Selective separation of methyl orange from water using magnetic ZIF-67 composites, Chem. Eng. J. 333 (2018) 49–57.
[122] R. Sahraei, Z. Sekhavat Pour, M. Ghaemy, Novel magnetic bio-sorbent hydrogel beads based on modified gum tragacanth/graphene oxide: removal of heavy metals and dyes from water, J. Clean. Prod. 142 (2016) 1–12.
[123] M. Ehyaee, F. Safa, S. Shariati, Magnetic nanocomposite of multi-walled carbon nanotube as effective adsorbent for methyl violet removal from aqueous solutions: response surface modeling and kinetic study, Korean J. Chem. Eng. 34 (2017) 1051–1061.
[124] N. Belachew, D.R. Devi, K. Basavaiah, Green synthesis and characterisation of L-Serine capped magnetite nanoparticles for removal of Rhodamine B from contaminated water, J. Exp. Nanosci. 12 (2017) 114–128.
[125] B. Yu, Y. Bai, Z. Ming, H. Yang, L. Chen, X. Hu, et al., Adsorption behaviors of tetracycline on magnetic graphene oxide sponge, Mater. Chem. Phys. 198 (2017) 283–290.
[126] B. Huanga, Y. Liu, B. Li, S. Liu, G. Zeng, Z. Zeng, et al., Effect of Cu(II) ions on the enhancement of tetracycline adsorption by Fe_3O_4@SiO_2-Chitosan/graphene oxide nanocomposite, Carbohydr. Polym. 157 (2017) 576–585.
[127] S. Aydin, M.E. Aydin, F. Beduk, A. Ulvi, Removal of antibiotics from aqueous solution by using magnetic Fe_3O_4/red mud-nanoparticles, Sci. Total Environ. 67020 (2019) 539–546.
[128] J. Miao, F. Wang, Y. Chen, Y. Zhu, Y. Zhou, S. Zhang, The adsorption performance of tetracyclines on magnetic graphene oxide: a novel antibiotics absorbent, Appl. Surf. Sci. 475 (2019) 549–558.
[129] X. Hu, Y. Zhao, H. Wang, X. Tan, Y. Yang, Y. Liu, Efficient removal of tetracycline from aqueous media with a Fe_3O_4 nanoparticles@graphene oxide nanosheets assembly, Int. J. Environ. Res. Public. Health 14 (2017) 495–509.

[130] M. Li, Y. Liu, G. Zeng, S. Liu, X. Hu, D. Shu, et al., Tetracycline absorbed onto nitrilotriacetic acid-functionalized magnetic graphene oxide: influencing factors and uptake mechanism, J. Colloid Interface Sci. 485 (2017) 269–279.
[131] M. Li, Y. Liu, S. Liu, D. Shu, G. Zeng, X. Hu, et al., Cu(II)-influenced adsorption of ciprofloxacin from aqueous solutions by magnetic graphene oxide/nitrilotriacetic acid nanocomposite: competition and enhancement mechanisms, Chem. Eng. J. 319 (2017) 219–228.
[132] Ö. Kerkez-Kuyumcu, S.S. Bayazit, M. Abdel Salam, Antibiotic amoxicillin removal from aqueous solution using magnetically modified graphene nanoplatelets, J. Ind. Eng. Chem. 36 (2016) 198–205.
[133] Y. Yang, X. Hu, Y. Zhao, L. Cui, Z. Huang, J. Long, et al., Decontamination of tetracycline by thiourea-dioxide–reduced magnetic graphene oxide: effects of pH, ionic strength, and humic acid concentration, J. Colloid Interface Sci. 495 (2017) 68–77.
[134] M. Li, Y. Liu, S. Liu, G. Zeng, X. Hu, X. Tan, et al., Performance of magnetic graphene oxide/diethylenetriaminepentaacetic acid nanocomposite for the tetracycline and ciprofloxacin adsorption in single and binary systems, J. Colloid Interface Sci. 521 (2018) 150–159.
[135] F. Tan, M. Liu, S. Ren, Preparation of polydopamine-coated graphene oxide/Fe_3O_4 imprinted nanoparticles for selective removal of fluoroquinolone antibiotics in water, Sci. Rep. 7 (2017) 5735.
[136] T. Fernandes, S.F. Soares, T. Trindade, A.L. Daniel-da-Silva, Magnetic hybrid nanosorbents for the uptake of paraquat from water, Nanomaterials 7 (2017) 68.
[137] Q. Zhou, Y. Wang, J. Xiao, H. Fan, C. Chen, Preparation and characterization of magnetic nanomaterial and its application for removal of polycyclic aromatic hydrocarbons, J. Hazard. Mater. 371 (2019) 323–331.
[138] Y. Li, C.-X. Yang, X.-P. Yan, Controllable preparation of core-shell magnetic covalent-organic framework nanospheres for efficient adsorption and removal of bisphenols in aqueous solution, Chem. Commun. 53 (2017) 2511–2514.
[139] W. Liu, J. Zha, X. Meng, J. Wu, C. Sun, P. Qiu, et al., Efficient removal and trace determination of chlorophenols from water by mixed hemi/ad-micelle ionic liquid-coated magnetic graphene oxide and adsorption mechanism, Anal. Methods 9 (2017) 4581–4591.
[140] A.M. Gutierrez, R. Bhandari, J. Weng, A. Stromberg, T.D. Dziubla, J.Z. Hilt, Novel magnetic core-shell nanoparticles for the removal of polychlorinated biphenyls from contaminated water sources, Mater. Chem. Phys. 223 (2019) 68–74.
[141] M. Wanga, P. Liu, Y. Wang, D. Zhou, C. Ma, D. Zhang, et al., Core–shell superparamagnetic Fe_3O_4@β-CD composites for host–guest adsorption of polychlorinated biphenyls (PCBs), J. Colloid Interface Sci. 447 (2015) 1–7.
[142] P. Strachowski, W. Kaszuwara, M. Bystrzejewski, Novel magnetic composite adsorbent of phenolic compounds based on waste poly (ethylene terephthalate) and carbon encapsulated magnetic nanoparticles, N. J. Chem. 41 (2017) 12617–12630.
[143] X. Xie, X. Ma, L. Guo, Y. Fan, G. Zeng, M. Zhang, et al., Novel magnetic multi-templates molecularly imprinted polymer for selective and rapid removal and detection of alkylphenols in water, Chem. Eng. J. 357 (2019) 56–65.
[144] Y. Zhou, J. Lu, Y. Zhou, Y. Liu, Recent advances for dyes removal using novel adsorbents: a review, Environ. Pollut. 252 (2019) 352–365.
[145] T. Estlander, Allergic dermatoses and respiratory diseases from reactive dyes, Contact Dermatitis 18 (1988) 290–297.

[146] C.A. Gonzales, E. Riboli, G. Lopez-Abente, Bladder cancer among workers in the textile industry: results of a Spanish case-control study, Am. J. Ind. Med. 14 (1988) 673–680.
[147] L.Q. Zhao, F.M. Xue, B.W. Yu, J.R. Xie, X.L. Zhang, R.H. Wu, et al., TiO_2-graphene sponge for the removal of tetracycline, J. Nanopart. Res. 17 (2015) 16.
[148] T. aus der Beek, F.A. Weber, A. Bergmann, S. Hickmann, I. Ebert, A. Hein, et al., Pharmaceuticals in the environment—global occurrences and perspectives, Environ. Toxicol. Chem. 35 (2016) 823–835.
[149] K.O. K'oreje, L. Vergeynst, D. Ombaka, P. De Wispelaere, M. Okoth, H. Van Langenhove, et al., Occurrence patterns of pharmaceutical residues in wastewater, surface water and groundwater of Nairobi and Kisumu city, Kenya, Chemosphere 149 (2016) 238–244.
[150] S. Snyder, C. Lue-Hing, J. Cotruvo, J.E. Drewes, A. Eaton, R.C. Pleus, et al., Pharmaceuticals in the Water Environment. https://www.dcwater.com/waterquality/PharmaceuticalsNACWA.pdf, pp. 1–34.
[151] A. Bhatnagar, I. Anastopoulos, Adsorptive removal of bisphenol A (BPA) from aqueous solution: a review, Chemosphere 168 (2017) 885–902.
[152] Innovative Waste Consulting Services, Gainesville, Florida, Endocrine Disrupting and Pharmaceutical Compounds in Municipal Landfill Leachate, 2007.
[153] T. Zacharewski, Identification and assessment of endocrine disruptors: limitations of in vivo and in vitro assays, Environ. Health Perspect. 106 (1998) 577–582.
[154] J.M. Teal, R.W. Howarth, Oil spill studies: a review of ecological effects, Environ. Manag. 8 (1984) 27–43.
[155] H. Neelamegan, D.-K. Yang, G.-J. Lee, S. Anandan, J.J. Wu, Synthesis of magnetite nanoparticles anchored cellulose and lignin-based carbon nanotube composites for rapid oil spill cleanup, Mater. Today Commun. 22 (2020) 100746.
[156] K. Qiao, W. Tian, J. Bai, L. Wang, J. Zhao, Z. Du, et al., Application of magnetic adsorbents based on iron oxide nanoparticles for oil spill remediation: a review, J. Taiwan. Inst. Chem. Eng. 97 (2019) 227–236.
[157] M. Talukdar, S.K. Behera, K. Bhattacharya, P. Deb, Surface modified mesoporous g-C_3N_4@$FeNi_3$ as prompt and proficient magnetic adsorbent for crude oil recovery, Appl. Surf. Sci. 473 (2019) 275–281.

Application of magnetic nanomaterials in forensic chemistry

8

Miguel Ángel Aguirre, Lorena Vidal and Antonio Canals
Department of Analytical Chemistry and Food Science and University Institute of Materials, Faculty of Science, University of Alicante, Alicante, Spain

8.1 Introduction

Nowadays, nanotechnology represents an expanding area, which involves almost every field of science and more recently, a new class of nanomaterial [i.e., magnetic nanomaterial (MNM)] has been applied in the fields of nanomedicine, food safety, biotechnology, biosensors, among other areas [1–4]. MNMs are unique because of their magnetic behaviors, showing responses under the application of an external magnetic field. Besides, MNMs have the potential to be modified by endless functional agents, and as a consequence, the modification of the MNMs enhances the ever-existing properties along with the addition of some desired characteristics.

In terms of forensic science, errors should be avoided in decision-making as they affect both reaching the truth and judicial credibility. Further, when it is to be deployed in serious crime investigations where punishment can be severe, then the error cannot be tolerated. Some fields of forensic science such as detection of fingerprints, DNA/RNA analysis, narcotics, and explosives demand high accuracy.

In forensic chemistry, MNMs have gained extremely high attention, since they present excellent properties. Among the different MNMs investigated to date, MNM based on magnetite nanoparticles (MNPs) is the most popular in forensic chemistry. Fig. 8.1 shows a summary of all MNPs addressed in this review for several applications. Throughout the chapter, the MNMs discussed are (1) MNPs coated with gold [5], silica [6], and platinum [7]; (2) MNPs embedded into a metal-organic framework (MOF) [8]; (3) carbon material decorated with MNPs [9,10]; (4) MNPs functionalized with polymer coating [11]; and (5) MNPs bioconjugated with antibody [12]. For the sake of simplicity, some examples are not included in Fig. 8.1 such as (1) magnetic alloys [13,14], (2) maghemite (γ-Fe_2O_3) or hematite (α-Fe_2O_3) core [15,16], or (3) bismuth-substituted yttrium iron garnet [17].

Magnetic Nanomaterials in Analytical Chemistry. DOI: https://doi.org/10.1016/B978-0-12-822131-0.00008-X

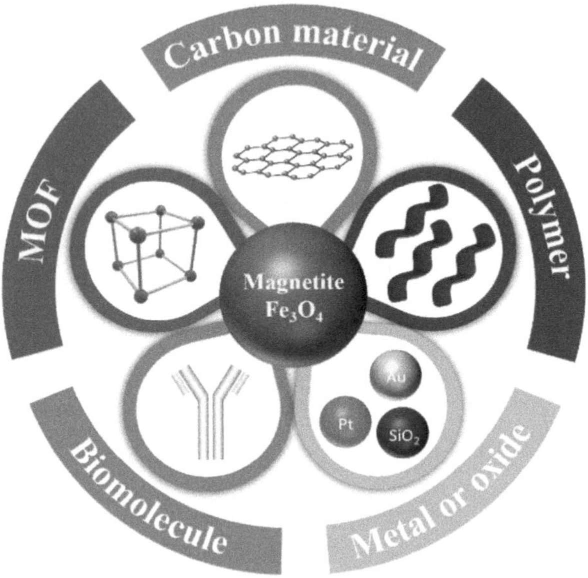

Figure 8.1 Diagram of different types of magnetite nanoparticles employed in forensic chemistry addressed in the chapter.

8.2 Applicability of magnetic nanomaterials in forensic chemistry

8.2.1 Fingerprint

Fingerprints are one of the oldest and most important pieces of evidence in forensic chemistry. The use of these ridge skin patterns on the end joint of the finger as a means of personal identification dates back many centuries. Consequently, a fingerprint match is widely accepted as certain evidence that identifies a specific person. The reasons why fingerprints are important in making unequivocal human identification are their two important features: (1) every fingerprint is unique to an individual and (2) fingerprints do not change throughout life unless damage has occurred to the dermal skin layer. This section not only addresses fingerprint visualization using MNM but also the analysis of fingerprints to obtain more information about the person.

8.2.1.1 Fingerprint visualization

The development of MNMs has also influenced the field of fingerprint visualization. During the last years, great efforts have been made to enhance the image of the fine structure of the fingerprint ridges on several surfaces with great results.

Lee et al. [11] reported the use of MNP-PDA (polydiacetylene) composite in fingerprint imaging. In this research, the addition of MNPs to appropriate diacetylene monomers produces powders that have unique advantages associated with their inherent magnetic properties as well as colorimetric responses. The authors demonstrated that when the new composite is deposited on solid substrates the resolution of the fingerprint image increases as a consequence of the fact that the powders align along with ridge patterns of the fingerprints when a magnetic field is applied. Also, the MNP-PDA composite improves the color change and after ultraviolet (UV) irradiation and subsequent heat treatment, the composite creates a red-colored and fluorescent image, achieving efficient visualization of fingerprints marked on papers.

Shahi et al. [13] reported a new lanthanide-based magnetic hybrid nanostructure composed of fluoride nanophosphor ($NaGd_{0.78}Er_{0.02}Yb_{0.2}F_4$) and β-diketones complex ($Eu[DBM]_3$ Phen) for fingerprint visualization (Fig. 8.2).

The MNM was found to be unique for the visualization of fingermarks since the surface-to-volume ratio is relatively high, which improves the attachment of particles with the fingerprint ridges. Furthermore, the unique structure of the MNM allows fingerprint imaging using dual-mode luminescence [i.e., UV and near-infrared (NIR) excitations, Fig. 8.2A]. Therefore the improvement of the luminescence quantum yield provides an opportunity to probe even very weak fingermarks. The authors demonstrated the sensitivity of the MNM for the visualization of fingerprints in glass slides and aluminum foils. On both surfaces, the MNM provides a good contrast image using UV and NIR light (Fig. 8.2B).

Boddis and Russell [18] focused their research in the visualization of fingermarks aged for up to 4 weeks stored on a glass substrate under four different conditions: at room temperature in the light; at room temperature in the dark; at an elevated temperature in the dark (55°C); and stored within a freezer (−20°C) in the dark. The authors concluded that the best quality images were obtained when the fingermarks were aged at 55°C. A possible explanation was the lack of oxidation or bacterial degradation of the fingermarks. Therefore the antibody−particle conjugates could be used in the detection of aged fingermarks in tropical climates, as pointed out by the authors.

Nevertheless, fingerprint images are difficult to visualize on rough surfaces such as clothes in comparison with smooth surfaces such as paper. Recently, Huang and Liu [7] synthesized Pt nanoclusters on the surface of MNPs modified with polyethyleneimine (PEI) to obtain a novel MNM with superparamagnetism and excellent fluorescence properties. The authors used the new composite on several complex surfaces: (1) a black plastic mouse, (2) a dark brown wood stool, (3) a piece of black paper, and (4) a black leather bag. They claimed that the fingerprint images show excellent ridge details with minimal background staining.

Huang and Tang [5] investigated several complex surfaces such as a window, a mouse, an iron railing, a printing paper, and a coin. The MNM developed by the authors is based on Au nanoclusters on the surface of MNPs. The magnetic composite with superparamagnetism and excellent fluorescence provides a new and efficient powder for the visualization of fingerprints. Additionally, the authors further investigated the visualization effect for aged fingerprint, the 1-day, 7-day, 14-day, 21-day fingerprints deposited on the abovementioned surfaces. The results showed

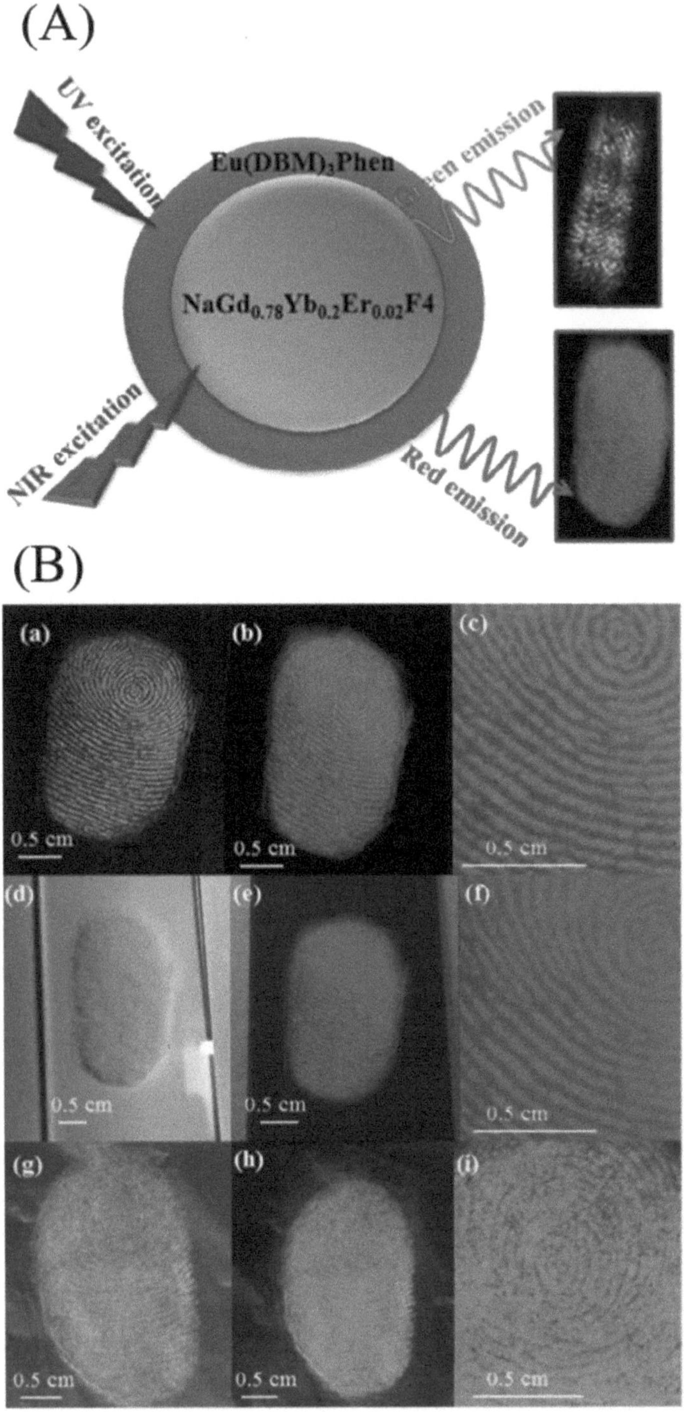
(A)
UV excitation
Eu(DBM)$_3$Phen
NaGd$_{0.78}$Yb$_{0.2}$Er$_{0.02}$F4
NIR excitation
Red emission
(B)
(a)
(b)
(c)
(d)
(e)
(f)
(g)
(h)
(i)
0.5 cm

that the MNM is detected in 1-day fresh fingerprints on all surfaces and the results also reveal that all fingerprints remained even for 21 days. Finally, among all the 21-day fingerprints deposited on different surfaces, the 21-day fingerprints deposited on the window and iron-railing surfaces obtain the biggest visualization rate.

8.2.1.2 Fingerprint analysis

The interest in obtaining—through the fingerprints—more information about the person of interest than pure identity pushed the forensic experts to elucidate on whether a person has taken narcotic drugs or has been in contact with explosive materials [19]. Focusing on the use of MNMs, the determination of chemical compounds is limited to the detection of the cotinine, a smoking biomarker. Cotinine is the major metabolite of nicotine and it is the most appropriate parameter to evaluate tobacco exposure and smoking status due to its higher stability and half-life when compared to nicotine.

Boddis and Russell [12] used MNPs functionalized with antibody (i.e., anticotinine) to distinguish between fingermarks from smoker and nonsmoker volunteers by detecting the presence of cotinine within the fingermarks. The authors demonstrated that there is a distinct difference in the appearance of the fluorescence images obtained from smoker and nonsmoker fingerprints. The image from the smoker showed definite binding on the ridges as pores, minutiae, and ridge shape were observed. In the image from a nonsmoker, the predominant binding is in the areas associated with the furrows, as determined by ridge counting. The authors concluded that the anticotinine antibody–particle conjugates can be used for both fingerprint visualization and distinguishing fingerprints deposited by smoker and nonsmokers.

Yang et al. [20] presented gold nanoparticle (AuNPs) embedded onto a magnetic network nanostructure based on IP_6 (inositol hexakisphosphate)-coated MNPs for detecting the same metabolite (i.e., cotinine) within fingerprints (Fig. 8.3). Fig. 8.3A shows the possible mechanism of the aggregation of the magnetic nanocomposite to create crevice structures between the AuNPs by applying a magnetic field. The magnetic nanocomposite was used as surface-enhanced Raman scattering (SERS) substrate to on-site monitor trace cotinine biomarkers in fingerprints. Due to the magnetic properties, such substrate presented a huge SERS activity (Fig. 8.3B), which has met the

Figure 8.2 Fingerprint visualization using the magnetic nanostructure composed of fluoride nanophosphor. (A) Scheme of the optical visualization of the fingerprint under UV excitation (i.e., red emission) and NIR excitation (i.e., green emission). (B) Fingerprint development on a glass slide in dark (a–c), on a glass slide in the light background (d–f), and on aluminum foil in the light background (g–i).
Source: Reprinted from P.K. Shahi, P. Singh, A.K. Singh, S.K. Singh, S.B. Rai, R. Prakash, A strategy to achieve efficient dual-mode luminescence in lanthanide-based magnetic hybrid nanostructure and its demonstration for the detection of latent fingerprints, J. Colloid Interface Sci. 491 (2017) 199–206. Copyright (2017), with permission from Elsevier.

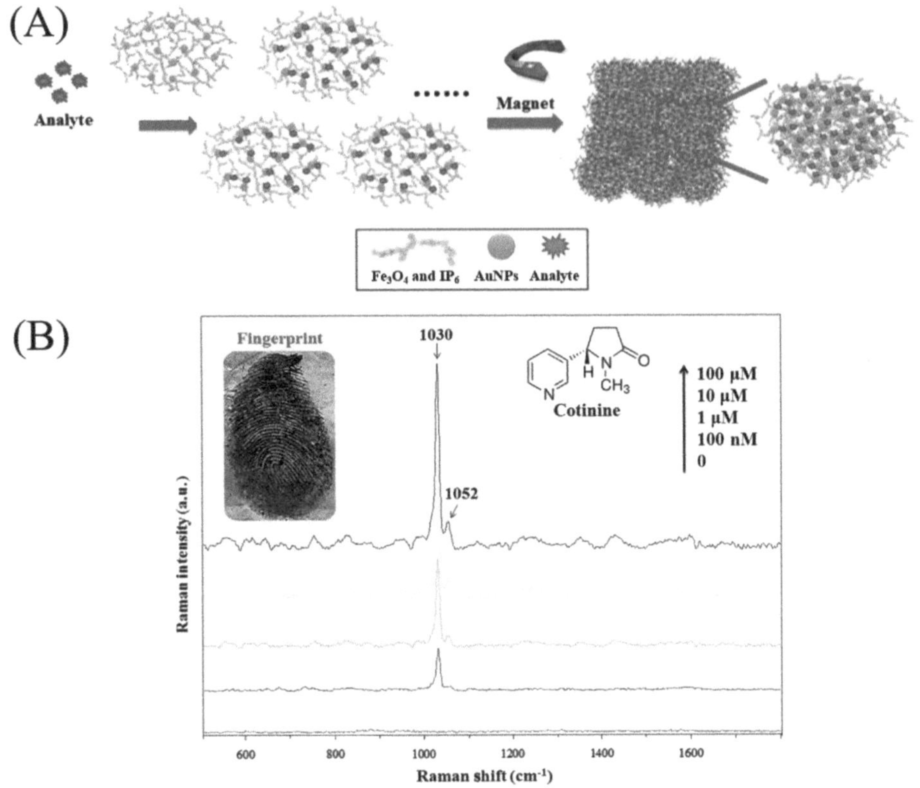

Figure 8.3 Magnetic SERS assay for rapid detection of cotinine in fingerprints. (A) Schematic illustration of the possible mechanism of the aggregation of the AuNPs onto a magnetic network nanostructure applying a magnetic field. (B) Raman spectra of the latent fingerprint formed with the magnetic nanocomposite that was doped by cotinine with concentrations varying from 100 nM to 100 μM.
Source: Adapted from T. Yang, X. Guo, H. Wang, S. Fu, Y. Wen, H. Yang, Magnetically optimized SERS assay for rapid detection of trace drug-related biomarkers in saliva and fingerprints, Biosens. Bioelectron. 68 (2015) 350–357. Copyright (2015), with permission from Elsevier.

sensitivity requirement for assay to detect the biomarkers in fingerprint from the substance abuse and mental health services administration (United States), and also the limit of detection (LOD) for the metabolite in fingerprint reached 100 nM (i.e., 17.6 ng mL^{-1}). The authors concluded that the proposed method could be employed as an on-site protocol to screen trace drugs of abuse.

8.2.2 Omics analysis

The term omics refers to a field of study in biological sciences that ends with -omics, such as genomics, transcriptomics, proteomics, metallomics, among others.

Omics analysis is now considered one of the most important and effective tools in the forensic analysis since it could provide critical pieces of evidence at crime scenes.

8.2.2.1 Genomics analysis

The genomics analysis is focused on the sequencing of the complete set of DNA in the single cell of an organism. Since its introduction in 1985 [21], DNA analysis in forensic chemistry has played a significant role in the justice process by preventing wrongful convictions and exoneration of the innocent. Recent advances in the development of new approaches to isolate DNA have employed MNMs and they have received more attention because of their easy manipulation and cost effectiveness.

The extraction of DNA with MNM is driven by hydrophobic, hydrogen bonding, and electrostatic interactions that are closely related to the surface properties [22]. Several researchers have focused on the development of new magnetic nanosorbents to increase the recovery efficiency of DNA, publishing a vast number of publications. Recently, MNMs have been grafted onto graphene oxide [23], and coated with silica [6,24], cobalt-zinc [6], and gold [15], whereas they have also modified their surface using functional groups (e.g., $-OH$ [25,26], $-NH_2$ [27,28], and $-CO_2H$ [29]), or ionic liquids [30].

Epigenomics is a relatively new and rapidly growing area of forensic chemistry and refers to analyze chemical modifications in DNA. DNA methylation varies with age, and the ability to predict the age of donor using DNA from evidence materials at a crime scene is of considerable value in forensic investigations. Concretely, the first publication proposing the forensic age-prediction model was in 2011 [31]. Hitherto, many studies have been reported and they seem very promising because of their high prediction accuracies (i.e., ± 4 years), which can be comparable to or even surpass those from, eyewitness reports [32]. Other applications of DNA methylation analysis in forensic genetics include (1) the ability to distinguish monozygotic twins due to their different DNA methylation patterns, (2) lifestyle prediction, such as smoking status, and (3) exposure to environmental factors, such as pollutants [33]. However, the incorporation of MNMs in epigenomics for forensic purpose is still not enough explored and only one manuscript have reported the use of magnetic microparticles [34].

8.2.2.2 Transcriptomics analysis

Transcriptomics is the study of the transcriptome, the set of all RNA transcripts in one cell, tissue, or organism. It includes the amount or concentration of each RNA molecule in addition to the molecular identities. Transcriptomics analysis is very much dependent on the quality of the RNA being assayed. Thus a major disadvantage of RNA profiling is its instability, as it is well known that RNA is unstable and rapidly degraded by endogenous and environmental ribonucleases. Because biological samples found at crime scenes are subject to a wide range of conditions and can be expected to be highly degraded, the quality and quantity of extracted RNA could be seriously impacted by the purification procedure compared to fresh biological samples. A purification procedure that minimizes or eliminates degradation and contamination should, therefore, be selected and meticulously performed by well-trained forensic examiners.

MNMs also played an important role in the separation of nucleic acids, proteins, and cells due to the convenience of magnetic separation and the diversity of surface modifications. However, in the transcriptomic field, only one manuscript has been published that employs magnetic bead–based purification procedures in forensic samples [35]. In this research, analysis of RNA from manual (silica column–based) and automated (magnetic bead–based) purification procedures in bloodstain samples was compared and magnetic bead–based purification can be as effective as silica column–based technology for RNA-based body fluid identification of forensic casework. Nonetheless, in highly degraded samples (i.e., saliva stain samples) magnetic beads can purify relatively large amounts of short-length RNA in comparison with manual purification. Therefore the automated purification procedure could be a more effective option for RNA-based body fluid identification of forensic samples.

8.2.2.3 Proteomics analysis

The term proteome refers to the sum of all the proteins in a cell, tissue, or organism. Proteomics is the science that studies those proteins as related to their biochemical properties and functional roles. Although the gold standard for identifying individuals in a forensic context is based on the amplification of fragments of DNA, proteomics in forensic chemistry is a new and powerful tool that can overcome some of the shortcomings of the conventional techniques since, unlike DNA, proteins are molecules that are more abundant, highly stable in physical and environmental conditions, and preserve well for prolonged periods [36]. Additionally, proteomic analysis allows the identification of insects/arthropods, helping forensic scientists to estimate the postmortem index, any change in position of the corpse as well as the cause of death [37,38]. Among these applications, one of the most extended applications of MNMs is their use in antidoping analysis. The use of drugs in sport has been named as doping and it can be defined as the employment of substance that could artificially improve athletic performance in sports. One of the most well-known substances of doping in sports is the use of erythropoietin, more commonly known as EPO. Increasing oxygen transport to the tissues and generating energy are the most effective ways of increasing performance. Thus the detection of EPO is an important task for antidoping. Immunochemical analyses are very frequently employed for its determination and to enhance the affinity toward the EPO, researchers conjugated MNMs with aptamers [39] and antibody [40,41]. Other banned performance-enhancing proteins, or peptides, commonly analyzed are insulin [42] and insulin-like growth factors [43,44]. In those cases, the analytical methodologies developed for these analyses employed commercial magnetic microparticles conjugated with specific antibodies.

8.2.3 Explosives detection

The detection and identification of explosive residues at crime scenes are of major importance in forensic chemistry. Explosives can be subdivided by compound sensitivity into primary, secondary, and tertiary explosives [45]. Primary explosives are highly sensitive, unstable materials and can ignite by shock, friction, heat, or impact. Triacetone triperoxide (TATP) is an example of a primary explosive. Secondary organic explosives typically contain a nitro group such as 2,4,6-trinitrotoluene (TNT) or 2,4-DNT (dinitrotoluene) and they are frequently used with a small number of primary explosives to initiate a detonation. On the other side, tertiary explosives, such as ammonium nitrate, are very insensitive to shock and require an explosive booster of secondary explosives. Any of the mentioned explosive types can be used as an improvised explosive device, and the identification of the components present can be used to determine the source and possibly provide a link to a suspect [45].

In recent years, several MNMs have been employed for explosive determination using nonportable instruments. Mahmoud and Zourob [46] employed magnetite/Au nanoparticle shells conjugated with aminated lignin for selective removal and TNT determination based on SERS detection. The presence of lignin and the highly packed magnetic composite enhanced the SERS signals allowing the detection of TNT below 1 pM (i.e., 0.2 ng L^{-1}) from contaminated soils.

Costa dos Reis et al. [9] developed an analytical method for the determination of TNT in water samples using graphene oxide and MNPs as a sorbent for magnetic solid-phase extraction (SPE) coupled with liquid chromatography ultraviolet-visible (UV-Vis) detection. The repeatability was evaluated extracting aqueous solutions of 10 $\mu g\, L^{-1}$ of TNT with a relative standard deviation value of 12% for the intra-day repeatability and 17% for the inter-day repeatability. The authors empirically determined the methodological LOD, progressively measuring more diluted concentrations of the analyte. The LOD value was found to be 0.3 $\mu g\, L^{-1}$. Also, the nanocomposite can be reused up to six times without losing extraction capacity, reducing cost and wastes.

Recently, Kgatitsoe et al. [47] biosynthesized MNPs functionalized with *Moringa oleifera* plant leaf extracts as potential nanosorbents in the extraction of nitroaromatic explosive compounds from aqueous solution. The biosynthesis approach can be considered as green, sustainable, and cost effective. Using plant extracts also introduces antioxidant and antibacterial properties thus resulting in nontoxic nanoparticles. The authors detected six explosive compounds employing a gas chromatography-time of flight-mass spectrometry: nitrobenzene, 1,3-dinitrobenzene, 2,6-DNT, 3-NT (nitrotoluene), 2-NT, and 4-NT, achieving LOD values of 0.90, 0.40, 0.44, 0.37, 1.03, and 4.47 $\mu g\, L^{-1}$, respectively.

The abovementioned instruments were relatively sensitive. However, those techniques were relatively bulky and they were not suitable for on-site testing. Without a shadow of a doubt, the development of on-site analytical methods is still on demand. Several methods have been developed, particularly, some researchers have focused on the combination of MNP as peroxidase mimetics for the determination

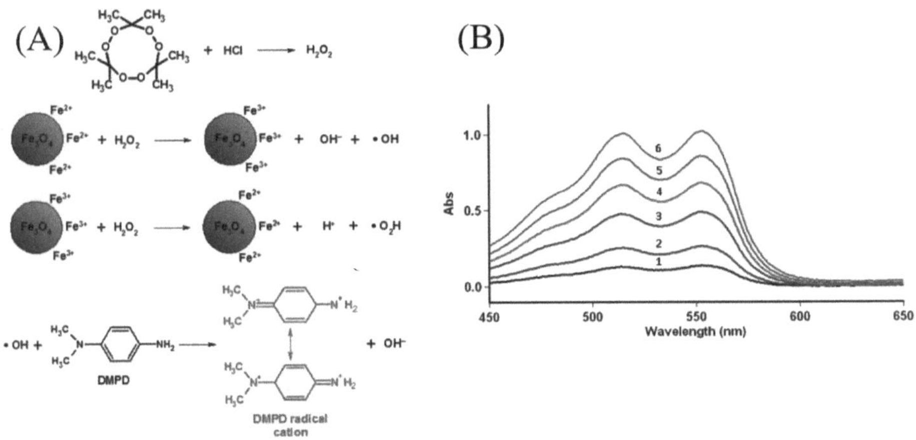

Figure 8.4 Basis of indirect spectrophotometric determination of TATP. (A) Schematic demonstration of H_2O_2 formation upon TATP hydrolysis and colored $DMPD^+$ radical formation. (B) Visible spectra of colored $DMPD^+$ compound at (1) 1 mg L^{-1}, (2) 2 mg L^{-1}, (3) 4 mg L^{-1}, (4) 6 mg L^{-1}, (5) 8 mg L^{-1}, and (6) 10 mg L^{-1} final concentrations. *Source*: Reprinted with the permission of the American Chemical Society (ACS) from Z. Can, A. Üzer, K. Türkekul, E. Erçağ, R. Apak, Determination of triacetone triperoxide with a *N,N*-dimethyl-*p*-phenylenediamine sensor on Nafion using Fe_3O_4 magnetic nanoparticles, Anal. Chem. 87 (2015) 9589–9594 (https://pubs.acs.org/doi/10.1021/acs.analchem.5b01775). Further permissions related to the material excerpted should be directed to the ACS.

of 2,4-DNT or TATP [48,49]. Can et al. [48] presented an interesting example of an on-site analytical method for the determination of the explosive TATP based on the peroxidase-like activity of MNPs (Fig. 8.4). This property has emerged as the next generation of enzyme mimics since the unexpected discovery in 2007 [50].

Fig. 8.4A illustrates the reaction mechanism. The first step is the H_2O_2 formation based on the acidic hydrolysis of TATP. Subsequently, the addition of MNPs to the medium produced hydroxyl radicals from H_2O_2. The formed radicals converted the *N,N*-dimethyl-*p*-phenylenediamine (DMPD) compound to the colored $DMPD^+$ radical cation, measuring its absorbance at a wavelength of 554 nm (Fig. 8.4B). The colored compound enabled its spectrophotometric measurement either in solution or on a Nafion membrane. The authors confirmed that the determination is free of possible interfering agents existing in explosive residues. In addition, the LOD values obtained for the proposed method were very suitable for field use, achieving 0.1 mg L^{-1} in the Nafion membrane. Finally, the authors claimed that the colorimetric sensor enables the development of a precise, sensitive, and selective method for the TATP assay that is suitable for both automated and on-site analysis.

One year later, Nie et al. [49] employed the same property of MNPs as peroxidase mimetics for the determination of 2,4-DNT in water samples. In this case, it was found that the MNPs catalyzed the oxidation of the 2,2′-azino-bis(3-ethylbenzothiazoline-6-sulfonate acid) (ABTS) by H_2O_2 to produce the colored product (i.e., oxidized ABTS) with maximum absorbance at 417 nm. The LOD value obtained was 0.15 μM (i.e., 0.027 mg L^{-1}), being sensitive enough for the determination of trace amount of 2,4-DNT that would have a great potential in the application of online monitoring of 2,4-DNT in environmental water samples.

8.2.4 Illicit drugs

Illicit drugs that are widely used include hallucinogens [e.g., lysergic acid diethylamide or tetrahydrocannabinol (THC)] and central nervous system agents: (1) inhibitors (e.g., heroin or ketamine) and (2) stimulants (e.g., cocaine or amphetamines). The illegal drug trade creates pernicious effects in the country population that puts the public at risk, not just from the substances being distributed, but also from the subsequent crimes committed by users, traffickers, and manufacturers. These crimes typically include burglary, assault, and fraud, but can also involve more serious crimes such as homicide, abduction, and human trafficking. Illicit drugs play a major role in forensic chemistry as they provide evidential links for drug trafficking. Among all the earlier mentioned illicit drugs, cocaine is one of the main targets in forensic chemistry using MNMs. In recent years, quantitative methods to analyze cocaine have been performed by high-performance liquid chromatography-mass spectrometry (HPLC-MS) [51–54], ion mobility spectrometry [55], and capillary electrophoresis with UV-Vis detector [56]. A promising alternative for the analysis of cocaine is the development of electrochemical [57], chemiluminescence [58], fluorescence [59], and colorimetric [60,61] sensors. Among all the cited analytical methods the electrochemical sensor fabricated by Hashemi et al. [57] showed the best analytical performance even in comparison with those obtained with HPLC-MS techniques [51]. The MNM based on reduced graphene oxide/MNPs/PA/AuNPs was employed for impedimetric determination of cocaine immobilizing a specific thiolated cocaine aptamer onto the surface of AuNPs. The aptamer was used for its excellent specificity and affinity ability for the determination of cocaine in urine and serum samples. The developed sensor was able to detect cocaine as low as 0.029 nM (i.e., 8.8 ng L^{-1}).

Nowadays, there is more awareness of the problems of illicit drugs and drug trafficking than ever before. As a result, considerable attention has been focused on the development of new methods for improving the analytical performance using MNMs. For instance, illicit drugs such as THC, heroin, ketamine, and amphetamine have been analyzed using different MNMs (i.e., MNPs bioconjugated with antibody [62,63], ceramic carbon-coated MNPs [64], polymeric MNPs [65,66], and magnetic nanographene oxide [67]).

8.2.5 Anticounterfeiting

Nowadays, the circulation of adulterated and counterfeit products provides a great challenge to the analytical chemists due to the increasing product diversity, the continuous development of new production technologies, and the emergence of new types of fraud. For instance, foodstuffs, fuels, medicines, inks, or banknotes are usually counterfeited, which seriously infringe legal rights and cause substantial economic damage [68]. Therefore the development of anticounterfeiting materials with improved security properties has become an important task.

In recent years, a wide diversity of anticounterfeiting technologies using MNMs has been developed. Some examples that demonstrate the potential of MNMs to be used in anticounterfeiting technologies are those based on security inks [69], barcodes [14], watermarks [70], and films [17]. These MNMs are made of stimuli-responsive molecules and a very evident change in their physical or optical properties can be triggered by a magnetic field. A step forward for information security is the use of DNA codes based on the determination of the order of their base sequence. A successful application of this anticounterfeiting approach has been developed by Puddu et al. [16] in three different samples (i.e., olive oil, cosmetic oil, and fuel) (Fig. 8.5). For encoding, DNA sequences were adsorbed on the surface of the positively charged maghemite nanoparticles. In the following step, the particles were encapsulated in silica by using *N*-trimethoxysilylpropyl-*N,N,N*-trimethylammonium chloride and tetraethoxysilane via sol–gel chemistry. The silica surface was further functionalized by hydrolysis of C_6 to achieve dispersibility in hydrophobic fluids. In order to demonstrate the long-term stability, the taggants showed good stability in decalin at room temperature, being stable for 2 years (Fig. 8.5A).

Nanoparticles containing DNA code are collected, cleaned, and dried, using a magnet when it is needed (Fig. 8.5B). Afterward, the DNA solution was incubated before DNA release in buffered oxide etch. Finally, they were analyzed by

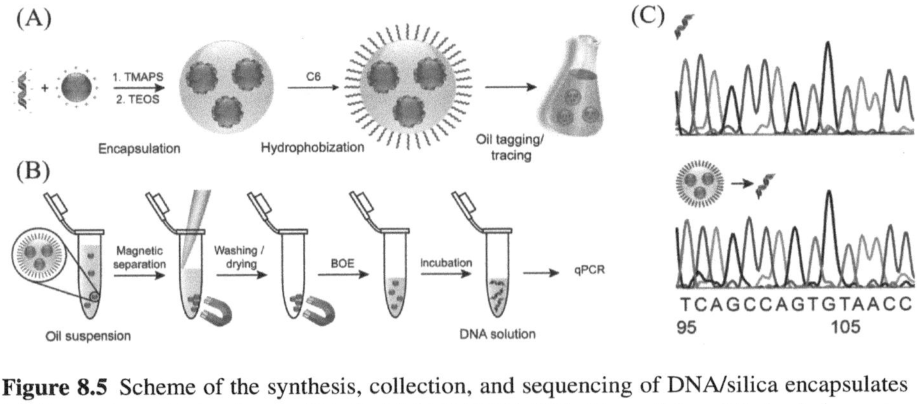

Figure 8.5 Scheme of the synthesis, collection, and sequencing of DNA/silica encapsulates in oil samples. (A) Synthesis of Fe_2O_3/TMAPS/DNA/SiO_2-C_6 particles. (B) Analytic route of DNA recovery from oil particle suspensions and subsequent quantification by qPCR. (C) Sequencing chromatograms of base 95–109 of unprocessed DNA sequence (top panel) and DNA encapsulated (bottom panel).
Source: Reprinted with the permission of the American Chemical Society (ACS) from M. Puddu, D. Paunescu, W.J. Stark, R.N. Grass, Magnetically recoverable, thermostable, hydrophobic DNA/silica encapsulates and their application as invisible oil tags, ACS Nano 8 (2014) 2677–2685 (https://pubs.acs.org/doi/10.1021/nn4063853). Further permissions related to the material excerpted should be directed to the ACS.

quantitative polymerase chain reaction (Fig. 8.5C). To show the applicability of the method, the authors utilized extra virgin olive oil as an example of an olive oil, bergamot oil as an example of a cosmetic oil, and gasoline as an example of a fuel. A concentration of 1 μg particles L^{-1} of the magnetic DNA/silica encapsulates was added to each sample. Afterward, they were retrieved, identified, and quantified from each sample. Finally, the samples containing the magnetic encapsulates were statistically discriminate from the negative control group (i.e., without tagging) to a certainty of >99.9%. Besides, the novel MNM can be utilized as low-cost (0.02 US Cent L^{-1}) tracers, and the method is universal because it does not require procedure optimization based on oil type.

8.2.6 Homeland security

New emerging challenges from organized crime and terrorist groups emphasize the necessity for sensitive, selective, and rapid identification of substances such as chemical and biological warfare (CBW) in the environment, public places, or workplaces to provide an early warning of sudden CBW contamination or potential terrorist attack. Nerve agents are considered the most noxious of chemical warfare in comparison with other chemical warfare agents (vesicants, vomit, pulmonary, lachrymating, or blood agents) because of their organophosphonate structure. In addition, the most dangerous nerve agents (e.g., Sarin, Cyclosarin, VX, Soman, and Tabun) are fairly easy to manufacture, and only low cost and simple technology are required for their synthesis [71]. Generally, much of the current researches on CBW quantification are focused on the use of MNMs as a sorbent in SPE. Perhaps, the most fruitful investigations that formed part of this field research are carried out by the Indian Defense Research & Development Establishment, since it is the only designated laboratory in India by the Organisation for the Prohibition of Chemical Weapons. This research group employed different MNMs, including MNPs coated with: (1) methacrylic acid and ethylene glycol dimethacrylate copolymer (MAA-*co*-EGDMA) [72], (2) divinylbenzene and *N*-vinylpyrrolidone copolymer (DVB-*co*-*N*-VP) [73,74], and (3) polystyrene divinylbenzene (PSDVB) [75]. Also, the same research group reported the use of multiwalled carbon nanotube filled with MNPs [10]. The lowest LOD values obtained was 0.5 ng mL^{-1} for Sarin, Cyclosarin, and Tabun [73]. Finally, the method was evaluated using a variety of water samples provided by the Organisation for the Prohibition of Chemical Weapons. The authors concluded that the developed materials and methods are advancement in facilitating the extraction of CBW from water samples that could aid to the forensic and verification analysis of CBW.

From a health point of view, the destruction of CBW is highly desired for proper handling of contamination. Chen et al. [8] reported the use of core–shell magnetic MOFs using MNPs and MOF-UiO-66-NH_2. The MOF was built of $[Zr_6O_4(OH)_4]^{12+}$ octahedron clusters and 2-aminoterephthalic acid ligands. The MNM was further tested in the catalytic hydrolysis of dimethyl-4-nitrophenyl phosphate (a nerve agent simulant) for adequate destruction capabilities. Finally, the authors claimed that this strategy could be used to prepare other types of magnetic core–shell composites and further combine them with other types of polymers to obtain higher catalytic activity.

Biodefense is another critical element in homeland security. A bioterrorism attack is the deliberate release of viruses, bacteria, or other germs to cause illness or death. These germs are often found in nature, but they can sometimes be bioactivated, or made more harmful by increasing their ability to cause disease, spread, or resist medical treatment. The dangerousness of these infectious diseases elevates the need for effective biodefense and surveillance systems that can detect and monitor the course of an outbreak. MNMs applicable for simultaneous pathogen detection and magnet-mediated separation have been employed for the first time to detect, quantify, and differentiate *Escherichia coli* cells, offering significant potential for biodefense purposes [76]. Baldrich and Muñoz [77] reported that magnetic beads modified with immunoaffinity (i.e., antibody) agents possess high specificity for *E. coli* cells and they can be recovered magnetically, but they are expensive, relatively unstable, and cannot be applied in situ for biodefense. One year later, Bromberg et al. [78] were able to capture germ-specific DNA from water resources using magnetite clusters modified by poly(hexamethylene biguanide) and polyethyleneimine copolymer (PHMBG-*co*-PEI). Afterward, the concentrated genetic material could be identified by a polymerase chain reaction. The levels of the DNA detection obtained in their experiments allow to reliably detect as low as 1 pg of bacteriophage DNA. The authors concluded that the biological material extracted from the magnetic clusters can be analyzed for the presence of several germs simultaneously and thus, the approach not only permits the detection of a specific germ but also allows for differentiating between genotypes to compare DNA isolates for epidemiological investigations. More recently, MNPs coated with polyaniline were used in an electrochemical biosensor for rapidly identifying influenza A virus in order to monitor animal influenza infections and to characterize the pandemic potential of strains that have transmitted from animals to humans [79].

8.3 Challenges and future outlook

Despite much information being shown in this review, the authors believe that there are still several challenges in forensic chemistry using MNMs to be studied, such as the following:

1. The development of new MNMs in the last decade has revolutionized the field of forensic chemistry. However, according to the reviewed bibliography, the authors found that some well-known MNMs have not been yet explored in this field area. For instance, magnetic molecularly imprinted nanomaterials have shown great potential for the recognition of a wide variety of analytes. Besides, their high selectivity and the advantages presented by their magnetic response for fast and easy isolation of the analyte from samples can increase both selectivity and sensitivity in comparison with those MNMs revisited in this review.

2. Among all the advantages that MNMs present, the fast sample manipulation shown would constitute an appreciable contribution to develop new field-based forensic analysis. Nowadays, field-based forensic analysis is an effective method of identifying body fluids, assessing the presence of DNA, or performing DNA profiling. However, the extensive employment of field-based forensic analysis is still an emerging field and there are large gaps in the research, as it is still necessary for many on-site tests to be used in combination with laboratory analysis due to their reasonable inferior performance. In recent years, there has been an effort to find an ideal candidate for use as an on-site forensic detection instrument. Researchers found that smartphones fulfill the requirements of the field-based systems as they are lightweight, handle, versatile, and reasonably cheap. Smartphones are also capable of use as analytical and data processing devices and they possess wi-fi capabilities and any data recorded using the application can be uploaded to a central "cloud" network for storage or downloaded to other devices.
3. Finally, although many challenges remain, the use of multiomics analysis for identifying and quantifying protein markers, combined with the use of DNA and RNA can be projected into a new stage of forensic chemistry, where the biological information provided by physical evidence enables to establish not only the identity of the subjects but also quantitative information on various events such as the health status of the individuals, their nutritional patterns, environmental conditions, migration events, and movement of individuals involved in a forensic scene. However, the number of studies using multiomics approaches in forensic samples is currently very limited. This is mainly because the acquisition and analysis of several multiomics data types simultaneously are very challenging. Although the cost of next-generation sequencing is decreasing, it is still beyond the reach of many small labs to perform integrated multiomics analysis of a complex forensic sample. However, these technologies are constantly evolving, and hopefully, soon, the forensic scientists will have appropriate multiomics tools to conduct a more comprehensive analysis of microbial communities and their relevance to forensic science. Similar to the Olympic motto "hendiatris citius, altius et fortius," which is Latin for faster, higher, and stronger, multiomics analysis in forensic chemistry can be expected to become more rapid, sensitive, and provides stronger evidence.

Acknowledgment

The authors would like to thank the Generalitat Valenciana (Spain) (project n. PROMETEO/2018/087) and the Spanish Ministry of Economy and Competitiveness and European Union (FEDER funds) (project n. CTQ2016-79991-R) for the financial support.

References

[1] S. Mukherjee, L. Liang, O. Veiseh, Recent advancements of magnetic nanomaterials in cancer therapy, Pharmaceutics 12 (2020) 147.
[2] Y. Zhang, G. Li, D. Wu, X. Li, Y. Yu, P. Luo, et al., Recent advances in emerging nanomaterials based food sample pretreatment methods for food safety screening, Trends Anal. Chem. 121 (2019) 115669.

[3] M. Pastucha, Z. Farka, K. Lacina, Z. Mikušová, P. Skládal, Magnetic nanoparticles for smart electrochemical immunoassays: a review on recent developments, Microchim. Acta 186 (2019) 312.
[4] F. Li, Y. Huang, K. Huang, J. Lin, P. Huang, Functional magnetic graphene composites for biosensing, Int. J. Mol. Sci. 21 (2020) 390.
[5] R. Huang, T. Tang, Assembly of magnetic nano-Fe_3O_4@GSH-Au NCs core-shell microspheres for the visualization of latent fingerprints, Nano 13 (2018) 1850128.
[6] J. Torres-Rodríguez, G. Soto, J. López-Medina, A. Portillo-López, E.L. Hernández-López, E. Vargas-Viveros, et al., Cobalt−zinc ferrite and magnetite SiO_2 nanocomposite powder for magnetic extraction of DNA, J. Sol-Gel Sci. Technol. 91 (2019) 33−43.
[7] R. Huang, R. Liu, Efficient in situ growth of platinum nanoclusters on the surface of Fe_3O_4 for the detection of latent fingermarks, J. Mater. Sci. 52 (2017) 13455−13465.
[8] R. Chen, C. Tao, Z. Zhang, X. Chen, Z. Liu, J. Wang, Layer-by-layer fabrication of core-shell Fe_3O_4@UiO-66-NH_2 with high catalytic reactivity toward the hydrolysis of chemical warfare agent simulants, ACS Appl. Mater. Interfaces 11 (2019) 43156−43165.
[9] L. Costa dos Reis, L. Vidal, A. Canals, Graphene oxide/Fe_3O_4 as sorbent for magnetic solid-phase extraction coupled with liquid chromatography to determine 2,4,6-trinitrotoluene in water samples, Anal. Bioanal. Chem. 409 (2017) 2665−2674.
[10] D. Pardasani, P.K. Kanaujia, A.K. Purohit, A.R. Shrivastava, D.K. Dubey, Magnetic multi-walled carbon nanotubes assisted dispersive solid phase extraction of nerve agents and their markers from muddy water, Talanta 86 (2011) 248−255.
[11] J. Lee, C.W. Lee, J.M. Kim, A magnetically responsive polydiacetylene precursor for latent fingerprint analysis, ACS Appl. Mater. Interfaces 8 (2016) 6245−6251.
[12] A.M. Boddis, D.A. Russell, Simultaneous development and detection of drug metabolites in latent fingermarks using antibody−magnetic particle conjugates, Anal. Methods 3 (2011) 519−523.
[13] P.K. Shahi, P. Singh, A.K. Singh, S.K. Singh, S.B. Rai, R. Prakash, A strategy to achieve efficient dual-mode luminescence in lanthanide-based magnetic hybrid nanostructure and its demonstration for the detection of latent fingerprints, J. Colloid Interface Sci. 491 (2017) 199−206.
[14] I.T. Jeon, S.J. Yoon, B.G. Kim, J.S. Lee, B.H. An, J.-S. Ju, et al., Magnetic NiFe/Au barcode nanowires with self-powered motion, J. Appl. Phys. 111 (2012) 07B513.
[15] K. Li, Y. Lai, W. Zhang, L. Jin, Fe_2O_3@Au core/shell nanoparticle-based electrochemical DNA biosensor for *Escherichia coli* detection, Talanta 84 (2011) 607−613.
[16] M. Puddu, D. Paunescu, W.J. Stark, R.N. Grass, Magnetically recoverable, thermostable, hydrophobic DNA/silica encapsulates and their application as invisible oil tags, ACS Nano 8 (2014) 2677−2685.
[17] T.Y. Kim, Y. Yamazaki, M. Abe, Preparation of hybrid coating film between magneto-optical and hologram layers, Phys. Status Solidi A. 201 (2004) 1759−1762.
[18] A.M. Boddis, D.A. Russell, Development of aged fingermarks using antibody-magnetic particle conjugates, Anal. Methods 4 (2012) 637−641.
[19] R.M. Connatser, S.M. Prokes, O.J. Glembocki, R.L. Schuler, C.W. Gardner, S.A. Lewis, et al., Toward surface-enhanced Raman imaging of latent fingerprints, J. Forensic Sci. 55 (2010) 1462−1470.
[20] T. Yang, X. Guo, H. Wang, S. Fu, Y. Wen, H. Yang, Magnetically optimized SERS assay for rapid detection of trace drug-related biomarkers in saliva and fingerprints, Biosens. Bioelectron. 68 (2015) 350−357.

[21] P. Gill, A.J. Jeffreys, D.J. Werrett, Forensic application of DNA 'fingerprints', Nature 318 (1985) 577–579.
[22] S.L. Sahoo, C.H. Liu, Adsorption behaviors of DNA by modified magnetic nanoparticles: effect of spacer and salt, Colloids Surf. A 482 (2015) 184–194.
[23] L. Qi, R. Xu, J. Gong, Monitoring DNA adducts in human blood samples using magnetic Fe_3O_4@graphene oxide as a nano-adsorbent and mass spectrometry, Talanta 209 (2020) 120523.
[24] Q. Fan, Y. Guan, Z. Zhang, G. Xu, Y. Yang, C. Guo, A new method of synthesis well-dispersion and dense Fe_3O_4@SiO_2 magnetic nanoparticles for DNA extraction, @Chem. Phys. Lett. 715 (2019) 7–13.
[25] X. Pan, S. Cheng, T. Su, G. Zuo, C. Zhang, L. Wu, et al., Poly (2-hydroxypropylene imines) functionalized magnetic polydopamine nanoparticles for high-efficiency DNA isolation, Appl. Surf. Sci. 498 (2019) 143888.
[26] M. Zhang, L. Li, B. Li, N. Tian, M. Yang, H. Zhang, et al., Adsorption of DNA by using polydopamine modified magnetic nanoparticles based on solid-phase extraction, Anal. Biochem. 579 (2019) 9–17.
[27] Y. Bai, D. Roncancio, Y. Suo, Y. Shao, D. Zhang, C. Zhou, A method based on amino-modified magnetic nanoparticles to extract DNA for PCR-based analysis, Colloids Surf. B 179 (2019) 87–93.
[28] R.J. da Silva, B.G. Maciel, J.C. Medina-Llamas, A.E. Chávez-Guajardo, J.J. Alcaraz-Espinoza, C. Pinto de Melo, Extraction of plasmid DNA by use of a magnetic maghemite-polyaniline nanocomposite, Anal. Biochem. 575 (2019) 27–35.
[29] C.Y. Wen, T.T. Liu, L.L. Wu, Y.M. Li, J.Y. Sun, J.B. Zeng, Magnetic nanospheres for convenient and efficient capture and release of hepatitis B virus DNA, Talanta 197 (2019) 605–611.
[30] M. Liu, X. Ding, X. Wang, J. Li, H. Yang, Y. Yin, Extraction of DNA from complex biological sample matrices using guanidinium ionic liquid modified magnetic nanocomposites, RSC Adv. 9 (2019) 23119–23128.
[31] C.M. Koch, W. Wagner, Epigenetic-aging-signature to determine age in different tissues, Aging 3 (2011) 1018–1027.
[32] W. Parson, Age estimation with DNA: from forensic DNA fingerprinting to forensic (epi)genomics: a mini-review, Gerontology 64 (2018) 1–7.
[33] A. Vidaki, M. Kayser, From forensic epigenetics to forensic epigenomics: broadening DNA investigative intelligence, Genome Biol. 18 (2017) 238.
[34] B.E. Tam, Y. Hao, H.D. Sikes, An examination of critical parameters in hybridization-based epigenotyping using magnetic microparticles, Biotechnol. Prog. 34 (2018) 1589–1595.
[35] T. Akutsu, T. Kitayama, K. Watanabe, K. Sakurada, Comparison of automated and manual purification of total RNA for mRNA-based identification of body fluids, Forensic Sci. Int. Gen. 14 (2015) 11–17.
[36] C. Wadsworth, M. Buckley, Proteome degradation in fossils: investigating the longevity of protein survival in ancient bone, Rapid Commun. Mass. Spectrom. 28 (2014) 605–615.
[37] C. Sánchez-Ovejero, F. Benito-López, P. Díez, A. Casulli, M. Siles-Lucas, M. Fuentes, et al., Sensing parasites: proteomic and advanced bio-detection alternatives, J. Proteom. 136 (2016) 145–156.
[38] R. Cardoso, H. Alves, J. Richter, M.C. Botelho, Parasites in forensic science: a historic perspective, Ann. Parasitol. 63 (2017) 235–241.

[39] Z. Chen, H. Li, Y. Zhao, M. Xu, D. Xu, Magnetic nanoparticles and polydopamine amplified FP aptasensor for the highly sensitive detection of rHuEPO-α, Talanta 189 (2018) 143–149.

[40] Y.S. Selbes, M.G. Caglayan, M. Eryilmaz, I.H. Boyaci, N. Saglam, A.A. Basaran, et al., Surface-enhanced Raman probe for rapid nanoextraction and detection of erythropoietin in urine, Anal. Bioanal. Chem. 408 (2016) 8447–8456.

[41] R. Agoston, E.L. Izake, A. Sivanesan, W.B. Lott, M. Sillence, R. Steel, Rapid isolation and detection of erythropoietin in blood plasma by magnetic core gold nanoparticles and portable Raman spectroscopy, Nanomedicine 12 (2016) 633–641.

[42] A. Thomas, W. Schanzer, P. Delahaut, M. Thevis, Sensitive and fast identification of urinary human, synthetic and animal insulin by means of nano-UPLC coupled with high-resolution/high-accuracy mass spectrometry, Drug Test. Anal. 1 (2009) 219–227.

[43] A. Thomas, M. Kohler, W. Schänzer, P. Delahaut, M. Thevis, Determination of IGF-1 and IGF-2, their degradation products and synthetic analogues in urine by LC-MS/MS, Analyst 136 (2011) 1003–1012.

[44] M. Thevis, A. Thomas, H. Geyer, W. Schänzer, Mass spectrometric characterization of a biotechnologically produced full-length mechano growth factor (MGF) relevant for doping controls, Growth Horm. IGF Res. 24 (2014) 276–280.

[45] L. Barron, E. Gilchrist, Ion chromatography-mass spectrometry: a review of recent technologies and applications in forensic and environmental explosives analysis, Anal. Chim. Acta 806 (2014) 27–54.

[46] K.A. Mahmoud, M. Zourob, Fe_3O_4/Au nanoparticles/lignin modified microspheres as effectual surface enhanced Raman scattering (SERS) substrates for highly selective and sensitive detection of 2,4,6-trinitrotoluene (TNT), Analyst 138 (2013) 2712–2719.

[47] M.M. Kgatitsoe, S. Ncube, H. Tutu, I.A. Nyambe, L. Chimuka, Synthesis and characterization of a magnetic nanosorbent modified with *Moringa oleifera* leaf extracts for removal of nitroaromatic explosive compounds in water samples, J. Environ. Chem. Eng. 7 (2019) 103128.

[48] Z. Can, A. Üzer, K. Türkekul, E. Erçağ, R. Apak, Determination of triacetone triperoxide with a *N,N*-dimethyl-*p*-phenylenediamine sensor on Nafion using Fe_3O_4 magnetic nanoparticles, Anal. Chem. 87 (2015) 9589–9594.

[49] D.X. Nie, G.Y. Shi, Y.Y. Yu, Fe_3O_4 magnetic nanoparticles as peroxidase mimetics used in colorimetric determination of 2,4-dinitrotoluene, Chin. J. Anal. Chem. 44 (2016) 179–185.

[50] L. Gao, J. Zhuang, L. Nie, J. Zhang, Y. Zhang, N. Gu, et al., Intrinsic peroxidase-like activity of ferromagnetic nanoparticles, Nat. Nanotechnol. 2 (2007) 577–583.

[51] Q. Lu, H. Guo, Y. Zhang, X. Tang, W. Lei, R. Qi, et al., Graphene oxide-Fe_3O_4 nanocomposite magnetic solid phase extraction followed by UHPLC-MS/MS for highly sensitive determination of eight psychoactive drugs in urine samples, Talanta 206 (2020) 120212.

[52] F. Yang, Y. Zou, C. Ni, R. Wang, M. Wu, C. Liang, et al., Magnetic dispersive solid-phase extraction based on modified magnetic nanoparticles for the detection of cocaine and cocaine metabolites in human urine by high-performance liquid chromatography–mass spectrometry, J. Sep. Sci. 40 (2017) 4234–4245.

[53] J. Sánchez-González, T. Barreiro-Grille, P. Cabarcos, M.J. Tabernero, P. Bermejo-Barrera, A. Moreda-Piñeiro, Magnetic molecularly imprinted polymer based-micro-solid phase extraction of cocaine and metabolites in plasma followed by high performance liquid chromatography-tandem mass spectrometry, Microchem. J. 127 (2016) 206–212.

[54] J. Sánchez-González, M.J. Tabernero, A.M. Bermejo, P. Bermejo-Barrera, A. Moreda-Piñeiro, Development of magnetic molecularly imprinted polymers for solid phase extraction of cocaine and metabolites in urine before high performance liquid chromatography-tandem mass spectrometry, Talanta 147 (2016) 641–649.
[55] A. Sorribes-Soriano, F.A. Esteve-Turrillas, S. Armenta, A. Montoya, J.M. Herrero-Martínez, M. de la Guardia, Magnetic molecularly imprinted polymers for the selective determination of cocaine by ion mobility spectrometry, J. Chromatogr. A 1545 (2018) 22–31.
[56] T. Baciu, F. Borrull, C. Neusüß, C. Aguilar, M. Calull, Capillary electrophoresis combined in-line with solid-phase extraction using magnetic particles as new adsorbents for the determination of drugs of abuse in human urine, Electrophoresis 37 (2016) 1232–1244.
[57] P. Hashemi, H. Bagheri, A. Afkhami, Y.H. Ardakani, T. Madrakian, Fabrication of a novel aptasensor based on three-dimensional reduced graphene oxide/polyaniline/gold nanoparticle composite as a novel platform for high sensitive and specific cocaine detection, Anal. Chim. Acta 996 (2017) 10–19.
[58] Y. Li, X. Ji, B. Liu, Chemiluminescence aptasensor for cocaine based on double-functionalized gold nanoprobes and functionalized magnetic microbeads, Anal. Bioanal. Chem. 401 (2011) 213–219.
[59] C. Ma, W. Wang, Q. Yang, C. Shi, L. Cao, Cocaine detection via rolling circle amplification of short DNA strand separated by magnetic beads, Biosens. Bioelectron. 26 (2011) 3309–3312.
[60] Y. Du, B. Li, S. Guo, Z. Zhou, M. Zhou, E. Wang, et al., G-Quadruplex-based DNAzyme for colorimetric detection of cocaine: using magnetic nanoparticles as the separation and amplification element, Analyst 136 (2011) 493–497.
[61] K. Mao, J. Ma, X. Li, Z. Yang, Rapid duplexed detection of illicit drugs in wastewater using gold nanoparticle conjugated aptamer sensors, Sci. Total. Environ. 688 (2019) 771–779.
[62] J.R. Lee, J. Choi, T.O. Shultz, S.X. Wang, Small molecule detection in saliva facilitates portable tests of marijuana abuse, Anal. Chem. 88 (2016) 7457–7461.
[63] N.V. Guteneva, S.L. Znoyko, A.V. Orlov, M.P. Nikitin, P.I. Nikitin, Rapid lateral flow assays based on the quantification of magnetic nanoparticle labels for multiplexed immunodetection of small molecules: application to the determination of drugs of abuse, Microchim. Acta 186 (2019) 621.
[64] A. Taghvimi, S. Dastmalchi, Y. Javadzadeh, Novel ceramic carbon-coated magnetic nanoparticles as stir bar sorptive extraction coating for simultaneous extraction of amphetamines from urine samples, Arab. J. Sci. Eng. 44 (2019) 6373–6380.
[65] M.L. Chen, L.L. Suo, Q.G. Yu, Q. Feng, Determination of eight illegal drugs in human urine by combination of magnetic solid-phase extraction with capillary zone electrophoresis, Electrophoresis 32 (2011) 2099–2106.
[66] A. Taghvimi, M. Ghorbani, H. Hamishehkar, Synthesis of a novel polymeric magnetic solid phase extraction adsorbent for selective extraction of amphetamine from urine samples coupled with high performance liquid chromatography, Drug Test. Anal. 10 (2018) 832–838.
[67] A. Taghvimi, H. Hamishehkar, M. Ebrahimi, Development and validation of a magnetic solid-phase extraction with high-performance liquid chromatography method for the simultaneous determination of amphetamine and methadone in urine, J. Sep. Sci. 39 (2016) 2307–2312.

[68] B. Yoon, J. Lee, I.S. Park, S. Jeon, J. Lee, J.-M. Kim, Recent functional material based approaches to prevent and detect counterfeiting, J. Mater. Chem. C 1 (2013) 2388–2403.
[69] B. Song, H. Wang, Y. Zhong, B. Chu, Y. Su, Y. He, Fluorescent and magnetic anti-counterfeiting realized by biocompatible multifunctional silicon nanoshuttle-based security ink, Nanoscale 10 (2018) 1617–1621.
[70] H. Hu, H. Zhong, C. Chen, Q. Chen, Magnetically responsive photonic watermarks on banknotes, J. Mater. Chem. C 2 (2014) 3695–3702.
[71] K. Kim, O.G. Tsay, D.A. Atwood, D.G. Churchill, Destruction and detection of chemical warfare agents, Chem. Rev. 111 (2011) 5345–5403.
[72] V. Singh, A.K. Purohit, S. Chinthakindi, R.D. Goud, V. Tak, D. Pardasani, et al., Analysis of chemical warfare agents in organic liquid samples with magnetic dispersive solid phase extraction and gas chromatography mass spectrometry for verification of the chemical weapons convention, J. Chromatogr. A 1448 (2016) 32–41.
[73] V. Singh, A.K. Purohit, S. Chinthakindi, R.D. Goud, V. Tak, D. Pardasani, et al., Magnetic hydrophilic–lipophilic balance sorbent for efficient extraction of chemical warfare agents from water samples, J. Chromatogr. A 1434 (2016) 39–49.
[74] V. Singh, S. Chinthakindi, A.K. Purohit, D. Pardasani, V. Tak, D.K. Dubey, Single vial sample preparation of markers of nerve agents by dispersive solid-phase extraction using magnetic strong anion exchange resins, J. Chromatogr. A 1395 (2015) 48–56.
[75] V. Singh, P. Garg, S. Chinthakindi, V. Tak, D.K. Dubey, Extraction and derivatization of chemical weapons convention relevant aminoalcohols on magnetic cation-exchange resins, J. Chromatogr. A 1329 (2014) 10–16.
[76] K. El-Boubbou, C. Gruden, X. Huang, Magnetic glyco-nanoparticles: a unique tool for rapid pathogen detection, decontamination, and strain differentiation, J. Am. Chem. Soc. 129 (2007) 13392–13393.
[77] E. Baldrich, F.X. Muñoz, Enzyme shadowing: using antibody-enzyme dually-labeled magnetic particles for fast bacterial detection, Analyst 133 (2008) 1009–1012.
[78] L. Bromberg, S. Raduyk, T.A. Hatton, Functional magnetic nanoparticles for biodefense and biological threat monitoring and surveillance, Anal. Chem. 81 (2009) 5637–5645.
[79] T.L. Kamikawa, M.G. Mikolajczyk, M. Kennedy, L. Zhong, P. Zhang, E.B. Setterington, et al., Pandemic influenza detection by electrically active magnetic nanoparticles and surface plasmon resonance, IEEE Trans. Nanotechnol. 11 (2012) 88–96.

Section III

Application of magnetic nanomaterials as sensors

Application of magnetic nanomaterials as colorimetric sensors

9

Farnaz Fazlali[1], *Pegah Hashemi*[1], *Ali Hajian*[2], *Hosein Khoshsafar*[1], *Tien Anh Nguyen*[3] *and Hasan Bagheri*[4]

[1]Research and Development Department, Farin Behbood Tashkhis LTD, Tehran, Iran, [2]Institute of Sensor and Actuator Systems, TU Wien, Vienna, Austria, [3]Department of Physics, Le Quy Don Technical University, Ha Noi, Vietnam, [4]Chemical Injuries Research Center, Systems Biology and Poisonings Institute, Baqiyatallah University of Medical Sciences, Tehran, Iran

9.1 Introduction

Nanotechnology is one of the most crucial investigations in current technology, permitting researchers in various fields of science to conduct studies at the molecular level and thereby generate significant improvements in life skills and health matters [1]. The application of particles at nanodimensions promises great benefits due to their exceptional dimension, shape, size, good biological compatibility—in other words, due to their physicochemical features and simple orientation to external magnetic fields. Given the extensive uses of nanomaterial with magnetic features in diverse fields such as engineering, environmental science biotechnology, biomedicine, and material science, considerable attention has been paid to the fabrication of diverse types of nanoparticles (NPs) with magnetic features [2]. The outstanding properties of these particles afford a promising future for their exploitation.

To fully comprehend a treatment of NPs with magnetic properties, it is essential to determine their physical and chemical factors, for instance, dimension, configuration, and magnetic features. Various kinds of materials with magnetic features can be characterized as alloys, oxides, and metallic metals. These display diverse magnetic features. The latest investigations concentrate on nanomaterial and sensors with magnetic features because they support utilization in biomedicine and detection through magnetic materials . For instance, several investigation and medical techniques are based on magnetic nanomaterial in hyperthermia behavior, drug delivery, and the exploiting of magnetic nanoparticles in MRI contrast agents [3].

Among the wide and diverse range of biosensors, it is safe to say that magnetic biosensors have attracted a lot of attention due to their numerous and useful features and properties. Most biomedical utilizations of NPs with magnetic properties are founded on the special properties of NPs with magnetic features and the generous

Magnetic Nanomaterials in Analytical Chemistry. DOI: https://doi.org/10.1016/B978-0-12-822131-0.00004-2

connections among magnetic fields and biological methods. For instance, when magnetic biosensors are compared with other methods like fluorescent-based techniques, biosensors with magnetic features have numerous benefits. First, biosensors with magnetic features or magnetic probes maintain their stability over a long period, and they have great stability. Fluorescent labels are classified as chemical composites that may lose their properties over time, but this is not the case with magnetic biosensors, which retain their properties over time [4]. This feature can be beneficial in tagging tests with tissue and organ construction for long periods. Another advantage of using magnetic biosensors is that they do not react to background noise, but this is very common in fluorescent-based techniques. Background fluorescence is a usual occurrence in biological materials due to the sticky features of tissue. The last advantage of magnetic biosensors is their sensitivity, which has been revealed to be superior to fluorescent analyses [5]. Their great sensitivity permits discovery with significantly minor quantities of protein in comparison with fluorescent-based methods. This chapter reviews magnetic nanomaterial as colorimetric sensors and presents a summary of improvements in and utilization of these magnetic sensors.

A sensor is described as an instrument or a module that helps sense every variation in body values such as pressure, power, or electrical quantity like current or any other type of energy. Also, a sensor can describe as a practical device that converts data and information about the physicochemical features of a system to a valuable and beneficial analytical signal based on possible actions and reactions between the medium and the probe conducting modification in optical features [5]. In still another explanation, conducting optical readable and interpretable signals is often done to generate optical probes. The colorimetric sensor has earned noticeable consideration due to its cost-effectiveness, facility, rapid response, and practicality. The crucial challenge for colorimetrics is all about altering recognition events into color modifications [6]. Colorimetric detecting and, from a wider viewpoint, detecting procedures based on adsorption spectroscopy exploit the photons in the visible light range (about 400–800 nm). However, in several samples, the adsorption spectra have been extended in range, usually UV–vis spectra (200–800 nm). A deep and high-quality insight into how colors change on the sensor can typically be attained from density functional theory (DFT) and various other theoretical techniques that have been demonstrated to be precise enough for theoretical research in the field of supramolecular chemistry. According to the research, the DFT method is a theoretical, logical, and accurate method that can be used in various software packages as well as on personal computers, meaning that you do not necessarily need Unix or a Unix-like environment, server, etc.

Since the changes in color can be seen even with the naked eye and without the use of advanced, complicated, and expensive instruments, it can be said that the colorimetric sensor is a cheap and economical device. An event that takes the form of assembly and disassembly, which happens among various molecules creating the appearance of an optical signal, is exploited in the design of colorimetric probes. It should be noted that one of the properties that make colorimetric sensors popular is their ability to recover and reuse. That's why, in the ideal design of colorimetric sensors, chemical interactions should be choosy but reversible. These actions and reactions contain metal

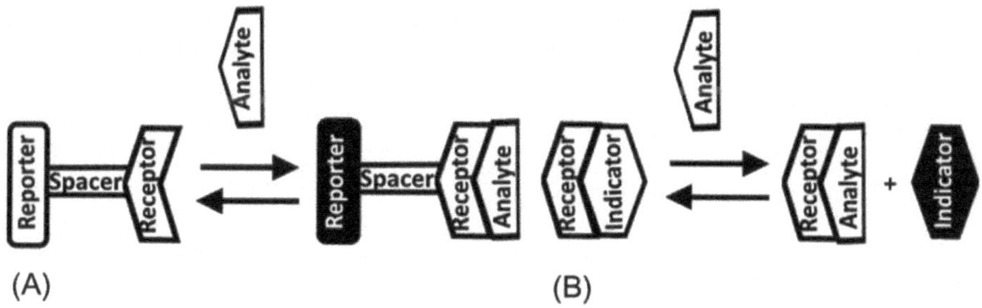

Figure 9.1 Illustration of the main approaches for the design of colorimetric chemical sensors: (A) receptor-spacer-reporter (RSR) and (B) indicator displacement assay (IDA). *Source*: I.I. Ebralidze, N.O. Laschuk, J. Poisson, O.V. Zenkina, Chapter 1-Colorimetric sensors and sensor arrays, in: Nanomaterials Design for Sensing Applications, 2019, pp. 1–39, reprinted by permission of the Elsevier.

coordination through ion–dipole connections and the creation of covalent bonds (for alkaline metal ions and transition metal complexes, respectively), as well as a variety of intermolecular actions and reactions associated to supramolecular actions and reactions. There are two main ways to design colorimetric sensors: The first method involves detecting based on a direct connection of an analyte, and the second method is about detecting based on a competitive connection of an analyte (see Fig. 9.1) [6].

Ideally, the receptor should be connected to the desired analyte in a very precise manner. Optimal physical or chemical stimulus changes such as pressure, pH, polarity, and concentration of the solvent and temperature can be used to optimize the selectivity of the receptor connected to the analyte [7].

Colorimetric sensors have the ability to quickly detect samples in a very short time and have the power to show outstanding color variation, which can be observed simply by the naked eye and which is valuable for in-arena analysis. The design of each probe is described through its actions and reactions with the exterior stimulus. The perfect specifications of probes based on the colorimetric method are noteworthy.

Primary sensors are complicated and bulky, requiring a variety of functional blocks, including a processing unit, a recognition unit, and transducer. All of these factors lead to delays in sensor response and consequently to lower sensor performance.

But current colorimetric methods are based on reducing size, dimension, and price, without any supplementary devices [8].

9.2 Magnetic nanomaterials as colorimetric sensors

Metal oxides are used in various medical and biomedical fields, in which applications include detection and therapeutic agents and cell tagging. It has been concluded that metal oxides are very widely used in the medical fields. Among these useful metal oxides, iron oxide is one of the most extensively used because it has more biocompatibility than other metal oxides [9].

Given the preceding explanations, nanotechnology has an extensive variety of uses in diverse areas of diagnostics. Nevertheless, the efficiency of some materials in nanodimensions is limited because of formidable isolation and low stability. Due to the problems sometimes encountered in the use of NPs, searching for materials with high stability, easy synthesis, easy and fast isolation, and the highest artificial catalytic activity seems essential for usefulness in an application [10]. Consequently, magnetic nanoparticles (MNPs) with significant properties such as high stability, dispersiveness, excellent catalytic properties, and simple isolation separation can be good candidates for the field of diagnosis. Researches demonstrate that MNPS with the size and dimensions of about 10–20 nm have the maximum super paramagnetism that leads to a speedy answer through a net arrangement of magnetic moments. When exterior magnetic fields are applied, the magnetic moments of MNPs arbitrarily adjust back to zero when the exterior magnetic field is detached [11].

In other words, MNPs—due to their unique properties such as exceptional magnetic features, great surface area and transmission of mass, extraordinary biocompatibility, simplicity of creation, and unique physicochemical features—are extremely important in the construction and development of colorimetric sensors. Using MNPs' catalytic activity, calorimetric sensors have been used to detect various analytes (Table 9.1) [12]. This section investigates new research and studies on the use of MNPs in the field of diagnosis.

Table 9.1 Advantages and disadvantages of the colorimetric sensors.

Advantages of colorimetric sensor arrays	Challenges
Fast	Reproducibility of printing
Inexpensive	Reproducibility of imaging
Lightweight and portable	Difficult to determine individual components of a mixture
Small sample size	Stability/shelf life
Minimal instrumentation and power needed	Large data sets of RGB values that need to be analyzed using chemometric methods
Ability to customize array for specific analytes	Sample applications may vary
Respond to a wide range of analytes	
Many potential sensors to choose from	
Can be used for identification and quantification	
Devices can be fabricated with simple methods such as inkjet printing	
With appropriate apps, data collection and analysis are user-friendly	
Potential to analyze vapor or liquid samples	
Customization possible for specific analytes	
Flexible array sizes	
Integration with smartphone technology	

Source: M.J. Kangas, R.M. Burks, J. Atwater, R.M. Lukowicz, P. Williams, A.E. Holmes, Detection and identification of chemical weapons and explosives, Crit. Rev. Anal. Chem. 47 (2) (2017) 138–153, reprinted by permission of Taylor & Francis Ltd, http://www.tandfonline.com.

9.2.1 Magnetic colorimetric sensors for heavy metal ions detection

Over time, the use of colorimetric sensors in the field of inorganic chemistry and organometallic chemistry has been increasing sharply [13]. Colorimetric sensors permit sample analysis without the need for very precise and complex, as well as very expensive, instruments. The basis of the performance of colorimetric sensors for the detection of heavy metals, as previously described, is that a change in color is observed in the presence of the metal species of interest. The color variation is preferably visible with the naked eye; nonetheless, it can also be detected with a modest spectrophotometer. Heavy metals affect the quality of human life because the human body desperately needs some of them to survive and wants a number of rare metals [14]. Although most heavy metals belong to the group of transfer metals, arsenic is a metalloid. Iron is the greatest plentiful heavy metal in the human body, found in many proteins that can carry oxygen, such as hemoglobin. Although the presence of iron is very important for the body, if there is too much, it is dangerous for the body due to the production of free radicals and toxins. Like iron and zinc, other metals in the body are dangerous to the body and produce toxins in excessive quantities.

There are several ways to identify heavy metals, each with its advantages and disadvantages: inductively coupled plasma mass spectrometry, electrochemistry, atomic absorption spectroscopy, electrochemistry, and high-performance liquid chromatography. While these methods are reliable, they also have problems [15]. They require precise, expensive, and time-consuming tools, and they need a complex pretreatment procedure. Access to diagnostic limits is difficult for complex samples due to the interposition of other coexisting ions, Working with them requires experienced and trained personnel, and samples must be measured in the laboratory. To remedy the disadvantages of these methods, a new method of diagnosis is needed that is simple, economical, and inexpensive and that can work with in situ samples. The colorimetric detection method promises to satisfy all these needs due to its simplicity, cost-effectiveness, and quick detection of the metal ion in a short period with notable color change that can be easily seen with the naked eye, ease of use, and usefulness for in-field analysis [12]. Detection using colorimetric sensors is one of the best, simplest, and most economical ways to quickly detect various analyses in a short period. It is a promising way to meet all these needs; by greatly simplifying diagnosis by changing color, which is clearly visible with the naked eye, it is useful for field examination.

Christus and coworkers [16] found an easy and practical colorimetric sensor that was established to sense As(V) and that is based on a Fenton-like response of Fe_3O_4 with methylene blue.

The Fenton-like catalytic response of Fe_3O_4 MNPs increases the indicator (methylene blue) oxidation reaction because the production of hydroxyl radicals (•OH) in the presence of hydrogen peroxide (H_2O_2) reduces the dark blue color of methylene blue and fades it. The dark blue fading is quite visible to the naked eye and also observed by UV–vis spectra. Then, with the addition of arsenate (As(V)) to the system under study, the blue color is recreated. This phenomenon can be due to the accumulation of arsenate on Fe_3O_4, which can cause the main and active sites of Fe_3O_4 to block, the result of which is a reduction in the inherent catalyst activity. In the optimum setting, the offered colorimetric probe shows a fast, accurate, and high-selectivity recognition of arsenate with the diagnosis restriction of 0.358 nM and a linear connection of $R^2 = 0.99$ in aqueous solution [16].

Christus and coworkers [17] demonstrated a colorimetric sensor for the detection of Hg(II) in aqueous solution, based on a MNPs–encapsulated metal–organic framework (MOF) or MOF-based magnetic nanocomposite (Fe_3O_4@ZIF-67). The strategy based on the approach is a highly efficient peroxidase mimetic reaction that can catalyze the oxidation of 3,3′,5,5′-tetramethylbenzidine (TMB) in the presence of hydrogen peroxide (H_2O_2). The result of this reaction was the production of the blue color in aqueous solution.

The blue color disappeared after the addition of antiradical biomolecules glutathione (GSH) to the solution, and consequently the oxidation of TMB was prevented due to the powerful restoration feature of cation radicals. When Hg^{2+} ions were added into the Fe_3O_4@ZIF-67/TMB/GSH system, the blue color of the solution was regained due to the strong affinity of Hg^{2+} with thiol groups (mercury–thiol complex) of antiradical biomolecules. Glutathione incites the oxidation of the substrates; in other words, GSH was exploited to neutralize the color and, on the other hand, the Hg^{2+} ions to regain the blue color. The mechanism proposed in this study significantly increased the sensitivity of the colorimetric sensor to identify mercury ions with the detection limit as low as 0.36 nM [17].

Jeong et al., [18] designed a reliable, fast, and sensitive colorimetric detector based on surface amine–modified iron oxide–silica core–shell NPs (APTES–Fe@SiO_2 NPs) to detect Co^{2+} ions. Core–shell NPs can be exploited as an effective and useful nominee sample for colorimetric detectors. In this method, an amine functional group was easily attached to the silica shell by the sol–gel selective method. The (3-aminopropyl) triethoxysilane (APTES) is nominated as the amine functional sample because it is an extensively exploited attaching agent for the specific recognition of toxic heavy metal ions. This inexpensive and speedy detection colorimetric sensor attained quick, naked-eye exposure of Co^{2+} ions at a detection limit as low as 0.08 mM. According to the strategy proposed in this study, APTES is conjugated with the magnetic core–shell and exploited as a colorimetric detector for Co^{2+}, Cu^{2+}, Fe^{2+}, and Hg^{2+}. The amine moiety was stabilized on the surface of Fe@SiO_2 NPs to create a complex of metal plus amine and to encourage a color alteration by accumulation among NPs. Various heavy metal ions were nominated to show colorimetric recognition. One of the most important reasons for using these magnetic core–shell NPs is their easy magnetic isolation for use in subsequent water treatment, thus reducing their environmental effects in the aqueous phase. A significant investigation is the ability to eliminate the NPs from the treated water to inhibit them from acting as pollutants or impurities themselves. The MNPs exploited in the colorimetric sensor of toxic heavy metal ions were effectively isolated in a very short time—a few minutes—by magnets [18].

9.2.2 Magnetic colorimetric sensors for antibiotics detection

Today, antibiotic pollution is considered a global concern due to its noteworthy influence on human health. Antibiotics are used effectively for both humans and animals to prevent and treat diseases, as well as to support the growth of animal husbandry. But the introduction of antibiotics into nature in various ways, due to their high resistance, poses many risks to human health. For these reasons, many methods have been created to identify these materials, one of the best and cheapest of which is to identify them using colorimetric sensors. Among these methods, colorimetric probe methods have the capacity for great material throughput, inexpensiveness, high sensitivity, and ease of use. In other words, the colorimetric method is a reliable method for the recognition of significant ionic or neutral species in medicine or the environment without any complicated and complex tool in qualitative and quantitative conditions [19]. Its exploitation as a trial kit for on-site recognition, With its very short time of recognition, ease of procedure, reversibility with ultraselectivity, high sensitivity, and no requirement for any pretreatment of material or physical skill, this method, as a trial kit, is a promising method among investigators.

Kwon and coworkers [20] demonstrated the quick and easy recognition of penicillin class (PEN) antibiotics, which includes several types of penicillin: penicillin G (PEN-G), penicillin V (PEN-V), nafcillin (NAF), oxacillin (OXA), cloxacillin (CLX), and dicloxacillin (DCX) in pork. This identification was based on the use of hybrid magnetic nanoparticle clusters (HMNCs), penicillin class-selective antibody (Ab)–functionalized platinum nanoparticles (Pt NPs), and TMB. Antibiotics were trapped by HMNCs due to the chemical bond formed between the amine groups in the HMNCs family and the carboxylic acid groups in the antibiotic family. After adding antibiotics to HMNCs, PENs-HMNC reacted with Ab-Pt, and the resulting Pt-antibiotic-HMNCs complexes were added to a TMB solution. As can be clearly seen in Fig. 9.2, because of TMB oxidation throughout the Pt NPs, the color of the

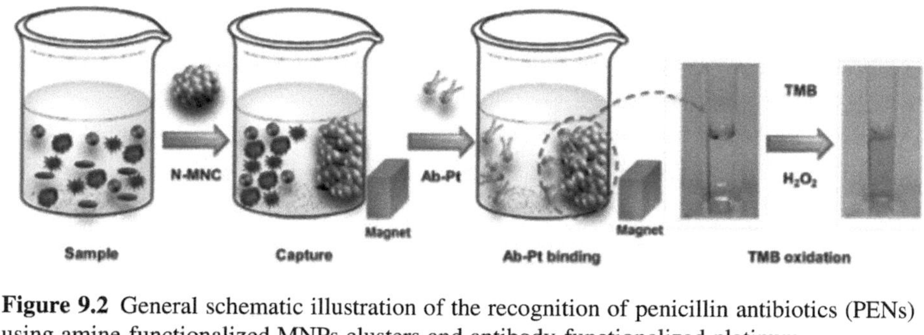

Figure 9.2 General schematic illustration of the recognition of penicillin antibiotics (PENs) using amine-functionalized MNPs clusters and antibody-functionalized platinum nanoparticles (Ab-functionalized Pt NPs).
Source: D. Kwon, W. Lee, W. Kim, H. Yoo, H.C. Shin, S. Jeon, Colorimetric detection of penicillin antibiotic residues in pork using hybrid MNPs and penicillin class-selective, antibody-functionalized platinum nanoparticles, Anal. Methods 7 (18) (2015) 7639–7645, reprinted by permission of the Royal Society of Chemistry.

solution changed from colorless to blue. The color change created during the oxidation reactions provided the easy detection of antibiotics (PENs) with the naked eye and quantified by calculating the absorption of light [20].

In recent studies, scientists added aptamers to the colorimetric immunoassays method to increase their selectivity to recognize diverse antibiotics. For instance, Miao and coworkers [21] found an ultraselective simple and sensitive colorimetric probe for recognizing the ultratrace level of antibiotics and exploiting the CAP (chloramphenicol) as the desired analyte. This research was based on triple signal-amplified and magnetic compound probes comprised of aptamer-HRP-Pt NPs nanotracer. Their results demonstrated that this method has suitable figures of merit. Due to the strong response based on color, CAP was visible at very low concentrations and in the range of 0.01 ng mL^{-1} with the naked eye, and it also could be detected at a concentration of 0.0003 ng mL^{-1} by the UV−vis spectrometer. The method presented in this study was successfully applied to real milk samples in a very short recognition time, about 30 minutes [21].

Also, Wang and coworkers [22] have been reported that the fast, applicable, and ultrasensitive colorimetry sensing of tetracyclines (TCs) based on Fe_3O_4 NPs. According to previous research, TCs are known to have a strong tendency to bind to metal ions, for instance, Fe(II) and Fe(III) on the surface of Fe_3O_4 . The reaction was accompanied by the oxidation of Fe_3O_4-TCs complex that catalyzed the TMB-H_2O_2 reaction. This oxidation was done by dissolving oxygen to produce more reactive oxygen species. The results showed that the hydroxyl radical created causes the color of the solution to deepen and deepen so that the color is recognizable with the unaided eye and also checked by UV−vis spectra. A colorimetric probe using Fe_3O_4 MNPs may afford a promising alternative recognition for tetracyclines in medical samples [22].

9.2.3 Magnetic colorimetric sensors for other analytes

Dong and coworkers used the colorimetric sensor based on graphene oxide−Fe_3O_4 magnetic nanocomposites to detect glucose, which showed a visible color change from colorless to blue in the existence of H_2O_2 [23]. The kinetic factors showed that the interaction between Fe_3O_4 and GO can enhance the catalytic attraction toward hydrogen peroxide. Unique peroxidase mimetics, GO−Fe_3O_4 nanocomposites, presented numerous benefits over natural enzymes and additionally exhibited features like simplicity of synthesis, inexpensiveness, superparamagnetic features, and improved sympathy. This strategy was based on the new peroxidase mimetic that could rapidly catalyze the oxidation reaction of the peroxidase substrate TMB in the presence of H_2O_2 [23].

Ma et al., [24] established a colorimetric nanosensor for the sensitive recognition of *Salmonella typhimurium* based on Fe_3O_4 MNPs and gold NPs (GNPs). The two balancing arrangements of the *S. typhimurium* aim DNA were matched to Fe_3O_4 MNPS and GNPs to synthesize capture detectors. Fe_3O_4 can attain the fast isolation and improvement of the aim DNA. By the addition of *S. typhimurium* aim DNA arrangements, the sandwich-like constructions were shaped with DNA hybridization detection technology. According to optical distance–dependent characteristics, the unique respectable spreading of GNPs was broken. GNPs presented different amounts of accumulation with varying degrees of *S. typhimurium* aim DNA. The color changed from red to purple to blue, which could be monitored by the unaided eyes. Finally, it can be said that this advanced technique was effectively exploited to examine samples of milk. There was no noteworthy variation among a traditional plate-counting technique and the advanced technique. Consequently, the method has the capability for usage in an extensive range in recognition of various food-borne pathogenic bacteria in nutrition examples [24].

Wu and coworkers [25] established a simple, high-sensitive, and ultraselective colorimetric sensor Fe_3O_4@MIL-100 (Fe) (MOF-based materials) based on the oxidation of TMB by H_2O_2 with great productivity. As a peroxidase-like catalyst, Fe_3O_4@MIL-100 (Fe) has numerous benefits such as simple fabrication and storage, high efficiency as a catalyst, inexpensiveness, and great stability. The offered technique was effectively exploited in the recognition of cholesterol amounts in clinical samples with a detection limit of 0.8 μM [25].

For another example, Liang and cowerkers [26] advanced a rapid and ultrasensitive recognition colorimetric probe using the Fe_3O_4 peroxidase assay, acetylcholinesterase (AChE), and choline oxidase (CHO) to detect organophosphorus (OP) pesticides and nerve agents in aqueous solution. The advanced MNPs peroxidase assay is more sensitive compared to other traditional enzyme activity–based techniques due to the high catalytic activity of MNPs. The strategy of this study was based on catalyzing the creation of hydrogen peroxide (H_2O_2) by the enzymes AChE and CHO existing in the enzyme acetylcholine. After this event and the production of hydrogen peroxide, the active MNPs catalyzed the oxidation of colorimetric substrates and caused a color reaction. After adding the OP neurotoxins, the enzymatic activity of AChE was prevented. As a result of the reduced enzymatic activity of AChE, the amount of hydrogen peroxide (H_2O_2) production decreased, causing a reduced catalytic oxidation of colorimetric substrates with MNPs peroxidase mimetics. Finally, the result was a sharp drop in the color intensity of the solution [26].

Chen et al., [27] designed a sensitive colorimetric sensor for L-cysteine recognition that was based on the greater peroxidase-like catalytic activity of superparamagnetic Fe_3O_4 nanofibers. The as-fabricated Fe_3O_4 nanofibers were revealed to have outstanding peroxidase-like catalytic activity in the oxidation of TMB in the existence of hydrogen peroxide. Due to the great catalytic activity of Fe_3O_4 nanofibers, a fast, ultraselective colorimetric recognition of L-cysteine with a detection limit as low as 0.028 μM has been established. The easy preparation method and excellent diagnostic ability showed that this strategy could provide promising programs for colorimetric sensor use in the medical and environmental sciences [27].

For more details, Li and coworkers [28] demonstrated a sensitive colorimetric detector based on a Pd/Fe_3O_4-PEI-RGO nanohybrid by dispersion of Pd and Fe_3O_4 NPs on polyetheylenimine (PEI) altered sheets of graphene oxide to detect hydrogen peroxide (H_2O_2). In this method, Pd/Fe_3O_4-PEI-RGO revealed improved peroxidase catalytic activity. This strategy was based on the high oxidation of the TMB material in the presence of hydrogen peroxide to create a suitable color response. Despite all the advantages and good properties of NPs, these compounds tend to agglomerate due to their small size and may lose their catalytic activity in the reaction due to this property. An effective strategy and solution to solve this problem of the NP is to use different solid substrates to deposit NPs with diverse forms and shapes on them.

Among the wide range of solid substrates for NPs use, graphene nanosheets and MNPs have attracted a lot of attention due to their great surface area, which can improve catalytic activity, and greater stability. MNPs have effective magnetic isolation ability for the simple isolation of the catalyst from the reaction solution. Also, the points highlighted in this study indicated that ternary nanohybrids Pd/Fe_3O_4-PEI-RGO demonstrated great stability, high biocompatibility, simple alteration, nontoxicity, easy maintenance, and quick isolation in comparison to traditional reaction situations. Consequently, it is predictable that Pd/Fe_3O_4-PEI-RGO ternary nanohybrids could usefully and powerfully sense H_2O_2 in various fields such as the medical, environmental, and industrialized areas. Pd/Fe_3O_4-PEI-RGO showed an extensive linear range from 0.5 to 150 mM H_2O_2 and a detection limit as low as 0.1 mM in aqueous solution [28].

Recently, Karami et al., [29] established a novel colorimetric sandwich-type immunosensor using a lateral surface plasmon resonance of triangular Au NPs and magnetic force separation of Fe_3O_4 NPs decorated with PSA- (prostate-specific antigen) specific antibodies to sense prostate-specific antigen (PSA). Consequently, the linear range for PSA recognition was exposed to be 0.01–20 ng mL^{-1} with the detection limit as low as 0.009 ng mL^{-1}. To achieve good and efficient isolation, the use of MNPs in conjunction with GNPs led to the approximation of such features as low detection limits and fast sensors in response. In the arena of isolation in the field of biology, MNPs are distinct as core nanomaterials; in particular, Fe_3O_4 are usually exploited to magnetically separate various samples such as mammalian cells, proteins, viruses, and DNA [29].

Zhou and coworkers [30] designed a novel and ultraselective colorimetric sensor based on ssDNA hybridization–controlled GO gathering, as a substitute for ssDNA discharge and with magnetic isolation to sense ochratoxin A (OTA). This colorimetric detector structure can be expediently exploited for quick and noticeable ochratoxin. A recognition using an easy magnetic isolation technique and an adjustment of the pH of the solution (see Fig. 9.3). Numerous colorimetric detectors are based on valuable and expensive metal nanomaterials as the indicator, like GNPs, gold nanorods, and silver NPs. Although these sensors perform well, the use of nonrenewable precious metals is not desirable due to their high cost and uneconomical cost. There is no doubt that carbon-based nanomaterials, obtained in a variety of ways, such as different organic and mineral resources, are economically better options for making low-cost sensors [30].

In other research, Le and coworkers [8] demonstrated a convenient and rapid colorimetric sensor to detect bacteria using chitosan-coated iron oxide MNPs (CS-MNPs). To improve the recognition techniques of bacteria that are chiefly appropriate for on-site recognition, colorimetric methods have acquired growing consideration

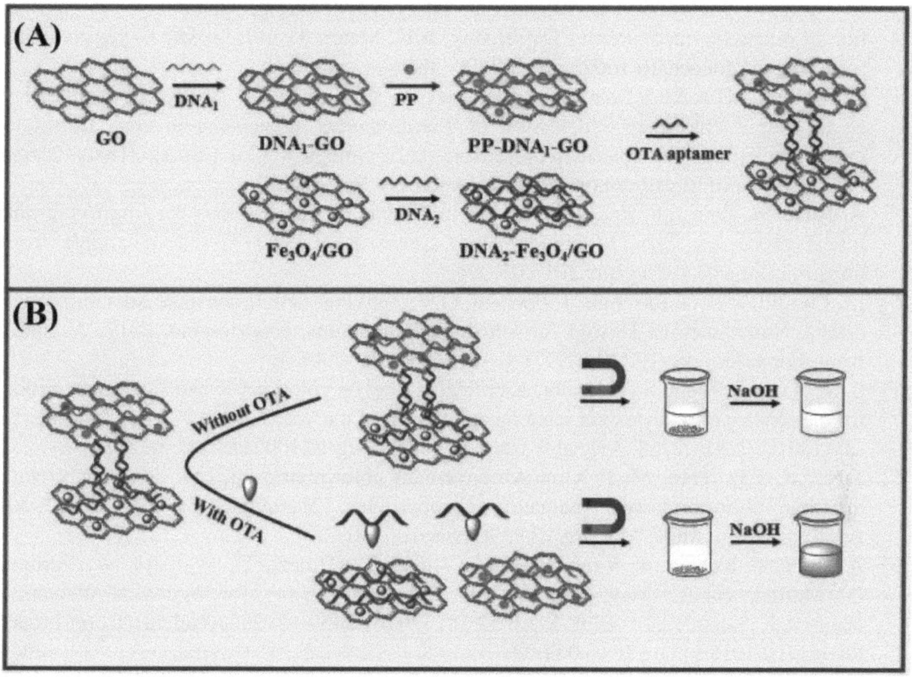

Figure 9.3 Schematic description of the synthesis procedure (A) and colorimetric recognition principle of the system to sense OTA (B).
Source: Z. Zhou, N. Hao, Y. Zhang, R. Hua, J. Qian, Q. Liu, et al., A novel universal colorimetric sensor for simultaneous dual-target detection through DNA-directed self-assembly of graphene oxide and magnetic separation, Chem. Commun. 53 (52) (2017) 7096–7099, reprinted by permission of the Royal Society of Chemistry.

because of their simple procedure and visual results without any complicated and expensive equipment and expertise This modest colorimetric tactic permits the fast recognition of bacterial cells to 10^4 CFU mL^{-1} by the naked eye and 102 CFU mL^{-1} by spectrophotometry, about 10 minutes. Based on the results of this study, it is predicted that the proposed method based on CS-MNPs will have a high potential for diagnosing bacterial infections in a restricted area or at point-of-care testing [8].

References

[1] T.A.P. Rocha-Santos, Sensors and biosensors based on magnetic nanoparticles, TrAC - Trends Anal. Chem. 62 (2014) 28–36. Available from: https://doi.org/10.1016/j.trac.2014.06.016.

[2] J.B. Haun, T.J. Yoon, H. Lee, R. Weissleder, Magnetic nanoparticle biosensors, Wiley Interdiscip. Rev. Nanomed. Nanobiotechnol. 2 (3) (2010) 291–304. Available from: https://doi.org/10.1002/wnan.84.

[3] L. Gloag, M. Mehdipour, D. Chen, R.D. Tilley, J.J. Gooding, Advances in the application of magnetic nanoparticles for sensing, Adv. Mater. 31 (48) (2019) 1–26. Available from: https://doi.org/10.1002/adma.201904385.

[4] S. Alhogail, G.A.R.Y. Suaifan, F.J. Bikker, W.E. Kaman, K. Weber, D. Cialla-May, et al., Rapid colorimetric detection of *Pseudomonas aeruginosa* in clinical isolates using a magnetic nanoparticle biosensor, ACS Omega 4 (26) (2019) 21684–21688. Available from: https://doi.org/10.1021/acsomega.9b02080.

[5] Á. Ríos, M. Zougagh, Recent advances in magnetic nanomaterials for improving analytical processes, TrAC - Trends Anal. Chem. 84 (2016) 72–83. Available from: https://doi.org/10.1016/j.trac.2016.03.001.

[6] I.I. Ebralidze, N.O. Laschuk, J. Poisson, O.V. Zenkina, Colorimetric sensors and sensor arrays, Nanomaterials Design for Sensing Applications, Elsevier Inc, 2019. Available from: https://doi.org/10.1016/B978-0-12-814505-0.00001-1.

[7] S. Suganya, S. Naha, S. Velmathi, A critical review on colorimetric and fluorescent probes for the sensing of analytes via relay recognition from the year 2012–17, ChemistrySelect 3 (25) (2018) 7231–7268. Available from: https://doi.org/10.1002/slct.201801222.

[8] T.N. Le, T.D. Tran, M. Il Kim, A convenient colorimetric bacteria detection method utilizing chitosan-coated magnetic nanoparticles, Nanomaterials 10 (1) (2020). Available from: https://doi.org/10.3390/nano10010092.

[9] A. Sobczak-Kupiec, J. Venkatesan, A. Alhathal AlAnezi, D. Walczyk, A. Farooqi, D. Malina, et al., Magnetic nanomaterials and sensors for biological detection, Nanomed. Nanotechnol. Biol. Med. 12 (8) (2016) 2459–2473. Available from: https://doi.org/10.1016/j.nano.2016.07.003.

[10] H. Maeda, P. Anzenbacher, Colorimetric sensors, Supramol. Chem. (2012). Available from: https://doi.org/10.1002/9780470661345.smc098.

[11] G. Liu, R.-W. Li, Y. Chen, Magnetic nanoparticle for biomedicine applications, Nanotechnol. Nanomed. Nanobiotechnol. 2 (1) (2015) 1–7. Available from: https://doi.org/10.24966/ntmb-2044/100003.

[12] M.J. Kangas, R.M. Burks, J. Atwater, R.M. Lukowicz, P. Williams, A.E. Holmes, Colorimetric sensor arrays for the detection and identification of chemical weapons and explosives, Crit. Rev. Anal. Chem. 47 (2) (2017) 138–153. Available from: https://doi.org/10.1080/10408347.2016.1233805.

[13] B. Jiang, L. Lian, Y. Xing, N. Zhang, Y. Chen, P. Lu, et al., Advances of magnetic nanoparticles in environmental application: environmental remediation and (bio)sensors as case studies, Environ. Sci. Pollut. Res. 25 (31) (2018) 30863–30879. Available from: https://doi.org/10.1007/s11356-018-3095-7.

[14] Z. Yang, L. Zhu, L. Chen, Selective adsorption and separation of dyes from aqueous solution by core-shell structured NH_2-functionalized UiO-66 magnetic composites, J. Colloid Interface Sci. 539 (2019) 76–86. Available from: https://doi.org/10.1016/j.jcis.2018.11.064.

[15] K. Wu, D. Su, J. Liu, R. Saha, J.-P. Wang, Magnetic Nanoparticles in Nanomedicine. (2018). <http://arxiv.org/abs/1811.01418>.

[16] A.A. Babu Christus, P. Panneerselvam, A. Ravikumar, Novel, sensitive and selective colorimetric detection of arsenate in aqueous solution by a Fenton-like reaction of Fe_3O_4 nanoparticles, Anal. Methods 10 (36) (2018) 4378–4386. Available from: https://doi.org/10.1039/c8ay01452c.

[17] A.A.B. Christus, P. Panneerselvam, A. Ravikumar, N. Morad, S. Sivanesan, Colorimetric determination of Hg(II) sensor based on magnetic nanocomposite (Fe_3O_4@ZIF-67) acting as peroxidase mimics, J. Photochem. Photobiol. A: Chem. 364 (Ii) (2018) 715–724. Available from: https://doi.org/10.1016/j.jphotochem.2018.07.009.

[18] U. Jeong, H.H. Shin, Y. Kim, Functionalized magnetic core-shell Fe@SiO_2 nanoparticles as recoverable colorimetric sensor for Co^{2+} ion, Chem. Eng. J. 281 (2015) 428–433. Available from: https://doi.org/10.1016/j.cej.2015.07.006.

[19] M.A. Abedalwafa, Y. Li, C. Ni, L. Wang, Colorimetric sensor arrays for the detection and identification of antibiotics, Anal. Methods 11 (22) (2019) 2836–2854. Available from: https://doi.org/10.1039/c9ay00371a.

[20] D. Kwon, W. Lee, W. Kim, H. Yoo, H.C. Shin, S. Jeon, Colorimetric detection of penicillin antibiotic residues in pork using hybrid magnetic nanoparticles and penicillin class-selective, antibody-functionalized platinum nanoparticles, Anal. Methods 7 (18) (2015) 7639–7645. Available from: https://doi.org/10.1039/c5ay01590a.

[21] Y. Miao, N. Gan, H.X. Ren, T. Li, Y. Cao, F. Hu, et al., A triple-amplification colorimetric assay for antibiotics based on magnetic aptamer-enzyme co-immobilized platinum nanoprobes and exonuclease-assisted target recycling, Analyst 140 (22) (2015) 7663–7671. Available from: https://doi.org/10.1039/c5an01142f.

[22] Y. Wang, Y. Sun, H. Dai, P. Ni, S. Jiang, W. Lu, et al., A colorimetric biosensor using Fe_3O_4 nanoparticles for highly sensitive and selective detection of tetracyclines, Sens. Actuators, B: Chem. 236 (2016) 621–626. Available from: https://doi.org/10.1016/j.snb.2016.06.029.

[23] Y.L. Dong, H.G. Zhang, Z.U. Rahman, L. Su, X.J. Chen, J. Hu, et al., Graphene oxide-Fe_3O_4 magnetic nanocomposites with peroxidase-like activity for colorimetric detection of glucose, Nanoscale 4 (13) (2012) 3969–3976. Available from: https://doi.org/10.1039/c2nr12109c.

[24] X. Ma, L. Song, Y. Xia, C. Jiang, Z. Wang, A novel colorimetric detection of *S. typhimurium* based on Fe_3O_4 magnetic nanoparticles and gold nanoparticles, Food Anal. Methods 10 (8) (2017) 2735–2742. Available from: https://doi.org/10.1007/s12161-017-0819-0.

[25] Y. Wu, Y. Ma, G. Xu, F. Wei, Y. Ma, Q. Song, et al., Metal-organic framework coated Fe_3O_4 magnetic nanoparticles with peroxidase-like activity for colorimetric sensing of cholesterol, Sensors Actuators B: Chem. 249 (2017) 195–202. Available from: https://doi.org/10.1016/j.snb.2017.03.145.

[26] M. Liang, K. Fan, Y. Pan, H. Jiang, F. Wang, D. Yang, et al., Fe_3O_4 magnetic nanoparticle peroxidase mimetic-based colorimetric assay for the rapid detection of organophosphorus pesticide and nerve agent, Anal. Chem. 85 (1) (2013) 308–312. Available from: https://doi.org/10.1021/ac302781r.

[27] S. Chen, M. Chi, Y. Zhu, M. Gao, C. Wang, X. Lu, A facile synthesis of superparamagnetic Fe_3O_4 nanofibers with superior peroxidase-like catalytic activity for sensitive colorimetric detection of L-cysteine, Appl. Surf. Sci. 440 (2018) 237–244. Available from: https://doi.org/10.1016/j.apsusc.2018.01.152.

[28] S. Li, H. Li, F.J. Chen, J. Liu, H.L. Zhang, Z.G. Yang, B.D. Wang, Strong coupled palladium nanoparticles decorated on magnetic graphene nanosheets as enhanced peroxidase mimetics for colorimetric detection of H_2O_2, Dyes Pigm. 125 (2016) 64–71. Available from: https://doi.org/10.1016/j.dyepig.2015.10.009.

[29] P. Karami, H. Khoshsafar, M. Johari-Ahar, F. Arduini, A. Afkhami, H. Bagheri, Colorimetric immunosensor for determination of prostate specific antigen using surface plasmon resonance band of colloidal triangular shape gold nanoparticles, Spectrochim. Acta - Part A: Mol. Biomol. Spectrosc. 222 (2019) 117218. Available from: https://doi.org/10.1016/j.saa.2019.117218.

[30] Z. Zhou, N. Hao, Y. Zhang, R. Hua, J. Qian, Q. Liu, et al., A novel universal colorimetric sensor for simultaneous dual target detection through DNA-directed self-assembly of graphene oxide and magnetic separation, Chem. Commun. 53 (52) (2017) 7096–7099. Available from: https://doi.org/10.1039/c7cc03914j.

Application of magnetic nanomaterials as resonance light scattering sensors

10

Shayessteh Dadfarnia, Roya Afsharipour and Ali Mohammad Haji Shabani
Department of Chemistry, Faculty of Science, Yazd University, Yazd, Iran

10.1 Introduction

Although we may not be familiar with the term "light scattering," it makes life enjoyable for us. Light scattering is a phenomenon that happened across human life in natural events such as exquisite sunrise and sunset glows, different colors of precious stones, and multicolored rainbows [1]. Any optically heterogeneous system consisting of a single molecule or self-aggregations of a few or a large number of molecules, when exposed to electromagnetic radiation, scattered light in all directions except in the direction of the incident beam [2,3]. Thus, except for vacuum, most of the system is somehow heterogeneous and so light scattering occurs widely in the universe. According to electromagnetic theory, the scattered light is radiated by electrons of the scattering system that are accelerated and oscillated by the excitation of an incident electromagnetic wave. In the following paragraphs, the principles of light scattering are discussed, then the role of nanoparticles (NPs) especially magnetic nanoparticles (MNPs) in the design of sensors by resonance light scattering (RLS) is considered, and the application of RLS signals from MNPs in various fields is reviewed.

10.2 Resonance light scattering theory and development

10.2.1 Light scattering theory

When an electromagnetic wave impinges on an object, the light is redirected causing light scattering to occur. Thus light scattering is a complex interaction between the incident electromagnetic wave and the atomic or molecular structure of the scattering object. The interaction of the electromagnetic wave with a discrete scattering particle results in the periodical perturbation of the electron orbits within the constituent's particles of the molecules with a frequency (ν_0) the same as the electric field of the incident beam. The perturbation or oscillation of the electron cloud induced dipole moment and consequently scattered light due to the periodic charge separation within the molecule. When the frequency of the majority of scattered light is the same as the incident light, the process is called elastic scattering.

Magnetic Nanomaterials in Analytical Chemistry. DOI: https://doi.org/10.1016/B978-0-12-822131-0.00002-9

In general, the theory of light scattering is categorized into two theoretical frames of Rayleigh and Mie scattering. The Rayleigh scattering applies to small, spherical, nonabsorbing (dielectric) particles, while the Mie scattering is suitable for absorbing or nonabsorbing spherical particles without the size limitation. Thus, the Mie theory can be used for describing most spherical scattering particles, including Rayleigh scattering. However, Rayleigh scattering theory is simpler and when applicable it is preferred.

The criteria of classification of light scattering are the dimensionless size parameter of α that is defined by the following equation:

$$\alpha = \frac{2\pi r}{\lambda} \tag{10.1}$$

In this equation, r is the radius of the particle, and λ is the relative scattering wavelength [4]. Thus if the size of the particle in comparison to wavelength is small, $\alpha << 1$, it is classified as Rayleigh scattering; when the particle size and wavelength are about the same, $\alpha \approx 1$ (or when the particle diameter is greater than about 1/20 of the wavelength of incident light), it is named Mie scattering; while if the particle size is much larger than the wavelength, $\alpha >> 1$, it is called geometric scattering [5]. The relative scattering wavelength λ is represented as:

$$\lambda = \frac{\lambda_0}{m_0} \tag{10.2}$$

where λ_0 and m_0 are the incident wavelength in the vacuum and the refractive index of the surrounding matrix, respectively. The refractive index of the scattering particle, m, is defined as:

$$m = n - ik \tag{10.3}$$

In this equation n denoted to the refraction of light (light speed in vacuum divided by the light speed in the material) and ik is the complex term related to light absorption. The absorption coefficient of material (cm^{-1}) and the complex part of the refractive index are related by the following relation:

$$\text{Absorption coefficient} = \frac{4\pi k}{\lambda} \tag{10.4}$$

Although the value of k for any material is not exactly zero, those with a k value close to zero are called dielectric. The magnitude of the refractive index $|m|$ for Rayleigh criteria is:

$$|m| = (n^2 + k^2)^{1/2} \tag{10.5}$$

Thus, according to the Rayleigh criteria, $\alpha << 1$ and $|m|\ \alpha << 1$, at any moment, the particle encounters a uniform electric field and the time for electric field penetration is significantly less than the electromagnetic wave oscillation period.

The scattering parameters may be defined for the Rayleigh and Mie theories based on the spherical coordinate scattering geometry for a single incident light ray on a single spherical particle (Fig. 10.1).

The intensities of the vertically and horizontally polarized scattered radiation (W cm^{-2}) at any scattering angle (ϕ, θ) concerning scattering plane are defined by the intensity of incident ray, I_0, and the scattered ray using the following equations:

$$I_{\varnothing} = I_0 \frac{\lambda^2}{4\pi^2 r^2} i_1 \sin^2\phi \tag{10.6}$$

$$I_{\varnothing} = I_0 \frac{\lambda^2}{4\pi^2 r^2} i_2 \cos^2\phi \tag{10.7}$$

As for perfectly spherical particles, polarized incident radiation and the produced polarized scattered radiation are similar, Eqs. (10.6) and (10.7) may be redefined and rewritten in terms of scattering plane and differential scattering cross-sections (cm^2 sr^{-1}):

$$I_{VV} = I_0 \frac{1}{r^2} \sigma'_{VV} \tag{10.8}$$

$$I_{HH} = I_0 \frac{1}{r^2} \sigma'_{HH} \tag{10.9}$$

where the subscripts refer to the state of the incident and scattered light polarization, respectively, and orientation is defined by the scattering plane. Thus, *VV* refers to vertically polarized incident and scattered light at scattering plane of $\phi = 90$ degrees and *HH* refers to horizontally polarized incident and scattered light at the

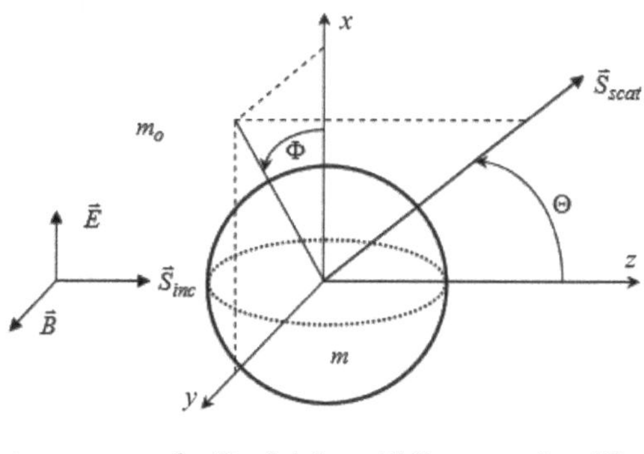

Figure 10.1 Coordinate geometry for Rayleigh and Mie scattering [4].

scattering plane of $\phi = 0$ degree. The scattering for unpolarized incident light is independent of ϕ and is represented by the following equation:

$$I_{scat} = I_0 \frac{1}{r^2} \sigma'_{scat} \quad (10.10)$$

In this equation σ'_{scat} is the average of σ'_{HH} and σ'_{VV}.

In addition to redistribution of incident radiation by light scattering represented by the equations 10.6-10.10, the particle may absorb the incident radiation too. The rate of the total amount of radiation energy of incident light removed because of its interaction with a single particle is directly determined from the extinction cross-section σ_{ext} (cm^2), that is:

$$\dot{E}_{removed} = I_0 \sigma_{ext} \quad (10.11)$$

where σ_{ext} denoted to the loss of energy of the incident beam because of the scattering and absorption and can be represented in term of absorption (σ_{abs}) and total scattering cross-section (σ_{scat}); thus:

$$\sigma_{ext} = \sigma_{abs} + \sigma_{scat} \quad (10.12)$$

The σ_{scat} is determined by integrating the differential cross-section over 4π steradians. Thus, based on the incident light intensity and the aforementioned equations, the scattering and extinction quantities can be determined if various cross-sections mentioned earlier are calculated.

The differential scattering cross-section of the small spherical particle can be calculated based on the Rayleigh theory using the following equations:

$$\sigma'_{VV} = \frac{\lambda^2}{4\pi^2} \alpha^6 \left| \frac{\overline{m}^2 - 1}{\overline{m}^2 + 2} \right|^2 \quad (10.13)$$

$$\sigma'_{HH} = \sigma'_{VV} \cos^2\theta \quad (10.14)$$

According to these equations, the differential scattering cross-sections are directly proportional to the sixth power of particle size (α) and inversely to the fourth power of λ. The total scattering and absorption cross-section (cm^2) are determined by:

$$\sigma_{scat} = \frac{2\lambda^2}{3\pi} \alpha^6 \left| \frac{\overline{m}^2 - 1}{\overline{m}^2 + 2} \right|^2 \quad (10.15)$$

$$\sigma_{abs} = \frac{-\lambda^2}{\pi} \alpha^3 \mathrm{Im} \left\{ \frac{\overline{m}^2 - 1}{\overline{m}^2 + 2} \right\} \quad (10.16)$$

The total extinction cross-section is then obtained by the addition of the absorption and scattering cross-section ($\sigma_{ext} = \sigma_{abs} + \sigma_{scat}$).

However, the differential scattering cross-section of particles that have a diameter similar to or larger than the wavelength of the incident light is determined by Mie theory. Based on Mie theory, the differential scattering cross-section is related to the angular intensity function i_1 and i_2 according to the following equations:

$$\sigma'_{VV} = \frac{\lambda^2}{4\pi^2} i_1 \tag{10.17}$$

$$\sigma'_{HH} = \frac{\lambda^2}{4\pi^2} i_2 \tag{10.18}$$

The average of these two equations provides the differential scattering cross-section of the unpolarized incident light, that is:

$$\sigma'_{scat} = \frac{\lambda^2}{8\pi^2}(i_1 + i_2) \tag{10.19}$$

where the intensity functions are determined using the following infinite series:

$$i_1 = \left| \sum_{n=1}^{\infty} \frac{2n+1}{n(n+1)} [\alpha_n \pi_n(\cos\theta) + b_n \tau_n(\cos\theta)] \right|^2 \tag{10.20}$$

$$i_2 = \left| \sum_{n=1}^{\infty} \frac{2n+1}{n(n+1)} [\alpha_n \tau_n(\cos\theta) + b_n \pi_n(\cos\theta)] \right|^2 \tag{10.21}$$

In these equations the angular dependent function π_n and τ_n is defined by Legendre polynomials:

$$\pi_n(\cos\theta) = \frac{P_n^{(1)}(\cos\theta)}{\sin\theta} \tag{10.22}$$

$$\tau_n(\cos\theta) = \frac{dP_n^{(1)}(\cos\theta)}{d\theta} \tag{10.23}$$

and the α_n and b_n parameters are expressed as:

$$a_n = \frac{\psi_n(\alpha)\psi'_n(m\alpha) - m\psi_n(m\alpha)\psi'_n(\alpha)}{\xi(\alpha)\psi'_n(m\alpha) - m\psi_n(m\alpha)\xi'_n(\alpha)} \tag{10.24}$$

$$b_n = \frac{m\psi_n(\alpha)\psi'_n(m\alpha) - \psi_n(m\alpha)\psi'_n(\alpha)}{m\xi(\alpha)\psi'_n(m\alpha) - \psi_n(m\alpha)\xi'_n(\alpha)} \tag{10.25}$$

The size parameter α is defined as before, that is $\alpha = (2\pi r m_0)/\lambda_0$

The functions of Riccati–Bessel, ψ and ξ are expressed by the function of half-integer-order Bessel of the first kind ($J_{n+1/2}(z)$) where:

$$\psi_n(z) = \left(\frac{\pi z}{2}\right)^{1/2} J_{n+1/2}(z) \tag{10.26}$$

and

$$\xi_n(z) = \left(\frac{\pi z}{2}\right)^{1/2} H_{n+1/2}(z) = \psi_n(z) + iX_n(z) \tag{10.27}$$

In these equations, $H_{n+1/2}(z)$ is the second kind function of the half-integer-order Hankel and the parameter X_n is expressed as the function of the half-integer-order Bessel of the second kind, $Y_{n+1/2}(z)$, thus:

$$X_n(z) = -\left(\frac{\pi z}{2}\right)^{1/2} Y_{n+1/2}(z) \tag{10.28}$$

The total extinction cross-section (σ_{ext}) and total scattering cross-section (σ_{scat}) can be determined by the following equations:

$$\sigma_{ext} = \frac{\lambda^2}{2\pi}\sum_{n=0}^{\infty}(2n+1)\mathrm{Re}\{a_n + b_n\} \tag{10.29}$$

$$\sigma_{scat} = \frac{\lambda^2}{2\pi}\sum_{n=0}^{\infty}(2n+1)\left(|a_n|^2 + |b_n|^2\right) \tag{10.30}$$

The total absorption cross-section can also be determined from Eqs. (10.29) and (10.30) [4,6,7].

10.2.2 Development of resonance light scattering

Pasternack and coworkers, in 1993, studied the light scattering of trans-bis(*N*-methylpyridinium-4-yl)-diphenylporphine and copper(II) derivatives through scanning both excitation and emission monochromators of a conventional fluorescence spectrophotometer (CFS) at the same time [8]. They observed an increase of several orders of magnitude in the light scattering signal of the system in the Soret region that was related to the self-assembling of components on the surface of DNA in a closed region. Furthermore, they found that the unusual enhancement in the light scattering signal occurs when the wavelength of the incident light beam is within or near the absorption envelopes of extended chromophores. So, the concept of RLS was established. Later they used the RLS technique for studying the aggregation of porphyrin diacid [9] and chlorophyll [10] and in 1995 the RLS technique was introduced by Pasternack and Collings [11]. The main advantages of the RLS technique in comparison to Raleigh scattering are good selectivity and convenience. The high selectivity is because in the RLS technique only the light scattering signals close to the absorbing region are considered resulting in an exhibition of the good selectivity of absorption spectra. The RLS technique is convincing because the signals can be monitored by a CFS without the need for a relatively complicated instrument due to the enhancement of several orders of magnitude in the light scattering [1]. The RLS spectra and intensity are influenced by molecular conformation, shape, size, and interfacial properties [12]. In these bases, RLS is introduced as an efficient technique with a high capability in generating a powerful signal and a good potential detection platform in analytical chemistry [1]. RLS has been employed for the detection of various species including proteins [13,14], bacteria [15], peptides [16], and, metal ions [11,17].

10.3 Light scattering by nanoparticles

In 1998 Yguerabide and coworker reported that when submicroscopic particle suspensions are irradiated with white light, scattering of colored light occurred and the color of scattered light is dependent on the particle size. They also observed that when the silver nanoparticle and gold nanoparticle (GNP) with the diameter of 40 nm are illuminated with a narrow beam of white light, they produced blue and green scatter light, respectively, while the 80 nm in diameter GNPs scatter yellow light [3]. Thus the light scattering of nanoparticles depends on size, homogeneity, shape, composition, and refractive index of the media. Based on these observations as well as others, they suggested that light scattering by nanoparticles is comparable to fluorescent traces and can be used in clinical, biochemical, biological, and cell imaging applications [3,18]. The main advantages of light scattering nanoparticles as a probe in trace analysis are their high scattering powers that allow their detection even with the naked eye as well as the ease and possibility of synthesizing nanomaterials with the ability to scatter various colors in the visible region of the electromagnetic spectrum. Furthermore, it should be noted that when the nanoparticles aggregate through interaction with other species including enzyme, antigen, antibody, and DNA, an enhancement in the light scattering signal is induced that can be employed for the quantification and determination of various target analytes [18,19]. Thus, in the last decade, great affords have been devoted to the application of nanomaterials as light scattering probes in the detection and quantification of various inorganic and organic species. For example, in 2001, Taton and coworkers designed a method for the identification of two DNA target sequences in one solution based on the difference in the light scattering of the 50 and 100 nm GNPs labeled with DNA [20]. Light scattering nanoparticles functionalized with proteins or other biomolecules have been employed for bioanalytical analysis. In this regard, Saviranta et al. reported a sandwich immunoassay in a microarray for the sensitive measurement of various analytes based on the generation of the light scattering signal by nanoparticles [21]. The enhancement of light scattering of nanomaterials (or biomodified nanoparticles) due to their aggregation during chemical interactions with various species such as DNA hybridization, antigen–antibody, and electrostatic attraction has been used in the biochemical determination of drugs, proteins, and DNA [22–34]. The RLS based on nanomaterials has also been used for the determination of inorganic pollutants in the environment. For example, Madrakian et al. developed an RLS technique for the determination of Cd(II) and Pb(II) in the presence of each other. They synthesized CdS hollow nanospheres functionalized with cysteamine as a probe for the quantification of selected analytes [35]. Graphene quantum dots (GQDs) contain various carboxyl and hydroxyl groups on their surface that makes its interaction with different inorganic, organic, biological, or polymeric species possible [36–38]. Kaewprom's group also designed a novel RLS sensor utilizing functionalized GQDs for the measurement of lead [39]. They modified the surface of GQDs with diethyldithiocarbamate as a chelating agent with remarkable affinity in the formation of a stable complex with heavy metals such as lead. It was observed that the intensity of RLS of diethyldithiocarbamate-doped GQDs increases linearly with an increase in lead concentration that was used as the analytical signal.

Although the enhancement in light scattering based on the aggregation of nanoparticles such as gold and silver had been used for the detection and determination of some important biomolecules and environmental pollutants, as the level of these species is usually very low, their accurate determination required a more sensitive method. In an attempt to alleviate this limitation, nanoscale magnetic materials had been employed in the establishment of sensing as well as a diagnosis based on RLS that is discussed further in the following paragraph.

10.4 Magnetic nanomaterials and their application in resonance light scattering

Nanoscale magnetic materials possess the advantages of the possibility of synthesis in a wide range of size of 10–100 nm with a defined structure for a particular application as well as exploiting by the external magnetic force. One of the unique characteristics of the nanoscale magnetic materials is their high surface-to-volume ratio that affect their chemical and physical properties and is strongly related to particle size. Thus, the surface area and the availability of the loading active site increase with a decrease in particle size [40]. The surface of nanoscale magnetic materials can be functionalized with inorganic and organic components to give it a specific affinity. For example, MNPs modified by biomolecules have the ability of specific binding to detect or purify biological entities. MNPs have been also extensively applied for designing biosensing devices due to their high protein loading capacity and great biocompatibility. Furthermore, the high molecular weight, as well as high refractive index, makes them a suitable candidate as an enhancer of RLS.

As was mentioned before to alleviate the limitation of determination of trace amounts of biomolecules and environmental pollutants by RLS based on the aggregation of nanoparticles such as gold and silver, nanoscale magnetic materials had been employed. Sensor based on MNPs possesses the unique advantages of high sensitivity, fast response, low limit of detection, and high signal to noise ratio. In these sensors, MNPs are directly used as a tagging support for sensing application and dispersion in the sample. The RLS sensor based on MNPs has been developed for the detection of species of interest in medicine, biotechnology, and the environment [41–45]. In this regard, for the first time, in 2013, Yue and coworkers designed a sensitive and reusable biosensor for the detection of the inorganic species of mercury(II) based on the turn-on RLS using MNPs as the probe [46]. Mercury(II) is a highly persistent element with hazardous effects on the environment and humans. They measured Hg(II) through functionalizing the MNPs by T-containing ssDNA with linkers of 5′-Biotin-TATTTTATTTTATA-3′ (linker 1) and 5′-Biotin-TATATTTTATTTTA-3′ (linker 2). Then, linker 1 and linker 2 are hybridized in the presence of Hg^{2+} causing the aggregation of MNPs as MNPs@linker 1-Hg^{2+}-MNPs@linker. The particle aggregation resulted in the enhancement of RLS that was proportional to Hg^{2+} concentration and used as the analytical signal for the quantification of Hg^{2+}. The designed sensor enjoyed the advantages of ease of magnetic separation of functionalized MNPs as well as excellent selectivity of the specific interaction of thymine and Hg^{2+} in the formation of T-Hg^{2+}-T complex. The proposed method provides a low detection limit of 500 fM with the applicability of Hg^{2+} determination in water samples containing excess amounts of other metal ions. Later, Moosavi et al. reported a sensing probe based on the plasmonic enhanced effect of silver MNPs for the determination of cyanide [47]. They reported that the addition of Ag/Fe_3O_4 to cyanide solution causes the etching of Ag NPs through specific interaction of cyanide ions and silver resulting in a decrease in Rayleigh scattering intensity that was proportional to cyanide concentration. However, the addition of nonmagnetic Ag NPs has no significant effect on the plasmon fluorescence intensity, proving the high molecular weight and refractive index of MNPs cause it to act as a powerful and efficient material for the enhancement of plasmonic responses.

Application of RLS combined with MNPs for the determination of organic species of interest in medicine and biotechnology has attracted more scientists. Lu and coworkers reported a sensitive double-bead sandwich immunoassay for alpha-fetoprotein (AFP) taking the advantages of GNPs, magnetic beads (MBs), and resonance Rayleigh scattering (RRS) technique [48]. AFP is a tumor marker that the National Academy of Clinical Biochemistry (NACB) and the European Association for the Study of the Liver (EASL) suggested its monitoring in serum for early detection and diagnosis of hepatocellular carcinoma [49,50]. They used the GNPs coated with antibody as a probe for the immunological detection of AFP and optical transduction. While the MBs conjugated with a second antibody were utilized for the formation of MBs-antigen-GNPs complexes in the presence of antigen as well as the separation of the complexes. The determination of AFP was based on monitoring the RLS intensity of unconjugated GNPs remained in solution after magnetic separation of the formed complexes (Fig. 10.2). They claimed that in comparison to the ELISA method, it has the advantages of higher sensitivity, lower analysis time, and rapid separation process.

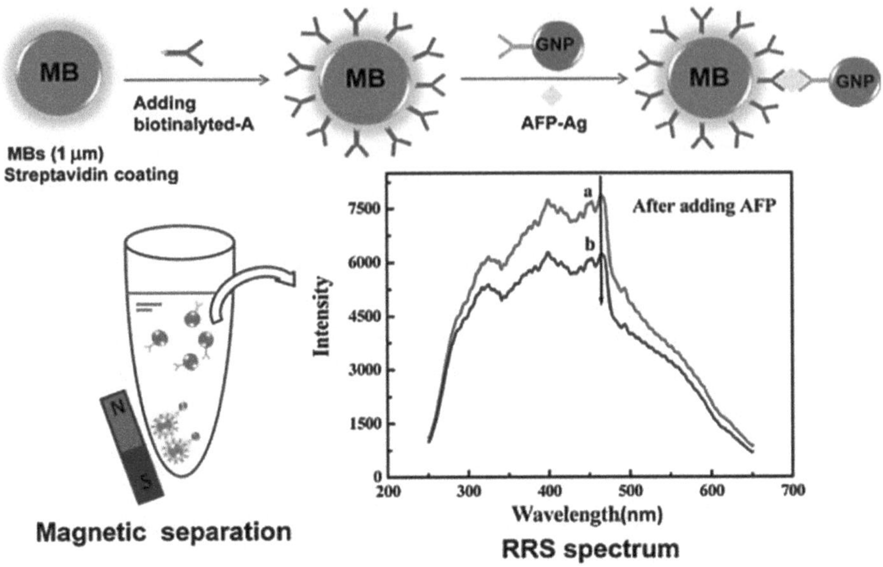

Figure 10.2 Schematic presentation of a double-bead sandwich immunoassay for alpha-fetoprotein (AFP) using gold nanoparticles (GNPs), magnetic beads (MBs), and resonance Rayleigh scattering (RRS) technique [48].

Yin and coworkers presented a novel RLS for sensitive determination of Alzheimer's amyloid-β peptide (Aβ) using Fe_3O_4@Au composites [51]. Alzheimer's disease is the most common cause of dementia, a continuous decline in thinking, behavioral, and social skills that disrupts a person's ability to function independently. The cure of Alzheimer's disease right now is not possible, but if diagnosed in the early stage it can be controlled. An indicator of Alzheimer's disease is the amount of Aβ, an amino acid containing the residue of the amyloid precursor protein, in blood, serum, plasma, and cerebrospinal fluid. Thus the determination of Aβ at a low level is important. RLS probes based on GNPs had been reported for the determination of Aβ at ng mL^{-1} level, but the level of Aβ in the blood of healthy and Alzheimer's disease patients is in pg mL^{-1}. To further enhance the sensitivity of RLS, Yu et al. combined the Fe_3O_4@Au composite probe with RLS and developed a highly sensitive method for quantification of Aβ. Thus the MNPs were synthesized by the hydrothermal method and its surface was coated with the amino groups through refluxing with (3-aminopropyl)triethoxysilane. Then, the amino-coated MNPs were covered by the gold shell with a thickness of ~9 nm using the classical Frens method. The Fe_3O_4@Au composites with an average diameter of ~320 nm possess high stability and biocompatibility and its RLS intensity significantly increased through interaction with Aβ. Furthermore, the enhancement in RLS intensity (I/I_0) at 463.0 nm was directly proportional to the logarithmic concentration of Aβ in the range of 5.0×10^{-15} to 5.56×10^{-9} M. The designed method has excellent sensitivity with a detection limit of 1.2×10^{-15} M, simplicity, label free, cost effective, and suitable for protein analysis as well as other various applications in medicine and biology [52–55].

Molecular-imprinted polymer–coated magnetic nanoparticles (MIP@MNPs) have been employed in the design of RLS nanosensors. MIP@MNPs have the advantages of specific recognition of MIP, low cost, environmental adaptability, excellent efficiency, as well as ease of magnetic separation. Ahmadi et al. synthesized a magnetic MIP for mefenamic acid (MFA) using suspension polymerization [56]. They found the RLS spectra of MNPs, SiO_2-coated MNPs (SCMNPs), and MIPSCMNPs to be similar, and the order of their intensities at the maximum wavelength of light scattering (454 nm) MIPSCMNPs > SCMNPs > MINPs followed the RLS theory that states the RLS intensity is proportional to r^6 (r is the radius of nanoparticles) [8]. Although the effect of the mechanism of the magnetic core of Fe_3O_4 on RLs is unclear, they found that the light scattering of MIPSCMNPs with an average diameter of 150 nm (greater than 1/20 of the incident light wavelengths) follows the Mie scattering. According to the theory of Mie scattering, the RLs properties are affected by molecular volume, the morphology of the particles, small magnetic spheres magnetism as well as small clusters magnetization of radius larger than 10 nm [57]. Thus, based on the Mie theory, upon the interaction of MFA with the NPs, aggregation of NPs with bigger complexes volume and higher hydrophobicity causes an increase in RLS intensities. The increase in intensity was proportional to MFA concentration in the range of 100.0–2000.0 ng L^{-1} with a detection limit of 50.0 ng L^{-1}. MIP@MNPs coupled to the RLS technique have also been used for recognition and quantification of viruses. Zhang et al. designed a magnetic MIP optical chemical sensor for recognition and determination of traces amount of the hepatitis A virus (HAV) [58]. The core–shell HAV-magnetic-MIPs were synthesized by surface imprinting technique using nanosized, thin, stable, and bioinspired polydopamine (PDA) as an imprinting layer. The HAV interacts with the hydroxyl and amine groups of the bioinspired PDA chain through noncovalent binding. Thus the imprinted sites of HAV-magnetic-MIPs bind with the model virus by hydrogen and p–p bonds resulting in an increase in its light scattering that followed the Mie scattering pattern. The increase in the RLS signal was proportional to virus concentration in the range of 0.02–1.40 nmol L^{-1} with a limit of detection of 6.2 pmol L^{-1}. They also found that the HAV-magnetic-MIPs showed excellent selectivity toward HAV in comparison to other viruses with different molecular weights. The high selectivity was related to the spatial interaction of the HAV-magnetic-MIPs with HAV having a specific shape, dimension, and functional groups. Thus the virus with different conformation or larger size may not interact strongly with the imprinted binding sites to be recognized, and those with smaller sizes might enter the cavity but they cannot increase the size of the HAV-magnetic-MIPs. However, the incubation time was relatively long and needs to be shortened by making the imprinting layer as thin as possible. To overcome this limitation, later, Luo and coworkers designed a magnetic surface molecularly imprinted-RLS sensor for sensitive and fast detection of Japanese encephalitis virus (JEV) [59]. In this study, silicon-coated Fe_3O_4 was chosen as a supporting material, and JEV as template virus was imprinted on its surface using aminopropyl-triethoxysilane

(APTES) as a functional monomer and tetraethyl-orthosilicate (TEOS) as a cross-linker. The synthesized magnetic MIP captures the target molecule JEV with great selectivity and rapidness, resulting in an enhancement of the RLS intensity. The combined advantages of RLS with magnetic MIP constructed an efficient sensor for detecting low levels of JEV virus within 20 minutes. Thus, the proposed strategy for detection and determination of virus is very selective, simple, cost effective, and fast, and offers a novel technique for virus detection and quantification in bioanalytical and clinical studies without the need for any complex operation and expensive equipment. The schematic of the developed method is represented in Fig. 10.3.

Chiral recognition and analysis are very important in science and technology due to the differences in the therapeutic and physiological properties of enantiomers. Ahmadi et al. combined RLS with modified magnetite nanospheres and developed a selective method for the determination of phenylglycine and naproxen enantiomers [60]. In this regard, they synthesized two types of chiral nanospheres, namely D-(+)-tryptophan-modified magnetite nanospheres (+MNSs) and L-(−)-tryptophan-modified magnetite nanospheres (−MNSs). Then, the light scattering behaviors of the prepared nanospheres in the presence and absence of (S)-(+)-naproxen (+NAP), (R)-(−)-naproxen (−NAP), L-(+)-α-phenylglycine (+PHY), and D-(−)-α-phenylglycine (−PHY) as the model chiral compound were studied by RLS method. They found that the RLS signal of +MNSs was greatly enhanced in the presence of −NAP or −PHY, while the RLS signal of −MNSs was enhanced when +PHY or +NAP was present. Thus the

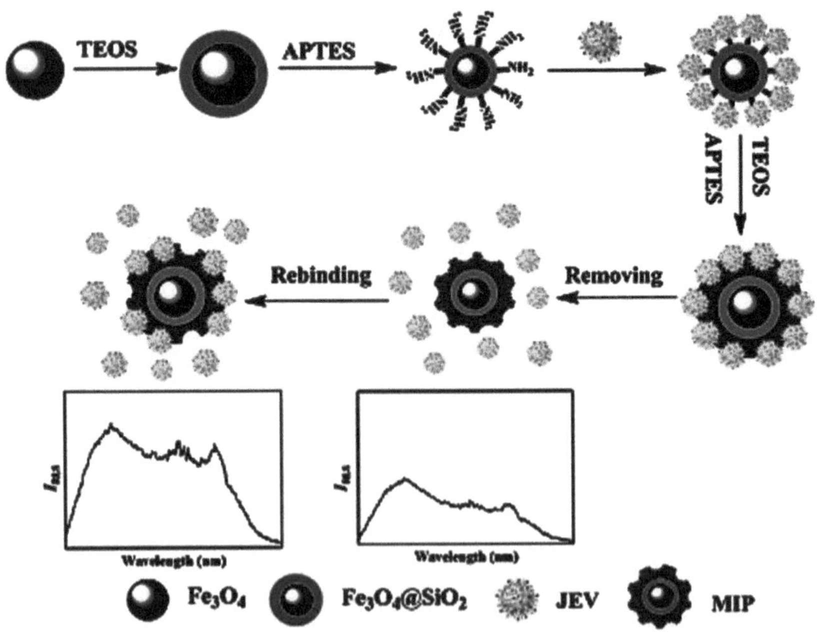

Figure 10.3 Schematic presentation of JEV virus based on a resonance light scattering sensor using magnetic molecular-imprinted polymer [59].

chiral nanospheres can act as a probe in the determination of enantiomers by the RLS method without the need for any separation step. The increase in the RLS signal was related to the enhancement of hydrophobicity due to the interaction between enantiomers and the nanospheres that counterbalance the charges of the enantiomers and the surface functional groups of the nanospheres. Thus the hydrophobic aryl groups of enantiomers act as the hydrophobic interface, causing the aggregation of the nanospheres and formation of larger complexes that result in an increase in RLS signal based on the Mie's formalism. It was also claimed that the designed method can be applied for the determination of purity of synthesized chiral materials and pharmaceutical formulation at ng L^{-1} level.

RLS based on the MNPs as the probe was employed in the fabrication of aptasensor for the determination of enzyme. Hou et al. designed a reusable biosensor for the detection of thrombin using MNPs as the RLS probe based on an extension–contraction movement of DNA interconversion [61]. Thrombin is the blood plasma enzyme that causes the clotting of blood by converting fibrinogen to fibrin and involves in the commencement and advancement of some diseases such as inflammatory cascades and atherosclerosis [62]. Thus, because of its remarkable clinical importance, the establishment of a reliable and quick method for its sensitive measurement is noticeably necessary. The 50-nm-sized MNPs coated with streptavidin (0.1 mg mL^{-1}) was treated with tris(hydroxy-methyl)-aminomethane (Tris), 4-(2-hydroxyethyl) piperazine-1-ethane-sulfonic acid (HEPES). In the next step, 5.0 nM of 15-mer thrombin aptamer labeled with biotin (5′-biotin-GGT TGG TGT GGT TGG-3′) (TBA1) and 29-mer aptamer labeled with biotin (5′-AGT CCG TGG TAG GGC AGG TTG GGG TGA CT-biotin-3′) (TBA2) were added to the dispersed MNPs and incubated at room temperature about 20 minutes resulting in the formation of MNPs and aptamers denoted as MNPs@TBA1 and MNPs@TBA2. Then, the addition of thrombin resulted in the binding between 15-mer TBA1 and the fibrinogen-binding site of thrombin (with a K_d of $\sim$100 nM), as well as the binding between 29-mer TBA2 and the heparin-binding site of thrombin ($K_d \sim$ 0.5 nM). Thus the interaction of thrombin with DNA sequence causes the MNPs aggregation through the formation of a sandwich structure resulting in the enhancement of the RLS signals as shown in Fig. 10.4. The ΔRLS increased linearly with thrombin concentration in the range of 60 pM–6.0 nM with a detection limit of 3.5 pM. The designed sensor was reused by destroying the sandwich structure of thrombin aptasensor through heating the solution and collecting the MNPs with an external magnetic field. The designed aptasensor takes the advantages of functionalized MNPs and the specific interaction between aptamers (TBA1 and TBA2) and thrombin in the formation of a sandwich structure. They also stated that the strategy has the potential of application to any other analyte by simply changing the labeling of MNPs surface with proper aptamers or ligands.

Yue et al. used the same idea and presented an aptasensor based on enhanced RLS by MNPs for assaying of miRNA [63]. They modified the surface of MNPs coated with streptavidin by two strands of biotin-labeled DNA probes that were complementary to miRNA, via the formation of a specific interaction between streptavidin and biotin and prepared the MNPs@probe 1 and MNPs@probe 2 with

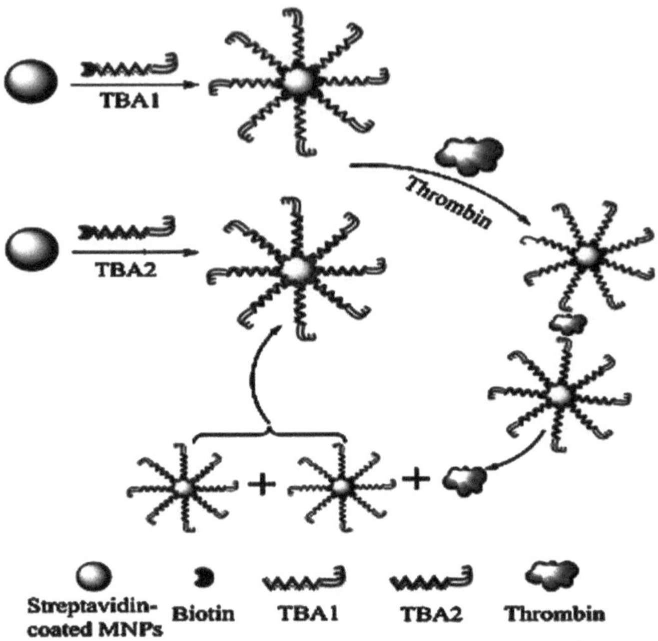

Figure 10.4 Schematic illustration of a reusable magnetic aptasensor for the detection of thrombin [61].

low RLS signal in homogeneous solution. The addition of miRNA then caused partial hybridization of DNA probes and miRNA followed by coagulation of MNPs in a dumbbell-like structure. The aggregation of MNPs produced the mild local magnetic fields that accelerate the further aggregation of MNPs as well as enhancement of the RLS signal. The increase in the ΔRLS signal was proportional to the miRNA concentration within the range of 10.0 pM–10.0 nM with a limit of detection of 0.9 pM. The method has high sensitivity and specificity and was capable of discriminating between the members of the miRNA family as well as a single-base mismatched sequence. The developed aptasensor was successfully applied in clinical analysis such as determination of miRNA in human cancer cells and human serum samples.

It should be noted that in the aforementioned RLS strategies the mechanism of detection was mostly an enhancement in RLS signal upon addition of analyte, which is called "turn-on mode." However, Tao et al. designed a "turn-off" RLS aptasensor using MNPs and aptamer for determination of 8-hydroxy-2′-deoxyguanosine (8-OHdG) [64]. 8-OHdG is a biomarker known as the indicator of the damaged nucleotide that causes the development and progression of cancer in humans [65] and its amount is increased in neurological diseases such as Parkinson's and Alzheimer's. In this study, they treated the MNPs modified with streptavidin directly with HEPES followed by placing the DNA probes on the MNPs surface that produce relatively low RLS signals. The hybridization of probe DNA against the 8-OHdG aptamer resulted in the formation of a sandwich structure MNPs@probe 1-aptamer-MNPs@probe 2 sensor for 8-OHdG and significant enhancement of RLS signal. The addition of 8-OHdG caused a decrease in RLS signal (turn-off mode) that was proportional to the 8-OHdG concentration in the range of 32 pM–12.0 nM. The reduction in RLS signal was related to the high affinity of the aptamer for 8-OHdG in comparison to probe DNA that resulted in a change in the conformation of the aptamer from a double-stranded to a G-quadruplex structure followed by the release of the labeled MNPs by probe DNA (Fig. 10.5). This method was utilized for the analysis of human urine samples containing 8-OHdG. The designed method has a limit of detection of 11 pM and was used for the determination of 8-OHdG in the human urine sample.

Despite the advantages of MNPs such as large surface area, low toxicity, biocompatibility, ease of preparation, and superparamagnetic property, they often aggregate into the large cluster that vanishes their specific properties related to single-domain magnetic nanostructures. Thus the chemical stability of MNPs as well as uniformity in their size and dispersive capability are important that are improved through surface functionalization. In this regard, carbon dots (CDs), zero-dimensional carbon-based materials containing numerous oxygen-containing functional groups such as –C=O, –COOH, and –OH are good candidates for coating the MNPs. In these bases, Maleki and coworkers applied polyacrylonitrile nanofibers decorated with magnetic carbon dots (MCDs@NFs) as a probe in RLS determination and designed a sensor for quantification of famotidine (FMD) in pharmaceutical preparation and human serum samples [66]. The main goal of their study was to synthesize the MNPs, CDs, and MCDs@NFs (~565 nm) in one step to reduce and simplify the synthesis stage. So, they utilized a novel approach in the synthesis of MCDs@NFs nanocomposite through the combination of electrospinning and the simple one-step hydrothermal method. The formation of MCDs inside

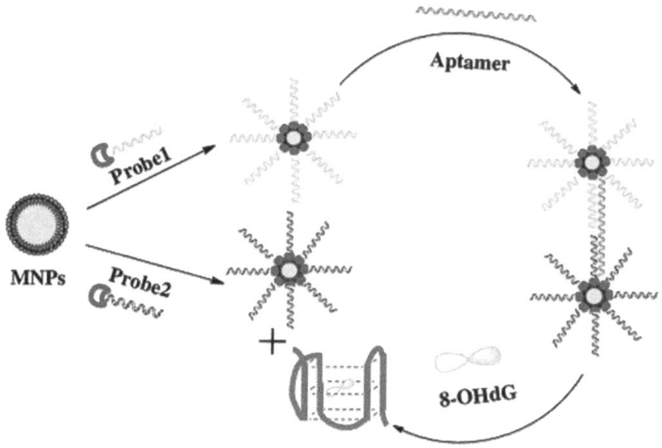

Figure 10.5 Schematic representation of resonance light scattering aptasensor for 8-hydroxy-2′-deoxyguanosine detection [64].

the nanofibers generates a composite with perfect physicochemical properties that prevent the aggregation of nanoparticles within the sample matrix. Then, they considered the effect of FMD on the RLS intensity of the MCDs@NFs nanocomposite and found a good linear relation between the ΔIRLS at 465 nm and FMD concentration in the range of 0.15–50.0 μmol L^{-1}. The method is schematically shown in Fig. 10.6. Furthermore, they considered the possible mechanism for the enhancement in RLS. As the average diameter of MCD@NFs nanocomposite (~565 nm) was greater than (~1.20) that of the incident light beam (465 nm), they concluded that the dominant mechanism is Mie scattering. Based on the theory of Mie scattering, there is a close relationship between the morphology of particles, the molecular volume, and the RLS intensity.

Recently, Pirdadeh-Beiranvand et al. introduced nickel-zinc ferrite ($Ni_{0.5}Zn_{0.5}Fe_2O_4$) nanoparticles modified by poly(vinyl alcohol) (PVA) nanofiber as the RLS probe for determination of sunitinib in real samples [67]. $Ni_{0.5}Zn_{0.5}Fe_2O_4$ nanoparticles–decorated PVA nanofibers with an average diameter of 200 nm were synthesized through the mixing of magnetic nanocomposite into the PVA matrix using the electrospinning method. The RLS technique was applied for investigation of the interaction between nickel-zinc ferrite nanoparticles–decorated PVA nanofibers and sunitinib. The presence of nickel-zinc ferrite in the fiber resulted in the strong interaction between the fiber and sunitinib and consequently a large quenching of scattered light by nickel-zinc ferrite nanoparticles–decorated PVA nanofibers. The decrease in the signal of scattered light was proportional to the analyte concentration in the range of 5.0×10^{-3} to 10.0 mg L^{-1} with a limit of detection of 1.0×10^{-3} mg L^{-1}. The developed method enjoyed the benefit of the possibility of direct and sensitive determination of sunitinib. Furthermore, the most important advantage of the method is the high stability of

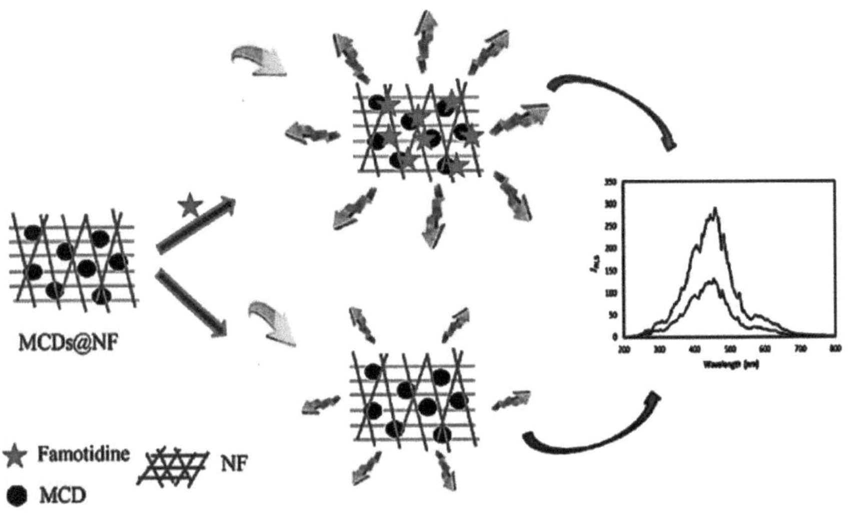

Figure 10.6 Fabrication procedure of famotidine detection based on an RLS sensor using MCDs@NFs nanocomposite [66].

synthesized nanofibers in the aqueous solution that results in a significant decrease in RSD of the RLS technique.

10.5 Conclusions

The RLS as an optical technique enjoys the benefit of high capability in generating a powerful signal, simplicity, facility, rapidness, availability, with a good potential detection platform in analytical chemistry. The application of nanoparticles as light scattering probes resulted in an enhancement of scattering signals that make trace analysis possible. Nanoscale magnetic materials with the advantages of the possibility of synthesis in a wide range of size, a high surface to volume ratio that can be functionalized with inorganic and organic components, high molecular weight, as well as high refractive index have been employed as a suitable candidate for enhancement of RLS. Sensor based on MNPs possesses the advantages of high sensitivity, fast response, low limit of detection, and high signal to noise ratio. However, the application of the RLS sensor based on MNPs with the excellent potential in the detection of species of interest in medicine, biotechnology, and the environment at trace level is still limited and needs to be extended.

References

[1] W. Lu, B.S.F. Band, Y. Yu, Q.G. Li, J.C. Shang, C. Wang, et al., Resonance light scattering and derived techniques in analytical chemistry: past, present, and future, Microchim. Acta 158 (2007) 29–58.

[2] C.F. Bohren, D.R. Huffman, Absorption and Scattering of Light by Small Particles, Wiley-VCH, Weinheim, 1998.

[3] J. Yguerabide, E.E. Yguerabide, Light-scattering submicroscopic particles as highly fluorescent analogs and their use as tracer labels in clinical and biological applications: II. Experimental characterization, Anal. Biochem. 262 (1998) 157–176.

[4] D.W. Hahn, Light Scattering Theory, Department of Mechanical and Aerospace Engineering, University of Florida, 2009.

[5] S.K. Brar, M. Verma, Measurement of nanoparticles by light-scattering techniques, TrAC Trends Anal. Chem. 30 (2011) 4–17.

[6] H.C. Hulst, H.C. van de Hulst, Light Scattering by Small Particles, Courier Corporation, 1981.

[7] C.F. Bohren, D.R. Huffman, Absorption and Scattering of Light by Small Particles, John Wiley & Sons, 2008.

[8] R.F. Pasternack, C. Bustamante, P.J. Collings, A. Giannetto, E.J. Gibbs, Porphyrin assemblies on DNA as studied by a resonance light-scattering technique, J. Am. Chem. Soc. 115 (1993) 5393–5399.

[9] R.F. Pasternack, K.F. Schaefer, P. Hambright, Resonance light-scattering studies of porphyrin diacid aggregates, Inorg. Chem. 33 (1994) 2062–2065.

[10] J.C. de Paula, J.H. Robblee, R.F. Pasternack, Aggregation of chlorophyll a probed by resonance light scattering spectroscopy, Biophys. J. 68 (1995) 335.

[11] R.F. Pasternack, P.J. Collings, Resonance light scattering: a new technique for studying chromophore aggregation, Science 269 (1995) 935–939.
[12] S. Liu, H. Luo, N. Li, Z. Liu, W. Zheng, Resonance Rayleigh scattering study of the interaction of heparin with some basic diphenyl naphthylmethane dyes, Anal. Chem. 73 (2001) 3907–3914.
[13] S. Liu, Z. Yang, Z. Liu, L. Kong, Resonance Rayleigh-scattering method for the determination of proteins with gold nanoparticle probe, Anal. Biochem. 353 (2006) 108–116.
[14] C.Z. Huang, S.F. Chen, Quantitation and differentiation of bioparticles based on the measurements of light-scattering signals with a common spectrofluorometer, J. Phys. Chem. B 112 (2008) 11785–11793.
[15] C. Wang, D. Liu, Z. Wang, Resonance light scattering as a powerful tool for sensitive detection ofβ-amyloid peptide by gold nanoparticle probes, Chem. Commun. 47 (2011) 9339–9341.
[16] Z.D. Liu, Y.F. Li, J. Ling, C.Z. Huang, A localized surface plasmon resonance light-scattering assay of mercury (II) on the basis of Hg^{2+} – DNA complex induced aggregation of gold nanoparticles, Environ. Sci. Technol. 43 (2009) 5022–5027.
[17] T. Wu, C. Liu, K.J. Tan, P.P. Hu, C.Z. Huang, Highly selective light scattering imaging of chromium (III) in living cells with silver nanoparticles, Anal. Bioanal. Chem. 397 (2010) 1273–1279.
[18] A. Akbarzadeh, M. Samiei, S. Davaran, Magnetic nanoparticles: preparation, physical properties, and applications in biomedicine, Nanoscale Res. Lett. 7 (2012) 144.
[19] J. Ling, C.Z. Huang, Y.F. Li, L. Zhang, L.Q. Chen, S.J. Zhen, Light-scattering signals from nanoparticles in biochemical assay, pharmaceutical analysis and biological imaging, TRAC. Trends Anal. Chem. 28 (2009) 447–453.
[20] T.A. Taton, G. Lu, C.A. Mirkin, Two-color labeling of oligonucleotide arrays via size-selective scattering of nanoparticle probes, J. Am. Chem. Soc. 123 (2001) 5164–5165.
[21] P. Saviranta, R. Okon, A. Brinker, M. Warashina, J. Eppinger, B.H. Geierstanger, Evaluating sandwich immunoassays in microarray format in terms of the ambient analyte regime, Clin. Chem. 50 (2004) 1907–1920.
[22] Z. Chen, Z. Wang, J. Chen, S. Wang, X. Huang, Sensitive and selective detection of glutathione based on resonance light scattering using sensitive gold nanoparticles as colorimetric probes, Analyst 137 (2012) 3132–3137.
[23] C. Li, J. Ma, Q. Fan, Y. Tao, G. Li, Dynamic light scattering (DLS)-based immunoassay for ultra-sensitive detection of tumor marker protein, Chem. Commun. 52 (2016) 7850–7853.
[24] B. Lorber, F. Fischer, M. Bailly, H. Roy, D. Kern, Protein analysis by dynamic light scattering: methods and techniques for students, Biochem. Mol. Biol. Educ. 40 (2012) 372–382.
[25] P. Chandra, E. Segal, Nanobiosensors for Personalized and Onsite Biomedical Diagnosis, The Institution of Engineering and Technology, 2016.
[26] N. Seow, Y.N. Tan, L.-Y.L. Yung, X. Su, DNA-directed assembly of nanogold dimers: a unique dynamic light scattering sensing probe for transcription factor detection, Sci. Rep. 5 (2015) 1–8.
[27] X. Liu, Q. Dai, L. Austin, J. Coutts, G. Knowles, J. Zou, et al., A one-step homogeneous immunoassay for cancer biomarker detection using gold nanoparticle probes coupled with dynamic light scattering, J. Am. Chem. Soc. 130 (2008) 2780–2782.
[28] H. Jans, X. Liu, L. Austin, G. Maes, Q. Huo, Dynamic light scattering as a powerful tool for gold nanoparticle bioconjugation and biomolecular binding studies, Anal. Chem. 81 (2009) 9425–9432.

[29] X. Miao, S. Zou, H. Zhang, L. Ling, Highly sensitive carcinoembryonic antigen detection using Ag@ Au core–shell nanoparticles and dynamic light scattering, Sens. Actuators B: Chem. 191 (2014) 396–400.
[30] X. Huang, Z. Xu, Y. Mao, Y. Ji, H. Xu, Y. Xiong, et al., Gold nanoparticle-based dynamic light scattering immunoassay for ultrasensitive detection of *Listeria monocytogenes* in lettuces, Biosens. Bioelectron. 66 (2015) 184–190.
[31] Q. Dai, X. Liu, J. Coutts, L. Austin, Q. Huo, A one-step highly sensitive method for DNA detection using dynamic light scattering, J. Am. Chem. Soc. 130 (2008) 8138–8139.
[32] H. Jans, Q. Huo, Gold nanoparticle-enabled biological and chemical detection and analysis, Chem. Soc. Rev. 41 (2012) 2849–2866.
[33] Y. Sun, H. Zhao, I. Boussouar, F. Zhang, D. Tian, H. Li, Highly sensitive chiral sensing by calix [4] arene-modified silver nanoparticles via dynamic light scattering, Sens. Actuators B: Chem. 216 (2015) 235–239.
[34] X.T. Zheng, W.L. Goh, P. Yeow, D.P. Lane, F.J. Ghadessy, Y.N. Tan, Ultrasensitive dynamic light scattering based nanobiosensor for rapid anticancer drug screening, Sens. Actuators B: Chem. 279 (2019) 79–86.
[35] T. Madrakian, M. Ahmadi, M. Ghadiri, A. Afkhami, Application of cysteamine functionalized CdS hollow nanospheres in determination of Cd (II) and Pb (II) in the presence of each other by resonance light scattering technique, J. Environ. Chem. Eng. 4 (2016) 3484–3491.
[36] R. Afsharipour, A.M.H. Shabani, S. Dadfarnia, E. Kazemi, Selective fluorometric determination of sulfadiazine based on the growth of silver nanoparticles on graphene quantum dots, Microchim. Acta 187 (2020) 54.
[37] J. Shen, Y. Zhu, C. Chen, X. Yang, C. Li, Facile preparation and upconversion luminescence of graphene quantum dots, Chem. Commun. 47 (2011) 2580–2582.
[38] J. Shen, Y. Zhu, X. Yang, J. Zong, J. Zhang, C. Li, One-pot hydrothermal synthesis of graphene quantum dots surface-passivated by polyethylene glycol and their photoelectric conversion under near-infrared light, N. J. Chem. 36 (2012) 97–101.
[39] C. Kaewprom, P. Sricharoen, N. Limchoowong, P. Nuengmatcha, S. Chanthai, Resonance light scattering sensor of the metal complex nanoparticles using diethyl dithiocarbamate doped graphene quantum dots for highly Pb (II)-sensitive detection in water sample, Spectrochim. Acta Part A: Mol. Biomol. Spectrosc. 207 (2019) 79–87.
[40] J. Lim, S.P. Yeap, H.X. Che, S.C. Low, Characterization of magnetic nanoparticle by dynamic light scattering, Nanoscale Res. Lett. 8 (2013) 381.
[41] A.K. Gupta, M. Gupta, Synthesis and surface engineering of iron oxide nanoparticles for biomedical applications, Biomaterials 26 (2005) 3995–4021.
[42] R. Hao, R. Xing, Z. Xu, Y. Hou, S. Gao, S. Sun, Synthesis, functionalization, and biomedical applications of multifunctional magnetic nanoparticles, Adv. Mater. 22 (2010) 2729–2742.
[43] L. Cumbal, J. Greenleaf, D. Leun, A.K. SenGupta, Polymer supported inorganic nanoparticles: characterization and environmental applications, Reactive Funct. Polym. 54 (2003) 167–180.
[44] W. Yantasee, C.L. Warner, T. Sangvanich, R.S. Addleman, T.G. Carter, R.J. Wiacek, et al., Removal of heavy metals from aqueous systems with thiol functionalized superparamagnetic nanoparticles, Environ. Sci. Technol. 41 (2007) 5114–5119.
[45] J. Hu, I.M. Lo, G. Chen, Comparative study of various magnetic nanoparticles for Cr (VI) removal, Sep. Purif. Technol. 56 (2007) 249–256.
[46] Q. Yue, T. Shen, J. Wang, L. Wang, S. Xu, H. Li, et al., A reusable biosensor for detecting mercury (II) at the subpicomolar level based on "turn-on" resonance light scattering, Chem. Commun. 49 (2013) 1750–1752.

[47] R. Moosavi, A. Afkhami, T. Madrakian, A Simple cyanide sensing probe based on Ag/Fe3O4 nanoparticles, RSC Adv. 5 (2015) 15886–15891.
[48] Y. Lu, X. Huang, J. Ren, Sandwich immunoassay for alpha-fetoprotein in human sera using gold nanoparticle and magnetic bead labels along with resonance Rayleigh scattering readout, Microchim. Acta 180 (2013) 635–642.
[49] G. Li, J. Zeng, H. Liu, P. Ding, J. Liang, X. Nie, et al., A fluorometric aptamer nanoprobe for alpha-fetoprotein by exploiting the FRET between 5-carboxyfluorescein and palladium nanoparticles, Microchim. Acta 186 (2019) 314.
[50] N. Ma, T. Zhang, D. Fan, X. Kuang, A. Ali, D. Wu, et al., Triple amplified ultrasensitive electrochemical immunosensor for alpha fetoprotein detection based on MoS_2@Cu_2O-Au nanoparticles, Sens. Actuators B: Chem. 297 (2019) 126821.
[51] H. Yin, C. Wang, H. Zhu, S.H. Overbury, S. Sun, S. Dai, Colloidal deposition synthesis of supported gold nanocatalysts based on Au–Fe_3O_4 dumbbell nanoparticles, Chem. Commun. (2008) 4357–4359.
[52] C. Yang, J. Wu, Y. Hou, Fe_3O_4 nanostructures: synthesis, growth mechanism, properties and applications, Chem. Commun. 47 (2011) 5130–5141.
[53] G.K. Kouassi, Magnetic and gold-coated magnetic iron oxide nanoparticles as detection tools: preparation, characterization, and biosensing applications, Curr. Nanosci. 7 (2011) 510–523.
[54] K.C.-F. Leung, S. Xuan, X. Zhu, D. Wang, C.-P. Chak, S.-F. Lee, et al., Gold and iron oxide hybrid nanocomposite materials, Chem. Soc. Rev. 41 (2012) 1911–1928.
[55] Y.-H. Bai, J.-Y. Li, J.-J. Xu, H.-Y. Chen, Ultrasensitive electrochemical detection of DNA hybridization using Au/Fe_3O_4 magnetic composites combined with silver enhancement, Analyst 135 (2010) 1672–1679.
[56] M. Ahmadi, T. Madrakian, A. Afkhami, Molecularly imprinted polymer coated magnetite nanoparticles as an efficient mefenamic acid resonance light scattering nanosensor, Anal. Chim. Acta 852 (2014) 250–256.
[57] W. Hergert, T. Wriedt, The Mie Theory: Basics and Applications, Springer, 2012.
[58] F. Zhang, L. Luo, H. Gong, C. Chen, C. Cai, A magnetic molecularly imprinted optical chemical sensor for specific recognition of trace quantities of virus, RSC Adv. 8 (2018) 32262–32268.
[59] L. Luo, J. Yang, K. Liang, C. Chen, X. Chen, C. Cai, Fast and sensitive detection of Japanese encephalitis virus based on a magnetic molecular imprinted polymer–resonance light scattering sensor, Talanta 202 (2019) 21–26.
[60] M. Ahmadi, T. Madrakian, A. Afkhami, Chiral magnetic nanospheres resonance light scattering properties studies for selective determination of naproxen and phenylglycine enantiomers, Sens. Actuators B: Chem. 210 (2015) 439–445.
[61] Y. Hou, J. Liu, M. Hong, X. Li, Y. Ma, Q. Yue, et al., A reusable aptasensor of thrombin based on DNA machine employing resonance light scattering technique, Biosens. Bioelectron. 92 (2017) 259–265.
[62] J.I. Borissoff, H.M. Spronk, S. Heeneman, H. ten Cate, Is thrombin a key player in the 'coagulation-atherogenesis' maze? Cardiovasc. Res. 82 (2009) 392–403.
[63] Q. Yue, L. Tao, Y. Hou, C. Zhang, Y. Wang, M. Hong, et al., Assay of miRNA in cell samples using enhanced resonance light scattering technique based on self aggregation of magnetic nanoparticles, Nanomedicine 13 (2018) 2301–2310.
[64] L. Tao, Q. Yue, Y. Hou, Y. Wang, C. Chen, C.-Z. Li, Resonance light scattering aptasensor for urinary 8-hydroxy-2′-deoxyguanosine based on magnetic nanoparticles: a preliminary study of oxidative stress association with air pollution, Microchim. Acta 185 (2018) 419.

[65] S.S. Wallace, Biological consequences of free radical-damaged DNA bases, Free Radic. Biol. Med. 33 (2002) 1–14.
[66] S. Maleki, T. Madrakian, A. Afkhami, Application of polyacrylonitrile nanofibers decorated with magnetic carbon dots as a resonance light scattering sensor to determine famotidine, Talanta 181 (2018) 286–295.
[67] M. Pirdadeh-Beiranvand, A. Afkhami, T. Madrakian, $Ni_{0.5}Zn_{0.5}Fe_2O_4$ nanoparticles-decorated poly (vinyl alcohol) nanofiber as resonance light scattering probe for determination of sunitinib in serum samples, Talanta 121190 (2020).

Application of magnetic nanomaterials in plasmonic sensors

11

Arash Ghoorchian[1], Mahdie Kamalabadi[1], Zahra Amouzegar[1], Samira Aghajani[2], Tayyebeh Madrakian[1], Abbas Afkhami[1] and Mazaher Ahmadi[1]
[1]Faculty of Chemistry, Bu-Ali Sina University, Hamedan, Iran, [2]Hamedan University of Technology, Hamedan, Iran

11.1 Introduction

Taking the advantages of both nanoscale size (such as high surface-to-volume ratios) and magnetic manipulability, magnetic nanomaterials (MNMs) have found a wide range of applications in different fields from chemistry to biomedicine. Their magnetic property differs significantly when bulk ferromagnetic materials are converted to nanoscale materials, mainly due to higher surface-to-volume ratios. Small enough ferromagnetic particles (typically smaller than 20 nm) exhibit a special magnetic property called superparamagnetism [1]. Unlike bulk materials, superparamagnetic nanoparticles exhibit single magnetic domains that align with the magnetic field quickly under an external magnetic field and then, when the external magnetic field is removed, exhibit no net magnetism, leading to easy manipulation of these materials using magnetic fields since the magnetic moment rapidly reverses [2].

MNMs can consist of either elemental magnetic materials such as iron, cobalt, and nickel or their oxides and mixed oxides. Among the magnetic materials, iron oxide nanoparticles, including magnetite (Fe_3O_4) and maghemite (γ-Fe_2O_3), have attracted much greater attention in various fields as drug delivery vehicles [3–6], MRI contrast agents [5,7,8], catalysts [9–12], sensors [13–18], and sorbents [13,19–22]. In analytical chemistry, MNMs can be utilized as sorbents and sensors. The superparamagnetism property and high surface-to-volume ratio make MNMs ideal sorbents since they can adsorb the target analyte with high adsorption capacities and can be easily separated from the medium using an external magnetic field within seconds or a few minutes. So far, MNMs have been utilized as sorbents for food analysis, bioanalysis, environmental monitoring, and forensic chemistry. On the other hand, these magnetic materials also can be utilized as sensing materials in colorimetric sensors, resonance light scattering sensors, electrochemical sensors, biosensors, and magnetoresistive sensors.

A surface plasmon, an optical phenomenon that deviates from what is standard, behavior was investigated on a metallic diffraction grating. For the first time, this

Magnetic Nanomaterials in Analytical Chemistry. DOI: https://doi.org/10.1016/B978-0-12-822131-0.00013-3

anomaly was reported by Wood in 1902 and was explained by Fano in 1941. In 1968, Otto illustrated that the decline in the intensity of reflected light in the attenuated total reflection (ATR) method arose as a result of the surface plasmon generation. In the same year, another arrangement of the ATR technique was proposed for surface plasmon excitation by Kretschmann and Raether. However, the first presentation of optical sensors based on surface plasmon resonance (SPR) occurred in 1980 for the study of surface processes at metal surfaces related to gas sensing. The marketing of an SPR device occurred in 1990 to detect interactions between proteins in the format of a flow system design [23–25]. Thus far, various plasmonic materials have been used in analytical chemistry for the construction of plasmonic sensors. Furthermore, various measurement strategies also have been developed. However, the use of MNMs for the construction of plasmonic nanosensors has been less investigated. Therefore this chapter will review the recent application of MNMs in plasmonic nanosensors.

11.2 Plasmonic sensors

Nanoplasmonic-based assay techniques mainly fall into three categories: SPR, localized surface plasmon resonance (LSPR), and surface-enhanced Raman scattering (SERS) (Fig. 11.1). Here, we describe the principles and the design of detection for each strategy.

11.2.1 Surface plasmon resonance sensors

Optical sensors based on SPR are among the different types of refractometers. This kind of sensor operates based on the excitation of surface plasmon [24]. The electrically conductive materials have a sea of electrons in their conduction band that can move freely throughout the space between the atomic nuclei. The interaction of plane-polarized light with conduction band electrons at the interface between the conductor and dielectric generates the collective oscillation of these charge carriers. This phenomenon triggers a wave of surface plasmon that can propagate along with the conductor–dielectric interface and cause SPR. Although generated along the surface, this propagation wave is limited to about 100–200 nm in the perpendicular direction. Metals are the best generator of surface plasmons because of their high density of free electrons [26].

When a monochromatic focused beam, plane-polarized light with an angle sufficient to cause total internal reflection, irradiates the interface, maximum light absorption occurs if the frequency of the incident light (as an electromagnetic wave) matches the frequency of surface plasmon (as a propagation wave). Therefore the intensity of the reflected light decreases, and it appears as a descent in the SPR spectrum. This is a function of incident angle along with the abscissa and reflection intensity along the ordinate. As the analyte enters the interface, the refractive index (RI) and the frequency of the surface plasmon wave change.

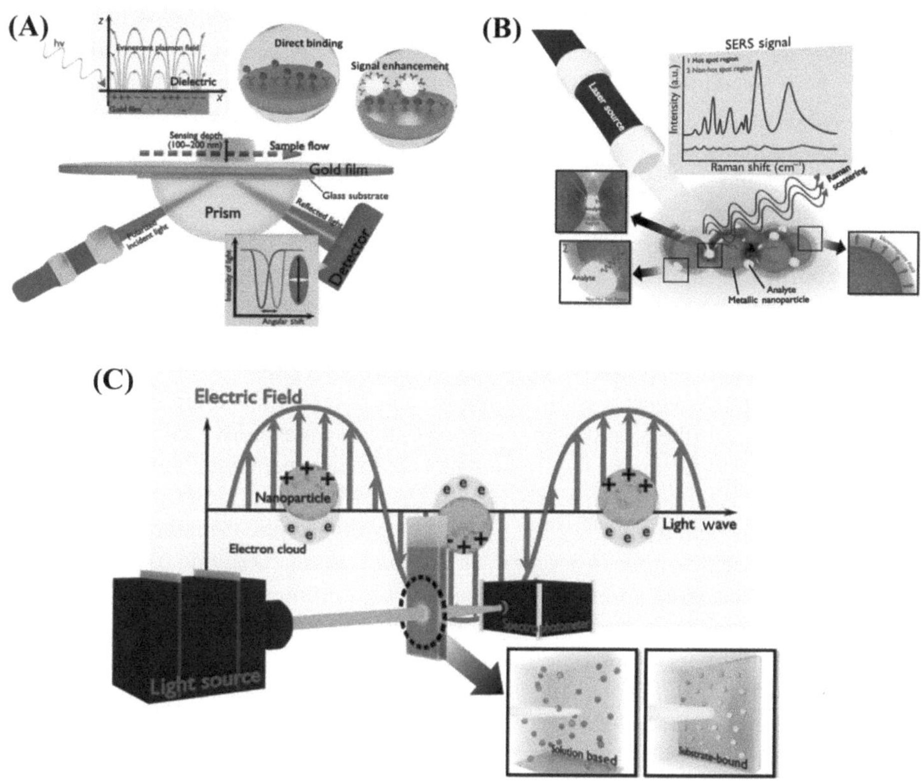

Figure 11.1 Frequently used measurement strategies using plasmonic sensors: (A) surface plasmon resonance, (B) surface-enhanced Raman scattering, and (C) localized surface plasmon resonance.
Source: Reproduced from A.R. Ferhan, J.A. Jackman, J.H. Park, N.-J. Cho, D.-H. Kim, Nanoplasmonic sensors for detecting circulating cancer biomarkers, Adv. Drug Deliv. Rev. 125 (2018) 48–77.

Accordingly, the amount of analyte can be determined. This can be done by either tracking the intensity of the reflected light at a fixed incident angle or tracking the incident angle shift while the reflection intensity is constant (Fig. 11.1A). The second method where the reflection intensity remains constant is more widely used because the appearance form of the SPR curve remains unchanged. In this way, easy signal analysis and measurement resolution are maintained because the characteristics such as the full width at half maximum do not change significantly.

In the traditional configuration of the SPR-based sensors, the prism-coupled form is applied. This configuration generally has two arrangements: Otto and Kretschmann. In both arrangements, the laser beam undergoes total internal reflection after passing through a glass prism. In the Otto configuration, a thin conductive film is placed at a distance near enough for the totally internally reflected light in the prism to interact with the plasmon on the interface, and the sample is placed in the notch between the prism and the conductive film. The Otto configuration is generally suitable for the analysis of solid materials (samples with relatively high refractive indices). It is less suitable to determining liquid materials (samples with relatively low refractive indices) because the mentioned notch reduces the efficiency of photo-n–plasmon coupling at the metal surface. The low efficiency of coupling the photo-n–plasmon would affect overall detection performance. In the Kretschmann configuration, the thin conductive film with a thickness of around 50 nm is put directly on the glass prism. The samples are placed on the opposite side of the total internally reflecting surface. This arrangement is suitable for both solid and liquid samples and provides maximum performance for analysis. Regardless of the configuration, the totally internally reflected light reaches a detector. Then the received signal is processed and converted to an SPR spectrum. Other diverse configurations are currently operational, which vary in different optical components employing diffraction gratings, optical fibers, and other types of optical waveguides [26].

In SPR-based techniques, the light beam does not penetrate the sample, and therefore the performance of analysis is not affected by the turbidity of the samples. This is a major advantage of the SPR method over other optical methods and allows for analysis of each sample in a clear, turbid, or colored medium. Another advantage of this method is its ability to directly analyze complex samples such as whole blood until the surface is inactivated by a proper strategy. This advantage is difficult to achieve by other optical methods such as absorption and scattering. It is also important to note that SPR methods essentially rely on RI variations. Therefore when the analyte binds to the surface, it can be traced with high precision, and this factor allows for a detailed analysis of the kinetics of binding and interaction [26].

11.2.2 Surface-enhanced Raman scattering sensors

SERS is a powerful technique used for the detection of molecules' fingerprint signals and the ultrasensitive exploration of surface chemical reactions [27,28]. The history of SERS goes back to 1974 when Fleischmann and coworkers investigated Raman scattering of pyridine on rough silver electrodes. They expressed that the observed amplification is related to the surface area. Later in 1977, two research groups separately encountered this phenomenon and suggested enhancement factors. Today, SERS is an exciting and growing field of research [29].

Typically, SERS relies on the utilization of metal nanostructures, either in solution or on planar support, which is illuminated by a laser source, and the enhancement of inelastic Raman scattering signals, which are generated from the surface-adsorbed molecule [29]. In this phenomenon, the amplification of scattered signals occurs due to the effect of two portions. The first portion is related to the contribution of the local electromagnetic field with a maximum SERS enhancement factor on the order of 10^{11}, which occurs when the laser beam interacts with the SERS-active materials (Fig. 11.1B). The second portion, identified as the chemical contribution, arises from the interaction between the surface-adsorbed molecules and the SERS-active surface and occurs due to the formation of charge-transfer states. It is believed that this factor has a minor role in scattering signal amplification, and it is effective with a factor of

$10-10^3$. Due to the short-range effect of the local amplified electromagnetic field, the SERS enhancement is maximized when the molecule and metal surface are in direct contact or in the immediate vicinity (around 10 nm) [26].

At the simplest level, four factors play a determining role in SERS measurements. Sample identity is the first factor. If the sample is known and the cross-section of its Raman scattering is well-defined, other factors combine to influence the signal. The second determining factor is the number of the surface-adsorbed molecules on the SERS-active surface being probed. So it depends on the amount of surface coverage and the area of the enhanced region being probed. Furthermore, the residence time of the analyte within the active surface is important, and the degree of its affinity to the surface determines this parameter [30]. The third factor is the amplification provided by the SERS substrate. The plasmonic enhancement is affected by a complex interaction of the material used, the geometry and arrangement of the nanostructure, and the coupling of the incident beam and the surface plasmon. Au, Ag, Cu, or Al as SERS-active metals [26,30] and TiO_2, ZnO, GaP, Fe_2O_3, Fe_3O_4, and etc. as SERS-active dielectrics [28,30] can be used in surface-enhanced spectroscopies. Finally, the instrumental parameters of the system such as laser power, focusing conditions, spectral resolution, and detection efficiency influence the generation and detection of the SERS signal. Although it seems that it is easy to control that number of factors, discovering an appropriate set of figures of merit in a SERS analysis is extremely challenging because each of these factors is determined by merging other variables, and controlling some of the subvariables is difficult [30].

However, although this method has been evaluated in terms of functional properties, the analytes used, instrumentation, and analytical matrix, generally, very few methods have been validated according to international guidelines since SERS-based methods are challenging for quantitative analysis and for comparison of the results obtained by different research groups around the world [27,30]. A recent study reported that usage of the quantitative SERS-based methods is still risky and that therefore standardizing SERS protocols is very necessary to define reproducible and accurate SERS methods [27]. In order to cancel out the intrinsic Raman scattering cross-sections of the analyte and the effects related to instrumental factors, an enhancement factor is defined. It acts as a scale for plasmonic amplification. Fundamentally, this presents a figure of merit for the usefulness of the substrate and can be used to compare the interlaboratory and intersample results [30].

11.2.3 Localized surface plasmon resonance sensors

As mentioned, the interaction between the light beam and the thin conductive film give rise to the SPR phenomenon. If the thin conductive film is replaced with discrete metallic nanoparticles, the incident beam excites the LSPR. The interaction of electromagnetic waves with the nanoparticles causes a nonpropagating collective oscillation for the free electrons of the nanoparticles. Thus resonant excitation generates a considerable electromagnetic field near the surface of the nanoparticles, called a local electromagnetic field. Under such conditions, the absorption and scattering of light occur at the plasmon resonance frequency by the nanoparticles with maximum intensity.

Spatially, the nonpropagating collective oscillation is confined to the close proximity of nanoparticle surface and therefore leads to a change in the local dielectric environment. Eventually, these changes vary the polarizability of the nanoparticles, followed by the resonance frequency of the plasmon and the optical extinction spectrum (Fig. 11.1C). In LSPR-based methods, the enhanced electromagnetic field is concentrated in a very small space near the nanoparticle surface, about 10–30 nm. Therefore this method is a very sensitive surface technique and is suitable for the determination of small biomolecules such as nucleic acids. Usually the status of nanoparticles in LSPR-based sensing is either freely suspended in a solution or deposited onto a solid support. In a typical configuration of LSPR, the light beam passes via the sensing platform (the package of nanoparticles). Then the overall optical extinction (involved absorbance and scattering spectra) is calculated according to the transmitted beam intensity [26].

According to Mie's theory, the LSPR of nanoparticles mainly depends on five factors: size, shape, composition, interparticle distance, and the dielectric constant (RI) of the surrounding medium. As can be seen, four of the factors mentioned are related to nanoparticles. The fifth factor, related to the surrounding medium of the nanoparticles, is the basis of the design of the LSPR sensors [31]. In general, LSPR-based sensors can operate with five strategies.

1. *Sensing the LSPR absorbance/scattering or color (known as labels).* This strategy is the simplest form of sensing and allows sensing through measurement of absorbance/light scattering or visual detection, taking advantage of the large extinction coefficient of nanoparticles. In this way, metal nanoparticles act as labels, and their color intensity determines the analyte concentration.
2. *Sensing the bulk RI of the medium by direct detection of the LSPR wavelength shift.* As mentioned previously, the LSPR absorbance (scattering) wavelength (intensity) of nanoparticles is dependent on the RI of the medium. Therefore the LSPR wavelength shift can be used to sense the RI changes. It is well accepted that the composition, size, and shape of nanostructures have a significant impact on RI sensitivity. Another important factor to enhance bulk RI sensitivity is the presence of a so-called hot spot, which can be fabricated by the dimerization of the nanoparticles. The area between nanoparticle dimers is considered a hot spot and presents strong electromagnetic field coupling.
3. *Detection of local analyte binding by direct detection of the LSPR wavelength shift (known as local binding sensitivity).* As the analyte molecules' RI is different from the surrounding medium's RI, the LSPR absorbance (scattering) wavelength (intensity) of metal nanoparticles is sensitive to the local binding of analyte molecules to the surface of nanoparticles. For this purpose, the surface of the nanoparticles is functionalized with a molecular receptor for an analyte (target) of interest, and the changes in the LSPR wavelength or absorbance/scattering intensity are monitored in the absence and presence of analyte molecules. Binding the analyte molecules to the receptor changes the local RI around the surface of the nanoparticles.
4. *Detection of an LSPR wavelength shift by analyte-induced metal nanoparticle aggregation.* The interparticle distance among adjacent nanoparticles strongly influences the optical properties of the plasmon. Decreasing the interparticle distance (small or large aggregates of metal nanoparticles compared to well-spaced individual nanoparticles) results in a strong overlap between the plasmon fields of adjacent nanoparticles and causes a shift toward longer wavelengths along with an increase in intensity.

5. *Detection of analyte-induced distance changes in linked metal nanoparticles (known as plasmonic rulers).* As described previously, the interparticle distance strongly influences the optical properties of plasmon. The last strategy is based on distance variations when the nanoparticles are linked together in any way and there are intervals between them. Analyte entry can change these distances and subsequently the optical properties of plasmon [31].

11.3 Application of magnetic nanomaterials in SPR sensors

Recently, MNMs have opened new avenues for SPR analysis. SPR sensors are unable to directly detect analytes at trace amounts or with small molecular mass because the attachment of these molecules induces only small changes in the RI of the surrounding medium [32,33]. One of the major approaches to overcome the challenge of sensitivity is employing MNMs because of their special properties. Low-cost production, high molecular weight, and a high RI are some of the main advantages of using MNMs for enhanced SPR detection [34,35]. MNMs show a high surface-to-volume ratio, which allows conjugation of the desired chemical and biomolecules on their surface. Also, they can directly capture analytes at the sensor surface and can concentrate and purify analytes from complex matrixes in an external magnetic due to their excellent magnetic properties [36].

Teramura et al. developed a magnetic nanobead (MNB)-SPR sensor for the determination of brain natriuretic peptide (BNP) [37]. Primary and secondary monoclonal antibodies were used for the detection of pictogram levels of BNP. A sensing surface was prepared by self-assembling monolayers of carboxyl group–terminated alkanethiol on a thin Au film. Then primary antibodies were covalently immobilized to the sensing surface. The sandwich-type immunoassay was made when the BNP solution and biotin-labeled secondary antibodies solution were injected. The signal of the SPR sensor was intensified by accumulating streptavidin-conjugated MNBs on the secondary antibody through bonding of biotin and streptavidin.

In another work, MNMs were used for the enhancement of the SPR signal in the detection of deltamethrin. Deltamethrin, which is a toxic insecticide present in food and agriculture, was detected in soybean with limit of detection (LOD) of 0.01 ng mL^{-1} and the linear concentration range of 0.01–1 ng mL^{-1} [38]. In the first step, carboxyl-coated Fe_3O_4 MNMs were functionalized with an antideltamethrin monoclonal antibody (Ab). Next, the sensing surface was constructed by the immobilization of chitosan on an Au chip modified with self-assembled mercaptoethylamine. The Fe_3O_4 MNMs antibody conjugates resulted in the enhancement of the RI change and rapid absorption of the analyte to the sensor surface.

Much attention has been focused on developing different biosensors for the earlier detection of diseases such as cancers [39]. Nanoparticle antibody label-based SPR is an emerging approach to detect biomarkers in biomedical samples. An SPR sensor based on MNMs was designed by Rusaling et al. for the determination of a prostate-specific antigen, which is a cancer biomarker, in the serum samples at the low level of 10 fg mL^{-1}. In this study, secondary antibodies (Ab_2) were conjugated with magnetic carboxyl–functionalized silica particles. Also, monoclonal antibodies (Ab_1) were attached on a carboxylate-functionalized Au chip surface. An enhanced SPR signal was achieved by the binding of MNMs-Ab_2-PSA complex to Ab_1.

Despite the fact that magnetic nanoparticles are desirable materials for sensitive SPR-based biosensors, they suffer from some disadvantages such as a disability to directly bind to biomolecules due to the lack of surface tenability and the loss of magnetic properties in complex biological systems. Magnetic nanoparticles are generally employed in the core–shell structure composed of Fe_3O_4 in the core and metals such as gold and silver in the shell to overcome these difficulties [40].

Liang et al. presented an SPR bioassay based on core–shell Fe_3O_4@Au MNMs for the determination of α-fetoprotein (AFP, a hepatocellular biomarker) [36]. A gold chip surface was modified with 3-mercapto-1-propanesulfonate (MPS), a positively charged chitosan-ferrocene (CS-Fc), and Au nanoparticles, respectively. Then a primary AFP antibody (Ab_1) was immobilized on the Au sensor surface. A sandwich immunoassay was formed by introducing Fe_3O_4@Au-AFP secondary antibody (Ab_2) conjugates as the signal enhancement agent to the MPS/CS-Fc/AuNPs/Ab_1 film. The Fe_3O_4@Au magnetic nanoparticles were used as the amplification agent because of their high RI and great mass. Also, they provided easy manipulation in the external field and high biocompatibility because of their magnetic cores and Au nanoparticles, respectively. The fabricated SPR sensor demonstrated high selectivity and sensitivity for the determination of AFP with LOD of 0.65 ng mL^{-1}. Fig. 11.2A SPR response curves for the detection of α-fetoprotein and the SPR signal amplification: (a) the free α-fetoprotein detection of in phosphate-buffered saline (PBS) in the absence of MNMs (b) signal enhancement with Ab_2, and (c) signal enhancement using Fe_3O_4@Au-Ab_2 conjugates. A schematic representation of the SPR sensor in different conditions for AFP detection is presented in Fig. 11.2B.

MNMs-enriched biomolecules have been used in SPR bioassays for the recognition of small molecules. In one study, MNM-antiadenosine aptamer conjugate–based indirect competitive inhibition assay was used for enhancement of SPR sensor, resulting in high selectivity and sensitivity in the detection of adenosine [41]. To easily functionalize MNMs with biomolecules, the surface of nanoparticles was modified by carboxylic groups. MNMs conjugated with antiadenosine aptamers were assembled on an amino group–modified gold sensing film. In the absence of adenosine, bioconjugated MNMs were adsorbed on the sensing surface through the DNA hybridization process, inducing a large change in the SPR signal because of the amplification effect of MNMs. Lastly, when adenosine with MNM-antiadenosine aptamer conjugates reacted, the antiadenosine aptamer (ss-DNA) on the surface of the Fe_3O_4 MNMs changed to a tertiary structure. So this structure cannot hybridize with its partially complementary ss-DNA bound on the Au sensing film. The decrease of signal change (the SPR angle shift) was proportional to the adenosine concentration. In this study, 14.51 and 32.82 nm nanoparticles were employed as amplification reagents. It was found that the SPR signal change resulting from using 32.82 nm MNMs was more significant than that of 14.51 nm MNMs. Also, the results showed that the signal change increased with an MNMs concentration increase from 0.016 to 1.6 nmol L^{-1}.

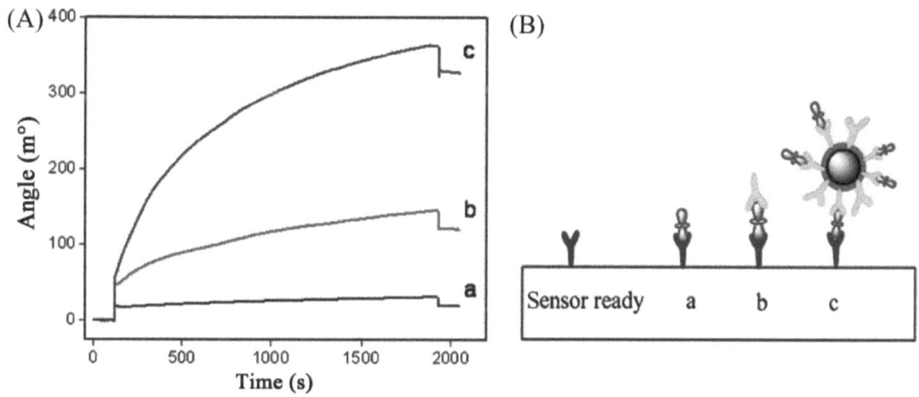

Figure 11.2 (A) The recorded SPR signal for determination of AFP (150 ng mL^{-1}): (a) the detection of AFP in PBS (in the absence of MNMs), (b) signal enhancement using Ab_2, and (c) signal enhancement using Fe_3O_4@Au-Ab_2 conjugates. (B) Schematic diagram of the sensor surface, showing the binding and amplified detection.
Source: Reproduced from R.P. Liang, G.H. Yao, L.X. Fan, J.D. Qiu, Magnetic Fe_3O_4@Au composite-enhanced surface plasmon resonance for ultrasensitive detection of magnetic nanoparticle-enriched α-fetoprotein, Anal. Chim. Acta 737 (2012) 22–28.

11.4 Application of magnetic nanomaterials in SERS sensors

The term MNM-based SERS-active substrate refers to combinations of magnetic and plasmonic nanoparticles to obtain ideal SERS substrates for analytical applications. Different MNMs can be applied to design a SERS, an active substrate that detects important targets. The employment of MNMs enables the capture of analytes and a sensitive identification using the Raman approach or its determination [42]. Taking advantage of the preconcentration ability and avoiding repeated centrifugation, many efforts have been devoted to the design and development of SERS-active substrates based on MNMs. Different MNMs used in the fabrication of SERS sensors can be grouped into four categorizations: core–shell MNMs, core–satellite MNMs, multifunctional MNMs, and special MNMs [43].

Core–shell MNMs like Fe_3O_4@Au, Fe_3O_4@SiO_2@Ag, and Fe_3O_4@SiO_2@Au have been successfully employed as SERS-active substrates for analytical applications. For example, a SERS sensor based on Fe_3O_4@Au core–shell MNMs with rough surfaces has been introduced for the detection of 4-aminothiophenol [44]. The as-prepared magnetic particles exhibited strong plasmon resonance absorption in the near-IR region.

In core–satellite MNMs, each of the large cores is surrounded by many nanosatellites. The active hot spots produced at the interface of different nanoparticles show an improvement in SERS sensitivity. The SERS detection of rhodamine 6G has been achieved using Fe_3O_4@SiO_2@Au seeds as core–shell magnetic particles

[45]. The superparamagnetic behavior of these MNMs has been used to form additional hot spots using magnetic means, improving a total enhancement by a factor of 10^3 with respect to a Raman analysis without magnetically inducing hot spot formation. An interesting paper has been published on the use of core−satellite Fe_3O_4-Au@TiO_2 nanoassemblies for the selective determination of Cr(VI) [46]. The reported sensor showed that Cr(VI) ions could be enriched using the TiO_2 layer close to Au nanoparticles.

In order to achieve the good performance of an MNMs-based SERS-active substrate, the functionalization of MNMs is essential. The use of functionalized MNMs provides an effective route to enhance the sensitivity as well as selectivity of the SERS method. Several functional materials have been employed to modify MNMs in SERS detection, for example, proteins, aptamers, polymers, etc. Multifunctional MNMs have been designed for a sensitive SERS immunosensor for the detection of carcinoembryonic antigen (CEA), a cancer biomarker widely applied for different types of cancers [47]. In this work, Fe_3O_4@Ag MNMs were coated with the capture antibody, and then the modification of the silver-coated gold nanorods (Au@Ag NRs) with detection antibody was carried out. The capture antibody−functionalized Fe_3O_4@Ag was employed to enrich the cancer biomarker magnetically. Afterward, the detection antibody−functionalized Au@Ag NRs probe was incubated (see Fig. 11.3). The developed SERS sensor exhibits detection of CEA in the linear range of 10 fg mL^{-1} to 100 ng mL^{-1}, with a detection limit of 4.75 fg mL^{-1}.

The analytical performance of MNMs-based SERS-active substrates strongly depends on the morphology and structure of MNMs. Therefore special plasmonic magnetic nanostructures with different morphologies, such as nanocubes [48], nanourchin [49], nanospindles [50], nano-Janus [51], nanostars [52], as well as nanoflowers [53] have been successfully applied as SERS substrates. The special MNMs present an astonishing class of MNMs that have attracted tremendous popularity due to high local curvature and relatively large aspect ratio. They can increase the enhancement factor field locally. In an interesting work, Reguera et al. utilized Janus-like MNMs comprising gold nanostars and iron oxide nanospheres as the superparamagnetic nanoparticles to improve SERS sensitivity [51]. The SERS signals of the as-prepared sensor to several dye molecules enhanced due to the binding of analytes on the Janus-like MNMs surface.

MNMs were also applied in the design and fabrication of SERS sensors to determine important analytical targets, including heavy metal ions, explosives, polycyclic aromatic hydrocarbons (PAHs), pesticides, antibiotics, pathogens, etc. Generally, SERS is an appropriate approach to determine organic substances because of its ultrahigh sensitivity.

PAHs are a class of potentially carcinogenic compounds found in the environment and are created from the incomplete combustion of various organic materials. The United States Environmental Protection Agency (USEPA) recommended 16 priority PAHs according to potential toxicity for human exposure. Therefore the determination of PAHs is very important and essential. The simultaneous employment of magnetic extraction and SERS technique considerably improves the sensitivity and selectivity of hybrid sensors toward PAHs. Du et al. developed the

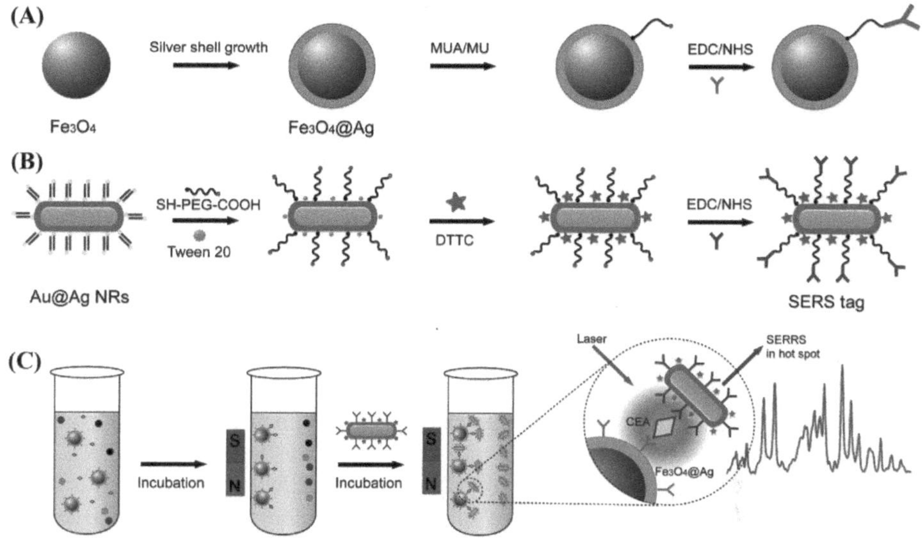

Figure 11.3 Schematic overview: (A) conjugation of Fe_3O_4@Ag MNMs to the capture antibody, (B) preparation of detection antibody-coated Au@Ag NRs, and (C) SERS analysis process to detect CEA.
Source: Reproduced from Z. Rong, C. Wang, J. Wang, D. Wang, R. Xiao, S. Wang, Magnetic immunoassay for cancer biomarker detection based on surface-enhanced resonance Raman scattering from coupled plasmonic nanostructures, Biosens. Bioelectron. 84 (2016) 15–21.

Fe_3O_4@Au SERS sensor, and the standard solutions of 16 PAHs as references were prepared and detected [54]. The proposed magnetic substrate was employed in SERS detection of PAHs in river water. The LODs of PAHs were $5-100$ nmol L^{-1}, indicating that the developed analytical method was applicable to detecting PAHs in real samples.

Pesticide residues from agricultural products have caused a series of problems in the environment and human health. The magnetic-based SERS sensor is an applicable strategy for the sensitive detection of pesticides. Wang et al. fabricated a polyethylenimine- (PEI-)interlayered core–shell–satellite MNM, Fe_3O_4@Ag-PEI-Au@Ag [55]. The prepared magnetic microsphere comprises three segments: (1) a Fe_3O_4@Ag core to supply SERS activity as well as magnetic property, (2) an Au@Ag satellite to provide additional hot spots, and (3) a PEI interlayer to form a precise nanogap between the Au@Ag shell and Fe_3O_4@Ag core. The substrate exhibited good performance for sensing paraquat and thiram, and the LODs were calculated to be 1×10^{-10} and 5×10^{-12} mol L^{-1}, respectively. In this work, effective hot spots were produced at the interface of satellites and core. Recent advances in MNMs-based SERS sensors enable portable SERS analysis [56,57], as well as online monitoring [58,59] of pesticides. Au NRs–coated Fe_3O_4 (Fe_3O_4@Au-NRs) microspheres were introduced as a magnetic SERS substrate for pesticide detection by portable Raman spectrometer with near infrared excitation [57]. Moreover, Ag/Fe_3O_4 magnetic nanocomposites have been successfully used for efficient online SERS detection of atrazine in real water samples [58].

Also, SERS-based methods for antibiotics have been proposed using MNMs as active substrates. A fast, selective, and sensitive SERS biosensor to determine chloramphenicol (CAP) was presented in which MNMs modified with the CAP antibody were applied as active substrate [60]. The main advantage of this approach was that the small molecule could be detected by sandwich immunosensor.

Developing rapid and accurate analytical methods for the detection of pathogens is of high importance for food safety and public health. The presentation of an efficient separation method for pathogens is a serious challenge, as in the real environmental samples they are at low concentrations. One candidate for the fast and accurate detection of pathogens is SERS substrates based on MNMs. The monitoring of pathogens with SERS sensors was performed in either a direct (label-free) or an indirect (labeled) way. The direct method is based on the intrinsic SERS spectrum of a pathogen as the analyte, according to the direct interaction of SERS-active substrate with the pathogen. The simultaneous preconcentration step, as well as SERS monitoring of bacterial cells, were achieved based on functionalized magnetic Fe_3O_4@Au MNMs [61]. Bacteria can be sensitively identified directly using SERS. In contrast, the indirect detection method relies on a known Raman probe to produce quantitative signals corresponding to the concentration of bound analytes. Wang et al. presented an MNMs-based SERS sensor using labeled Au@Ag NRs as SERS probe that could detect *Staphylococcus aureus* based on aptamer recognition [62]. Later, they improved their work by replacing Ag MNMs for Au MNMs and the aptamer to the antibody as substrates and reorganization elements, respectively [63]. In these works, the LODs have obtained about 10 CFU mL^{-1}. In addition to pathogens, SERS-active substrates based on MNMs can also be robust for the indirect detection of small targets [64]. The preparation of Fe_3O_4@Ag NPs modified with chitosan was reported to detect hydrogen peroxide, cyanide, and nitrite. The developed approach involves the covalent binding of myoglobin (Mb) to chitosan, and then its SERS spectrum was changed due to the attachment of small analytes to Mb. The detection limits of as-prepared magnetic SERS-active substrate toward nitrite, cyanide, and hydrogen peroxide were estimated to be 1 nmol L^{-1}, 0.2 μmol L^{-1}, and 10 nmol L^{-1}, respectively.

11.5 Application of magnetic nanomaterials in LSPR sensors

The fabrication of LSPR assay for small molecules detection is a challenge because of the occurrence of small changes in the RI. Moreover, the nonspecific binding and biofouling of real biomarkers constitute a serious obstacle to their detection via LSPR sensors. Therefore it is essential to discover new LSPR strategies to overcome these drawbacks [65]. MNMs are a potential tool for improving the sensitivity and selectivity of LSPR nanosensors. It is known that LSPR is very sensitive to the amount of analyte bound to the surface [66]. Hence, the application of MNMs-based LSPR sensors in analytical chemistry amplifies the LSPR spectral shift due to the relatively high molecular weight and RI of MNMs, such as Fe_3O_4. Tang's research group introduced a label-free LSPR nanosensor based on Fe_3O_4 MNMs for the quantification of cardiac troponin I (cTnI) in blood plasma for the first time [67]. Further to this, Tang et al., in 2016, developed MNMs-based LSPR sensors [68]. The anti-cTnI-immobilized Fe_3O_4 MNMs could selectively capture cTnI in blood plasma, while the anti-cTnI-coated on the surface of a gold nanorod (GNR) was used as a LSPR sensing platform (Fig. 11.4A). The preconcentration of cTnI, as a model analyte, occurred when MNMs were used, and consequently the sensing performance of the MNM-based LSPR sensor was significantly improved compared to a common LSPR sensor (Fig. 11.4B). The magnetic behavior and large surface-to-volume ratio of the modified Fe_3O_4 MNMs enable the enrichment and sensitive detection of cTnI, respectively. Additionally, to obtain more information about the mechanism of the MNM increment of the GNR sensing signal, COMSOL simulation was employed. According to Fig. 11.4C, the surface plasmon intensity is strongly dependent on the distance between the Fe_3O_4 nanoparticles and GNRs. The coupling effect increased with decreasing the distance between Fe_3O_4 nanoparticle and GNR surface, and a surface plasmon intensity of LSPR assay was achieved.

Today, the development of fast and reliable methods for providing quantitative information in the clinics is an ongoing need. To address this issue, an analytical strategy for label-free immunoassay based on the magnetic-assisted LSPR detection of four cytokines in a complex biological sample was reported [69]. In this work, a Fe_3O_4/Au core–shell MNMs sensor array was fabricated and employed for the precise detection of leukemia tumor markers. The as-prepared MNMs-based LSPR sensor array enabled new capabilities for analysis thanks to its plasmonic and magnetic characteristics. The application of this microarray sensor into an optofluidic system provided the online detection of cytokines with an excellent detection limit of 20 pg mL^{-1}. Finally, the advantage of the developed sensor was the detection of traces at low sample volume (c. 1 μL).

11.6 Conclusions

By far, various plasmonic materials have been used in analytical chemistry for the construction of plasmonic sensors. However, the use of MNMs for the construction of plasmonic nanosensors has been less investigated. The MNMs can consist of either elemental magnetic materials such as iron, cobalt, and nickel or their oxides and mixed oxides. Magnetic iron oxide nanoparticles, including magnetite and maghemite, have attracted very great attention in various fields including plasmonic nanosensors. Nanoplasmonic-based assay techniques mainly fall into three categories: SPR, LSPR, and SERS.

SPR sensors are unable to directly detect analytes at trace amounts or with small molecular mass because the attachment of these molecules induces only small

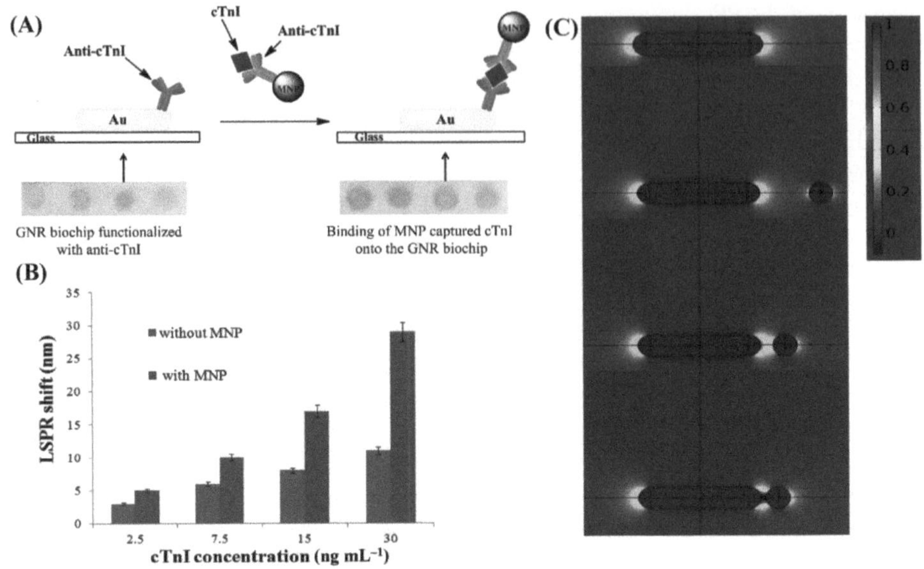

Figure 11.4 Schematic overview: (A) anti-cTnI-immobilized Fe_3O_4 MNMs—enhanced LSPR sensor on GNRs; (B) comparison of the longitudinal LSPR shift of the GNRs toward different concentrations of cTnI with (*red-right column*) and without (*blue-left column*) modified Fe_3O_4 MNMs, respectively; (C) result of the electric field simulation of the GNR-based LSPR sensor in close to Fe_3O_4 MNMs.
Source: Reproduced from Z. Mei, A. Dhanale, A. Gangaharan, D.K. Sardar, L. Tang, Water dispersion of magnetic nanoparticles with selective biofunctionality for enhanced plasmonic biosensing, Talanta 151 (2016) 23–29.

changes in the RI of the surrounding medium. One of the major approaches to overcome the challenge of sensitivity is employing MNMs because of their special properties such as low-cost production and high RI. Due to their high surface-to-volume ratio, MNMs can be easily conjugated with desired chemical and biomolecules. Furthermore, they can directly capture analytes at the sensor surface and can concentrate and purify analytes from complex matrixes in an external magnetic due to their excellent magnetic properties. Much attention has focused on developing different biosensors for the earlier detection of diseases such as cancers using SPR sensors. Despite the fact that MNMs are desirable materials for sensitive SPR-based biosensors, they suffer from some disadvantages such as the inability to directly bind to biomolecules due to the lack of surface tenability and loss of magnetic properties in complex biological systems. To address these drawbacks, magnetic nanoparticles are generally employed in core—shell structures composed usually of Fe_3O_4 in the core and metals such as gold and silver in the shell.

Various MNMs can be applied to design a SERS, an active substrate that detects important targets. The employment of MNMs enables the capture of analytes and sensitive identification using the Raman approach or its determination. Taking advantage of the preconcentration ability and avoiding repeated centrifugation, many efforts have been devoted to the design and development of SERS-active substrates based on MNMs. Different MNMs are used in the fabrication of SERS sensors and can be grouped into four categorizations: core–shell MNMs, core–satellite MNMs, multifunctional MNMs, and special MNMs. In order to achieve good performance of MNMs-based SERS-active substrate, the functionalization of MNMs is essential. The use of functionalized MNMs provides an effective route to enhance the sensitivity as well as selectivity of the SERS method. The analytical performance of MNMs-based SERS-active substrates strongly depends on the morphology and structure of MNMs. Therefore special plasmonic magnetic nanostructures with different morphologies, such as nanocubes, nanourchin, nanospindles, nano-Janus, nanostars, as well as nanoflowers have been successfully applied as SERS substrates.

The fabrication of LSPR assay for small molecules detection is a challenge because of the occurrence of small changes in the RI. Moreover, the nonspecific binding and biofouling of real biomarkers constitute a serious obstacle to their detection via LSPR sensors. Therefore it is essential to discover new LSPR strategies to overcome these drawbacks. MNMs are a potential tool for improving the sensitivity and selectivity of LSPR nanosensors. It is known that LSPR is very sensitive to the amount of analyte bound to the surface. Hence the application of MNMs-based LSPR sensors in analytical chemistry amplifies the LSPR spectral shift due to the relatively high molecular weight and RI of MNMs.

References

[1] Wahajuddin, S. Arora, Superparamagnetic iron oxide nanoparticles: magnetic nanoplatforms as drug carriers, Int. J. Nanomed. 7 (2012) 3445–3471.

[2] M.N. Alves, M. Miró, M.C. Breadmore, M. Macka, Trends in analytical separations of magnetic (nano)particles, TrAC Trends Anal. Chem. 114 (2019) 89–97.

[3] B.K. Purushothaman, P. Uma Maheswari, K.M. Meera Sheriffa Begum, Magnetic casein-$CaFe_2O_4$ nanohybrid carrier conjugated with progesterone for enhanced cytotoxicity of citrus peel derived hesperidin drug towards breast and ovarian cancer, Int. J. Biol. Macromol. 151 (2020) 293–304.

[4] X.J. Lee, H.N. Lim, N.S.K. Gowthaman, M.B.A. Rahman, C.A. Che Abdullah, K. Muthoosamy, In-situ surface functionalization of superparamagnetic reduced graphene oxide – Fe_3O_4 nanocomposite via *Ganoderma lucidum* extract for targeted cancer therapy application, Appl. Surf. Sci. 512 (2020) 145738.

[5] J. Zhao, Z. Zhang, X. Wang, Fabrication of pH-responsive PAA-$NaMnF_3$@DOX hybrid nanostructures for magnetic resonance imaging and drug delivery, J. Alloy Compd. 820 (2020) 153142.

[6] R. Ghafelehbashi, M. Tavakkoli Yaraki, L. Heidarpoor Saremi, A. Lajevardi, M. Haratian, B. Astinchap, et al., A pH-responsive citric-acid/α-cyclodextrin-functionalized Fe_3O_4 nanoparticles as a nanocarrier for quercetin: an experimental and DFT study, Mater. Sci. Eng. C 109 (2020) 110597.

[7] Y. Huang, L. Li, D. Zhang, L. Gan, P. Zhao, Y. Zhang, et al., Gadolinium-doped carbon quantum dots loaded magnetite nanoparticles as a bimodal nanoprobe for both fluorescence and magnetic resonance imaging, Magn. Reson. Imaging 68 (2020) 113−120.

[8] Q. Ren, K. Yang, R. Zou, Z. Wan, Z. Shen, G. Wu, et al., Biodegradable hollow manganese/cobalt oxide nanoparticles for tumor theranostics, Nanoscale 11 (2019) 23021−23026.

[9] P.L. Liang, L.Y. Yuan, H. Deng, X.C. Wang, L. Wang, Z.J. Li, et al., Photocatalytic reduction of uranium(VI) by magnetic $ZnFe_2O_4$ under visible light, Appl. Catal. B: Environ. 267 (2020).

[10] N. Nasseh, T.J. Al-Musawi, M.R. Miri, S. Rodriguez-Couto, A. Hossein Panahi, A comprehensive study on the application of $FeNi_3$@SiO_2@ZnO magnetic nanocomposites as a novel photo-catalyst for degradation of tamoxifen in the presence of simulated sunlight, Environ. Pollut. 261 (2020).

[11] T. Touqeer, M.W. Mumtaz, H. Mukhtar, A. Irfan, S. Akram, A. Shabbir, et al., Fe_3O_4-PDA-lipase as surface functionalized nano biocatalyst for the production of biodiesel using waste cooking oil as feedstock: characterization and process optimization, Energies 13 (2020).

[12] H. Veisi, T. Ozturk, B. Karmakar, T. Tamoradi, S. Hemmati, In situ decorated Pd NPs on chitosan-encapsulated Fe_3O_4/SiO_2-NH_2 as magnetic catalyst in Suzuki-Miyaura coupling and 4-nitrophenol reduction, Carbohydr. Polym. 235 (2020).

[13] A. Afkhami, S. Aghajani, M. Mohseni, T. Madrakian, Effectiveness of $Ni_{0.5}Zn_{0.5}Fe_2O_4$ for the removal and preconcentration of Cr(VI), Mo(VI), V(V) and W(VI) oxyanions from water and wastewater samples, J. Iran. Chem. Soc. 12 (2015) 2007−2013.

[14] M. Ghapanvari, T. Madrakian, A. Afkhami, A. Ghoorchian, A modified carbon paste electrode based on Fe_3O_4@multi-walled carbon nanotubes@polyacrylonitrile nanofibers for determination of imatinib anticancer drug, J. Appl. Electrochem. 50 (2020) 281−294.

[15] N.R. Jalal, T. Madrakian, A. Afkhami, M. Ghamsari, Polyethylenimine@ Fe_3O_4@carbon nanotubes nanocomposite as a modifier in glassy carbon electrode for sensitive determination of ciprofloxacin in biological samples, J. Electroanal. Chem. 833 (2019) 281−289.

[16] S.M. Khoshfetrat, H. Khoshsafar, A. Afkhami, M.A. Mehrgardi, H. Bagheri, Enhanced visual wireless electrochemiluminescence immunosensing of prostate-specific antigen based on the luminol loaded into MIL-53(Fe)-NH_2 accelerator and hydrogen evolution reaction mediation, Anal. Chem. 91 (2019) 6383−6390.

[17] S. Maleki, T. Madrakian, A. Afkhami, Application of polyacrylonitrile nanofibers decorated with magnetic carbon dots as a resonance light scattering sensor to determine famotidine, Talanta 181 (2018) 286−295.

[18] M. Pirdadeh-Beiranvand, A. Afkhami, T. Madrakian, Ag nanoparticles for determination of bisphenol A by resonance light-scattering technique, J. Iran. Chem. Soc. 15 (2018) 1527−1534.

[19] A. Afkhami, S. Aghajani, T. Madrakian, H. Bagheri, Simultaneous preconcentration and determination of trace quantities of inorganic arsenic species in water using $Ni_{0.5}Zn_{0.5}Fe_2O_4$ magnetic nanoparticles, Chem. Pap. 74 (2020).

[20] E. Bavel, A. Afkhami, T. Madrakian, Removal and preconcentration of Pb(II) heavy metal ion from water and waste-water samples onto poly (vinyl alcohol)/polyethyleneimine/Fe_3O_4 microfibers nanocomposite, J. Polym. Environ. 28 (2020) 614−623.

[21] M. Pirdadeh-Beiranvand, A. Afkhami, T. Madrakian, Magnetic molecularly imprinted electrospun nanofibers for selective extraction of nilotinib from human serum, Anal. Bioanal. Chem. 412 (2020) 1629−1637.

[22] M. Soleimani, M. Ahmadi, T. Madrakian, A. Afkhami, Magnetic solid phase extraction of rizatriptan in human urine samples prior to its spectrofluorimetric determination, Sens. Actuators, B: Chem. 254 (2018) 1225–1233.

[23] C.M. Miyazaki, F.M. Shimizu, M. Ferreira, 6-Surface plasmon resonance (SPR) for sensors and biosensors, in: A.L. Da Róz, M. Ferreira, F. de Lima Leite, O.N. Oliveira (Eds.), Nanocharacterization Techniques, William Andrew Publishing, 2017, pp. 183–200.

[24] J. Homola, Surface plasmon resonance sensors for detection of chemical and biological species, Chem. Rev. 108 (2008) 462–493.

[25] R. Karlsson, SPR for molecular interaction analysis: a review of emerging application areas, J. Mol. Recognit. 17 (2004) 151–161.

[26] A.R. Ferhan, J.A. Jackman, J.H. Park, N.-J. Cho, D.-H. Kim, Nanoplasmonic sensors for detecting circulating cancer biomarkers, Adv. Drug. Deliv. Rev. 125 (2018) 48–77.

[27] S. Fornasaro, F. Alsamad, M. Baia, L.A.E. Batista de Carvalho, C. Beleites, H.J. Byrne, et al., Surface enhanced Raman spectroscopy for quantitative analysis: results of a large-scale European multi-instrument interlaboratory study, Anal. Chem. 92 (2020).

[28] I. Alessandri, J.R. Lombardi, Enhanced Raman scattering with dielectrics, Chem. Rev. 116 (2016) 14921–14981.

[29] J. Langer, D. Jimenez de Aberasturi, J. Aizpurua, R.A. Alvarez-Puebla, B. Auguié, J.J. Baumberg, et al., Present and future of surface-enhanced Raman scattering, ACS Nano 14 (2020) 28–117.

[30] S.E.J. Bell, G. Charron, E. Cortés, J. Kneipp, M.L. de la Chapelle, J. Langer, et al., Towards reliable and quantitative surface-enhanced Raman Scattering (SERS): from key parameters to good analytical practice, Angew. Chem. Int. Ed. 59 (14) (2020) 5454–5462.

[31] F.P. Zamborini, L. Bao, R. Dasari, Nanoparticles in measurement science, Anal. Chem. 84 (2012) 541–576.

[32] W.-C. Law, K.-T. Yong, A. Baev, P.N. Prasad, Sensitivity improved surface plasmon resonance biosensor for cancer biomarker detection based on plasmonic enhancement, ACS Nano 5 (2011) 4858–4864.

[33] W.-C. Law, K.-T. Yong, A. Baev, R. Hu, P.N. Prasad, Nanoparticle enhanced surface plasmon resonance biosensing: application of gold nanorods, Opt. Express 17 (2009) 19041–19046.

[34] S. Zeng, D. Baillargeat, H.-P. Ho, K.-T. Yong, Nanomaterials enhanced surface plasmon resonance for biological and chemical sensing applications, Chem. Soc. Rev. 43 (2014) 3426–3452.

[35] S. Mohammadzadeh-Asl, A. Keshtkar, J. Ezzati Nazhad Dolatabadi, M. de la Guardia, Nanomaterials and phase sensitive based signal enhancment in surface plasmon resonance, Biosens. Bioelectron. 110 (2018) 118–131.

[36] R.P. Liang, G.H. Yao, L.X. Fan, J.D. Qiu, Magnetic Fe_3O_4@Au composite-enhanced surface plasmon resonance for ultrasensitive detection of magnetic nanoparticle-enriched α-fetoprotein, Anal. Chim. Acta 737 (2012) 22–28.

[37] Y. Teramura, Y. Arima, H. Iwata, Surface plasmon resonance-based highly sensitive immunosensing for brain natriuretic peptide using nanobeads for signal amplification, Anal. Biochem. 357 (2006) 208–215.

[38] X. Liu, L. Li, Y.-Q. Liu, X.-B. Shi, W.-J. Li, Y. Yang, et al., Ultrasensitive detection of deltamethrin by immune magnetic nanoparticles separation coupled with surface plasmon resonance sensor, Biosens. Bioelectron. 59 (2014) 328–334.

[39] S. Krishnan, V. Mani, D. Wasalathanthri, C.V. Kumar, J.F. Rusling, Attomolar detection of a cancer biomarker protein in serum by surface plasmon resonance using superparamagnetic particle labels, Angew. Chem. Int. Ed. 50 (2011) 1175–1178.
[40] F. Fathi, M.-R. Rashidi, Y. Omidi, Ultra-sensitive detection by metal nanoparticles-mediated enhanced SPR biosensors, Talanta 192 (2019) 118–127.
[41] J. Wang, A. Munir, Z. Zhu, H.S. Zhou, Magnetic nanoparticle enhanced surface plasmon resonance sensing and its application for the ultrasensitive detection of magnetic nanoparticle-enriched small molecules, Anal. Chem. 82 (2010) 6782–6789.
[42] P.C. Pinheiro, A.L. Daniel-da-Silva, H.I.S. Nogueira, T. Trindade, Functionalized inorganic nanoparticles for magnetic separation and SERS detection of water pollutants, Eur. J. Inorg. Chem. 2018 (2018) 3443–3461.
[43] D. Song, R. Yang, F. Long, A. Zhu, Applications of magnetic nanoparticles in surface-enhanced Raman scattering (SERS) detection of environmental pollutants, J. Environ. Sci. (China) 80 (2019) 14–34.
[44] Y. Zhai, J. Zhai, Y. Wang, S. Guo, W. Ren, S. Dong, Fabrication of iron oxide core/gold shell submicrometer spheres with nanoscale surface roughness for efficient surface-enhanced Raman scattering, J. Phys. Chem. C 113 (2009) 7009–7014.
[45] Y. Hu, Y. Sun, Stable magnetic hot spots for simultaneous concentration and ultrasensitive surface-enhanced Raman scattering detection of solution analytes, J. Phys. Chem. C 116 (2012) 13329–13335.
[46] B. Lv, Z. Sun, J. Zhang, C. Jing, Multifunctional satellite Fe_3O_4-Au@ TiO_2 nanostructure for SERS detection and photo-reduction of Cr (VI), Colloids Surf. A: Physicochem. Eng. Asp. 513 (2017) 234–240.
[47] Z. Rong, C. Wang, J. Wang, D. Wang, R. Xiao, S. Wang, Magnetic immunoassay for cancer biomarker detection based on surface-enhanced resonance Raman scattering from coupled plasmonic nanostructures, Biosens. Bioelectron. 84 (2016) 15–21.
[48] W. Cai, X. Wang, Y. Yan, Controllable fabrication and sensitive detection based on SERS substrates with Au nanocubes coated Fe_3O_4, Mater. Res. Bull. 52 (2014) 1–5.
[49] Y. Ye, J. Chen, Q. Ding, D. Lin, R. Dong, L. Yang, et al., Sea-urchin-like Fe_3O_4@ C@ Ag particles: an efficient SERS substrate for detection of organic pollutants, Nanoscale 5 (2013) 5887–5895.
[50] Q. He, A. Zhao, L. Li, H. Sun, D. Wang, H. Guo, et al., Fabrication of Fe_3O_4@ SiO_2@ Ag magnetic–plasmonic nanospindles as highly efficient SERS active substrates for label-free detection of pesticides, N. J. Chem. 41 (2017) 1582–1590.
[51] J. Reguera, D.J. de Aberasturi, N. Winckelmans, J. Langer, S. Bals, L.M. Liz-Marzán, Synthesis of Janus plasmonic–magnetic, star–sphere nanoparticles, and their application in SERS detection, Faraday Discuss. 191 (2016) 47–59.
[52] P. Quaresma, I. Osório, G. Dória, P.A. Carvalho, A. Pereira, J. Langer, et al., Star-shaped magnetite@gold nanoparticles for protein magnetic separation and SERS detection, RSC Adv. 4 (2014) 3659–3667.
[53] C. Wang, J. Wang, P. Li, Z. Rong, X. Jia, Q. Ma, et al., Sonochemical synthesis of highly branched flower-like Fe_3O_4@ SiO_2@ Ag microcomposites and their application as versatile SERS substrates, Nanoscale 8 (2016) 19816–19828.
[54] J. Du, J. Xu, Z. Sun, C. Jing, Au nanoparticles grafted on Fe_3O_4 as effective SERS substrates for label-free detection of the 16 EPA priority polycyclic aromatic hydrocarbons, Anal. Chim. Acta 915 (2016) 81–89.
[55] C. Wang, P. Li, J. Wang, Z. Rong, Y. Pang, J. Xu, et al., Polyethylenimine-interlayered core–shell–satellite 3D magnetic microspheres as versatile SERS substrates, Nanoscale 7 (2015) 18694–18707.

[56] Z. Liu, Y. Wang, R. Deng, L. Yang, S. Yu, S. Xu, et al., Fe_3O_4@Graphene Oxide@Ag particles for surface magnet solid-phase extraction surface-enhanced Raman scattering (SMSPE-SERS): from sample pretreatment to detection all-in-one, ACS Appl. Mater. Interfaces 8 (2016) 14160–14168.
[57] X. Tang, R. Dong, L. Yang, J. Liu, Fabrication of Au nanorod-coated Fe_3O_4 microspheres as SERS substrate for pesticide analysis by near-infrared excitation, J. Raman Spectrosc. 46 (2015) 470–475.
[58] Y. Shan, Y. Yang, Y. Cao, Z. Huang, Facile solvothermal synthesis of Ag/Fe_3O_4 nanocomposites and their SERS applications in on-line monitoring of pesticide contaminated water, RSC Adv. 5 (2015) 102610–102618.
[59] W. Song, Z. Yang, F. Ma, M. Chi, B. Zhao, X. Lu, Electrospun magnetic $CoFe_2O_4$/Ag hybrid nanotubes for sensitive SERS detection and monitoring of the catalytic degradation of organic pollutants, RSC Adv. 7 (2017) 40334–40341.
[60] K. Yang, Y. Hu, N. Dong, A novel biosensor based on competitive SERS immunoassay and magnetic separation for accurate and sensitive detection of chloramphenicol, Biosens. Bioelectron. 80 (2016) 373–377.
[61] L. Zhang, J. Xu, L. Mi, H. Gong, S. Jiang, Q. Yu, Multifunctional magnetic–plasmonic nanoparticles for fast concentration and sensitive detection of bacteria using SERS, Biosens. Bioelectron. 31 (2012) 130–136.
[62] J. Wang, X. Wu, C. Wang, N. Shao, P. Dong, R. Xiao, et al., Magnetically assisted surface-enhanced Raman spectroscopy for the detection of *Staphylococcus aureus* based on aptamer recognition, ACS Appl. Mater. Interfaces 7 (2015) 20919–20929.
[63] J. Wang, X. Wu, C. Wang, Z. Rong, H. Ding, H. Li, et al., Facile synthesis of Au-coated magnetic nanoparticles and their application in bacteria detection via a SERS method, ACS Appl. Mater. Interfaces 8 (2016) 19958–19967.
[64] X.X. Han, A.M. Schmidt, G. Marten, A. Fischer, I.M. Weidinger, P. Hildebrandt, Magnetic silver hybrid nanoparticles for surface-enhanced resonance Raman spectroscopic detection and decontamination of small toxic molecules, ACS Nano 7 (2013) 3212–3220.
[65] L. Tang, H.J. Kwon, K.A. Kang, Theoretical and experimental analysis of analyte transport in a fiber-optic, protein C immuno-biosensor, Biotechnol. Bioeng. 88 (2004) 869–879.
[66] D. Grigoriev, D. Gorin, G.B. Sukhorukov, A. Yashchenok, E. Maltseva, H. Möhwald, Polyelectrolyte/magnetite nanoparticle multilayers: preparation and structure characterization, Langmuir 23 (2007) 12388–12396.
[67] L. Tang, J. Casas, M. Venkataramasubramani, Magnetic nanoparticle mediated enhancement of localized surface plasmon resonance for ultrasensitive bioanalytical assay in human blood plasma, Anal. Chem. 85 (2013) 1431–1439.
[68] Z. Mei, A. Dhanale, A. Gangaharan, D.K. Sardar, L. Tang, Water dispersion of magnetic nanoparticles with selective biofunctionality for enhanced plasmonic biosensing, Talanta 151 (2016) 23–29.
[69] Y. Cai, J. Zhu, J. He, W. Yang, C. Ma, F. Xiong, et al., Magnet patterned superparamagnetic Fe_3O_4/Au core–shell nanoplasmonic sensing array for label-free high throughput cytokine immunoassay, Adv. Healthc. Mater. 8 (2019) 1801478.

Application of magnetic nanomaterials as electrochemical sensors

12

Hadi Beitollahi[1], *Fariba Garkani Nejad*[1] *and Somayeh Tajik*[2]
[1]Environment Department, Institute of Science and High Technology and Environmental Sciences, Graduate University of Advanced Technology, Kerman, Iran, [2]Research Center for Tropical and Infectious Diseases, Kerman University of Medical Sciences, Kerman, Iran

12.1 Introduction

Interest in nanotechnologies and nanoscale materials, particularly magnetic nanomaterials (MNMs) have demonstrated great potency in catalysts [1], absorbents [2], sensors [3], wastewater treatments [4], semiconductors [5], pigments, magnetic resonance imaging [6], magnetic data storage devices [7], bioseparation [8], as well as drugs [9]. In particular, introducing MNMs into conventional sensing procedures has largely simplified determination operations and improved the sensing function that made the MNMs-based sensing procedures an interesting research topic [10].

In comparison to other sensing procedures like fluorescence, chromatography, mass spectrometry, as well as the surface-enhanced Raman scattering, electrochemistry displayed specific features of high sensitivity, fingerprint specificity, short analysis time, and nondestructive detection [11]. By electrochemical methods, miniaturization is possible, which offers the possibility of online monitoring for detection and can be directly operated in real samples. Electrochemical sensors have widespread utilization, such as environmental monitoring, controlling food quality, medical diagnostics, drug compounds analysis, as well as chemical treatment determinations. Sensitivity and selectivity have been considered to be prominent elements for consideration in the optimization of the performance and applicability of the electrochemical assay [12]. These electrochemical sensors usually demand a three-electrode setup, such as the working, counter and reference with a potentiostat. The operation mode for electrochemical sensors involved the production of the electrical signal by reaction with the desired analyte. Therefore the signal produced is proportionate to the analyte concentration. The majority of electrochemical sensors may be categorized into potentiometric, impedimetric, conductometric, and amperometric sensors [13]. The commonly applied electroanalytical sensor is voltammetry that was introduced as a subclass of amperometry, which measures the currents at diverse potential points in the current−voltage curve as opposed to the constant potential points in the amperometric procedure [14].

Magnetic Nanomaterials in Analytical Chemistry. DOI: https://doi.org/10.1016/B978-0-12-822131-0.00006-6

Unmodified electrodes suffer from several problems like nonrepeatability of surface behavior, slower kinetics of the electrochemical reaction of some compounds on the electrode surface, and overpotential requirements. These restrict their ability and applications to analytical goals; therefore, eliminating such concerns would probably take place via physicochemical modifications in the electrodes' surface and their modification. Moreover, electrode modification was considered a process wherein the surface of an unmodified electrode is covered by a biological or chemical compound known as the modifier, or this compound can be introduced into the electrode matrix to alters its electrochemical features [15–19]. Increases in the stability, selectivity, and sensitivity of the electrode, improvement of the electrocatalytic features, the prevention of unfavorable reactions, the increased rate of electron transfer, greater currents, and reduction of overpotential have been considered as some of the benefits of the modified electrode [20,21].

Hence, in the course of the last 10 years, experts in the field have actively investigated diverse compositions and phases of MNPs like pure metals of Co, Ni, and Fe; metal oxides like Fe_3O_4 and γ-Fe_2O_3; ferrites like MFe_2O_4 (M = Cu, Ni, Mn, Mg, etc.); and metal alloys like FePt, CoPt as novel electrocatalytic modifiers due to their low cost, nontoxic behavior, facile functionalization, higher rate of electron transfer, very high ratio of the surface areas to the volume, and very good catalytic behavior [22,23].

Concerning the wide utilization of the MNMs in electrochemical sensing systems, the present chapter describes recent applications of MNMs in electrochemical sensors for clinical diagnosis, bioanalysis and pharmaceutical compounds analysis, and food analysis and environmental monitoring.

12.2 MNMs-based electrochemical sensors

12.2.1 Application in clinical diagnosis

Clinical diagnostics makes up the first and most important step in the treatment of a disease. Current situations in clinical diagnostics and monitoring require fast as well as precise analysis. Significant parameters that complicate such procedures have been considered to be the sophisticated laboratory and the cost of the analytical tools for whole measurements [24]. For this purpose, MNMs-based electrochemical sensors can serve as an option to solve such problems or to become part of the solution.

12.2.1.1 Glucose

It is well-known that diabetes is a chronic clinical condition resulting from the greater level of glucose due to the malfunction of the pancreatic β cells responsible for producing the insulin that closely monitors levels of glucose in the blood. The heightened levels of glucose create unpleasant and even life-threatening circumstances like nervous, cardiac, ocular, renal, peripheral vascular, and cerebral illnesses in patients with diabetes [25]. Although there is no treatment for diabetes, the levels of the blood glucose of diabetic patients must be closely monitored to avoid additional consequences. Because of the ever increasing concerns of the disease, continual monitoring of glucose in the blood is a potent procedure that leads to the construction of precise as well as sensitive blood sugar detection sensors for clinical diagnoses and personal care.

Shahnavaz and coworkers designed $ZnFe_2O_4$ magnetic NPs/reduced graphene oxide (rGO) nanosheets modified glassy carbon electrode (GCE) ($ZnFe_2O_4$ MNPs/rGO/GCE) as a new system for electrochemical glucose sensing [26]. They used an easy in situ hydrothermal path for simultaneous GO reduction as well as $ZnFe_2O_4$ MNPs formation, which enabled $ZnFe_2O_4$ MNPs dispersion on the reduced graphene (Gr) sheet. Also, electrochemical examinations demonstrated very good electrocatalytic activity of $ZnFe_2O_4$/rGO/GCE for glucose oxidation in concentration ranges between 0.1 and 12.5 mM with 1689.6 $\mu A\ mM^{-1}\ cm^{-2}$ sensitivity. Moreover, the glucose enzyme-free sensor showed good repeatability, reproducibility, selectivity, and stability.

In another recent work, Vennila and coworkers devised a strongly selective as well as sensitive enzyme-free glucose sensor based on the nickel-cobalt (Ni-Co)/iron oxide (Fe_3O_4) flower-like nanocomposite with the use of a simplified chemical phase procedure [27]. The electrocatalytic function of the as-prepared nanostructures for the oxidation of glucose was examined with cyclic voltammetry (CV) and amperometric procedures. The Ni-Co/Fe_3O_4 nanocomposite showed a rapid response toward the determination of nonenzymatic glucose, and the achieved amperometric signals have been linearly proportionate to the glucose concentration in ranges between 0.001 and 11.0 mM, reflecting a lower limit of detection (LOD) of 0.19 μM and great sensitivity equal to 2171 $\mu A\ mM^{-1}\ cm^{-2}$. Moreover, real-time utilization in the human serum specimens with usable recoveries showed very good functions concerning reliability, selectivity, acceptable accuracy as well as reproducibility that reflects potent examination of the as-prepared catalysts in the strongly selective and sensitive nonenzymatic glucose sensor utilizations.

12.2.1.2 Progesterone

As far as we know, progesterone has been proposed as one of the natural steroids and essential hormones for stabilizing and maintaining pregnancy [28–30]. According to studies on mammals, progesterone is formed from cholesterol that exists basically in the ovaries, adrenal glands, and placenta wherein it is synthesized following ovulation [31]. Additionally, it is considered a precursor of other essential hormones, that is, estrogen and testosterone in the adrenal glands that influence their activities. Moreover, premature birth has presented challenges in perinatology because it is a key cause of infant death or risk of severe medical conditions. Furthermore, universal statistics show that the number of premature births ranged from 5% to 10%; premature births have recently increased principally because of pregnancy in older women, utilization of the auxiliary reproductive procedures, greater body weight index, and diverse infections. Hence, progesterone contributes importantly to maintaining a pregnancy by suppressing the activities of the uterus muscles, otherwise actuated via prostaglandin and oxytocin, which may stimulate a premature birth. Investigations have indicated that control of the progesterone level in the pregnant women provides invaluable data to prognosticate the development of the normal embryo early in the pregnancy [32]. Hence, developing new analytical procedures for progesterone detection is of great importance in clinical applications.

A simplified, stable, sensitive, and reproducible electrochemical sensing platform for progesterone based on GCE modified with graphene quantum dots (GQDs), Fe_3O_4 nanoparticles (NPs), as well as functionalized multiwalled carbon nanotubes (f-MWCNTs) (Fe_3O_4 NPs@GQDs/f-MWCNTs/GCE) was constructed for electrochemical analysis [33]. Electrochemical examinations indicated that the as-prepared sensor has several benefits concerning highly efficient surface areas, more reactive sites, and very good electrochemical catalytic activities for oxidizing progesterone. Upon analytical condition optimization, the peak current for progesterone linearly varied with its concentration in a broad range from 0.01 to 0.5 and 0.5 to 3.0 μM. Moreover, the sensitivity and estimated LOD of the electrode equaled 16.84 $\mu A\ \mu M^{-1}$ and 2.18 nM. The sensor was satisfactorily utilized to determine progesterone in humans' sera specimens and drug products with very good recovery from any interference in the coexisting materials.

12.2.1.3 Creatinine

Out of the several health indicating biomarkers, creatinine has been regarded as a significant diagnostic substance in biological fluids. Therefore creatinine analysis may be utilized to diagnose thyroid, renal, and muscle malfunction [34], in biomedical diagnoses of acute myocardial infarction, as well as for quantitative characterization of hemodialysis treatment [35]. According to the analysis, normal physiological concentration ranges for creatinine in human blood equaled 40–150 μM, and a physiological value of 27 μM suggests severe renal injuries, which eventually result in dialysis or transplantation [36]. Therefore, in clinical diagnosis, it is very important to monitor creatinine levels in the human body.

Wen's group fabricated a new electrochemical sensing platform to sensitively detect creatinine in biological fluids based on magnetic field–induced self-assembly of Fe_3O_4@Polyaniline NPs (Fe_3O_4@PANI NPs) [37]. The template molecule, creatinine, was self-assembled over the surface of Fe_3O_4@PANI NPs with the functional monomer aniline by the formation of N–H hydrogen bonds. Following preassembly via the magnetic-induction of GCE, the ordered structure of the molecularly imprinted polymers (MIPs) were established by electropolymerization and assembled on the GCE surface with magnetic fields by a simplified one-step strategy. Moreover, the structural controllability of MIPs film fixed by the magnetic field–induced self-assembly was examined, and prepared MIEs showed acceptable reproducibility and stability to determine creatinine with 0.35 nM LOD. Fig. 12.1 demonstrates the construction path of a molecularly imprinted electrochemical sensor (MIES). Additionally, the strongly selective and sensitive MIES were substantially utilized to clinically detect creatinine in human plasma and urine specimens. Finally, average recovery equaled 90.8% to 104.9% with a relative standard deviation (RSD) of $<2.7\%$.

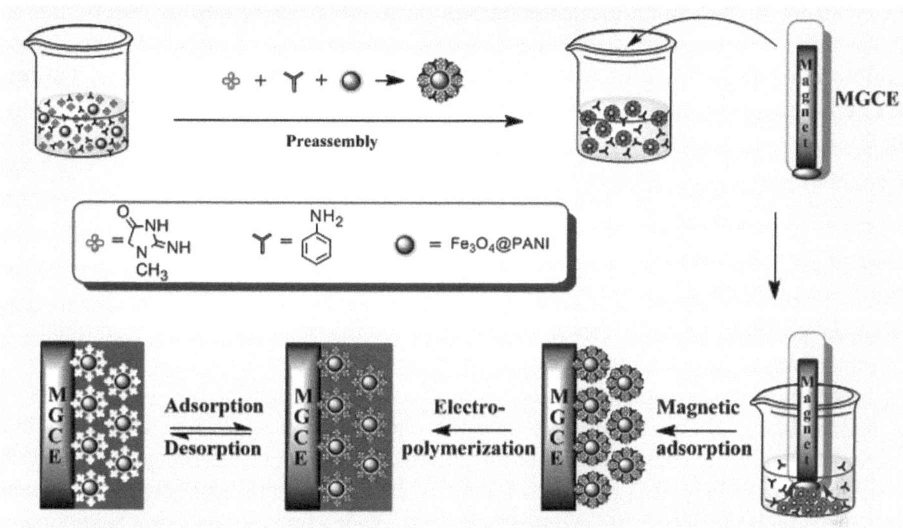

Figure 12.1 Schema to prepare MIES.
Source: Reused with permission from T. Wen, W. Zhu, C. Xue, J. Wu, Q. Han, X. Wang, et al., Novel electrochemical sensing platform based on magnetic field-induced self-assembly of Fe_3O_4@ Polyaniline nanoparticles for clinical detection of creatinine, Biosens. Bioelectron. 56 (2014) 180–185.

12.2.1.4 Uric acid

According to the studies, 2,6,8-trihydroxypurine or uric acid has been proposed as the main end product of purine metabolism that occurs in specimens of the blood and urine of humans, bird, and other animals. Therefore the human kidneys and intestinal tract excrete the generated uric acid with no extra metabolism. Moreover, the generation/excretion balance controls serum uric acid concentrations. Abnormal levels of uric acid essentially indicate the necessity for conventional clinical determination of diverse conditions such as renal insufficiency, cardiovascular illnesses, hyperuricemia or Lesch–Nyhan syndrome, arthritis, neurological diseases, hypertension, gout, and insulin resistance [38], stressing the clinical values of an exact detection of uric acid in diverse biological specimens.

Ni-doped Fe_2O_3 nanorods were synthesized by coprecipitation followed by the thermal decomposition method. Also, Ni–Fe_2O_3 nanorods have been utilized for fabricating an electrochemical sensor (Ni–Fe_2O_3/GCE) to detect uric acid in the biological condition using CV and chronoamperometry (CA) [39]. Results indicated the best responses of 5% Ni–Fe_2O_3/GCE to uric acid with less positive potential and greater current responses. Moreover, this new sensor gave acceptable linear current responses in concentration ranges between 6.6 and 112.4 μM with greater sensitivity equal to 0.849 mA mM^{-1} cm^{-2}.

In another work, Arvand et al. prepared a novel type of nanocomposite based on multiwalled carbon nanotubes (MWCNTs) decorated with the magnetic core–shell Fe_3O_4@SiO_2 NPs (Fe_3O_4@SiO_2/MWCNTs). The Fe_3O_4@SiO_2/MWCNTs were employed for fabricating a modified carbon paste electrode (CPE) for electrochemical sensing of uric acid [40]. Electrochemical behaviors of uric acid have been examined on the $Fe_3O_4@SiO_2$/MWCNTs-CPE via CV and square wave voltammetry (SWV) in phosphate buffer solution (PBS) of pH = 6.0. Based on the optimal condition, peak current linearly augmented with uric acid concentration ranges between 0.6 and 100.0 μM, with 0.13 μM LOD. Finally, this new sensor has been satisfactorily utilized to determine uric acid in biological fluids.

12.2.1.5 Hydrogen peroxide

Hydrogen peroxide (H_2O_2), one of the frequent intermediate substances, contributes to several enzymatic procedures, involves essentially in diverse biological procedures so that it is very necessary for investigations on the clinical diagnosis and life sciences. In the human body, H_2O_2 is produced by endogenous enzymes, as well as the electron transfer system of the cells for fighting infections via removal of the detrimental bacteria, fungi, and viruses [41]. It is also used for the regulation of body procedures like protein turnover, protein metabolism, carbohydrate, cell cycle, fats, and so on. Additionally, H_2O_2 has been considered to be a key by-product of diverse oxidative metabolic paths in animal cells. However, at a lower concentration, it is partly poisonous and contributes significantly as a biochemistry mediator of cellular pathology in the signaling processes associated with progressive neurodegenerative illnesses like Parkinson's disease. Nonetheless, at increased concentrations, H_2O_2 can be hazardous. It can easily lead to damage of cellular constituents, deoxyribonucleic acid (DNA), proteins, carbohydrates, as well as lipids and lead to gene mutation, diverse diseases, and possibly cancers. Moreover, H_2O_2 dangers largely are the result of fast conversion into the indiscriminately reactive hydroxyl radical (OH°) via interaction with some transition metal ions in vivo, for example, iron, or by exposure to the ultraviolet light [42].

A new electrochemical procedure to detect H_2O_2 based on Fe_3O_4 NPs with graphene oxide (GO) and polyamidoamine (PAMAM) dendrimer (Fe_3O_4/GO/PAMAM) was developed by Yang and coworkers [43]. The Fe_3O_4/GO/PAMAM was modified over the Au electrode surface as the working electrode. Therefore the detection of H_2O_2 was studied by the method of amperometric *i*/*t* curve. At the optimum experimental condition, it was possible to determine H_2O_2 in the linear calibration ranges between 2.0×10^{-5} and 1.0×10^{-3} M with the correlation coefficient equal to 0.9950 ($n = 10$), LOD equaled 2.0×10^{-6} M, and recovery ratio equaled 96.9%−108.1%, indicating that the method has very good accuracy. Moreover, the modified electrodes displayed very good electrochemical function, higher sensitivity, acceptable reproducibility, and prolonged stability.

Benvidi's group designed a strongly sensitive voltammetric sensor based on a CPE with a nanocomposite of rGO and $CuFe_2O_4$ NPs (rGO/$CuFe_2O_4$/CPE) to detect H_2O_2 [44]. The construction procedure of the rGO/$CuFe_2O_4$/CPE sensor is shown in Fig. 12.2. Based on the optimal condition of pH = 5, the modified electrode revealed a quick amperometric response of <2 seconds, acceptable linear ranges between 2.0 and 200.0 μM, lower LOD equal to 0.52 μM to determine H_2O_2. Moreover, the peak current of H_2O_2 differential pulse voltammetry (DPV) increased linearly with its concentration in

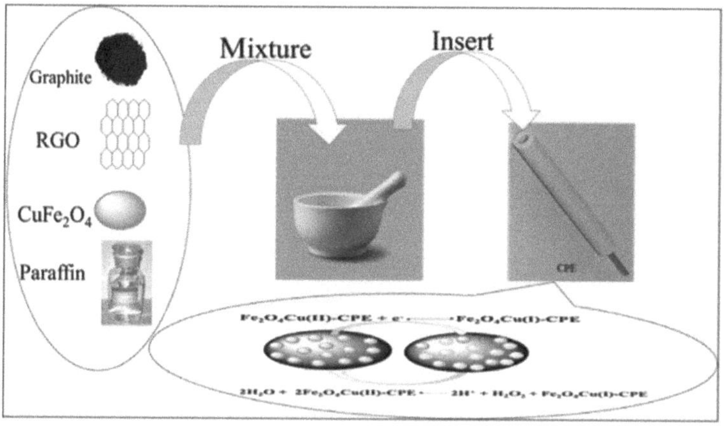

Figure 12.2 Schema of the modified electrochemical sensor based on the rGO/$CuFe_2O_4$/CPE platform.
Source: Reused with permission from A. Benvidi, M.T. Nafar, S. Jahanbani, M.D. Tezerjani, M. Rezaeinasab, S. Dalirnasab, Developing an electrochemical sensor based on a carbon paste electrode modified with nano-composite of reduced graphene oxide and $CuFe_2O_4$ nanoparticles for determination of hydrogen peroxide, Mater. Sci. Eng. C 75 (2017) 1435–1447.

a range between 2.0 and 10.0 μM, as well as between 10 and 1000 μM. Also, H_2O_2 LOD has been assessed as 0.064 μM by DPV. Finally, this new sensor has been satisfactorily utilized to assay H_2O_2 in the drug and biological specimens like milk, hair dye cream, mouthwash solution, and green tea.

Mollarasouli and coworkers introduced a new strategy for developing a nonenzymatic H_2O_2 sensor based on $CoFe_2O_4$@CdSe quantum dots (QDs) magnetic nanocomposite and rifampicin (RIF) mediator–modified GCE ($CoFe_2O_4$@CdSeQDs/RIF/GCE). $CoFe_2O_4$@CdSeQD magnetic nanocomposite was synthesized by a chemical coprecipitation procedure and is described by various techniques [45]. To prepare the nonenzymatic H_2O_2 sensor, first, the GCE surface was modified via dropping 10 μL of 5 mg mL^{-1} $CoFe_2O_4$@CdSeQD magnetic nanocomposite. In the next step, RIF was electrodeposited on the actuated $CoFe_2O_4$@CdSeQDs/GCE via using −0.7 V potential for 400 seconds in a pH 2.0 PBS consisting of 190 μM of RIF. Also, CV and electrochemical impedance spectroscopy (EIS) have been utilized for investigating the electrochemical behaviors of the sensor and have been utilized to reduce H_2O_2. Calibration plots for H_2O_2 have been drawn by an amperometric procedure (−0.2 V vs Ag/AgCl) at the modified electrode. Also, two linearity ranges have been found between 7.0 and 145.0 and 145.0 μM to 1.43 mM with 143.01 and 28.67 μA mM^{-1} sensitivities for the first and second linearity ranges. Moreover, LOD equaled 0.38 μM, and the reliability of the nanosensor verified the real sample analyses in diverse beverages like juice and milk with reasonable recovery outputs. Table 12.1 lists applications of MNMs-based electrochemical sensors for clinical diagnosis.

Table 12.1 Summary of the utilization of MNMs-based electrochemical sensors for clinical diagnosis.

Electrochemical sensor	Method	Analyte	LOD	Linear range	Ref.
$ZnFe_2O_4$/rGO/GCE	Amperometry	Glucose	1.2 μM	0.1–12.5 mM	[26]
Ni-Co/Fe_3O_4/GCE	Amperometry	Glucose	0.19 μM	0.001–11.0 mM	[27]
Fe_3O_4@GQDs/f-MWCNTs/GCE	DPV	Progesterone	2.18 nM	0.01–3.0 μM	[33]
Fe_3O_4@PANI/GCE	DPV	Creatinine	0.35 nM	0.02–1.0 μM	[37]
Ni–Fe_2O_3/GCE	Chronoamperometry	Uric acid	3.1 mM	6.6–112.4 μM	[39]
Fe_3O_4@SiO_2/MWCNTs-CPE	SWV	Uric acid	0.13 μM	0.6–100.0 μM	[40]
Fe_3O_4/GO–PAMAM/Au	Amperometry	H_2O_2	2.0 μM	20–1000 μM	[43]
rGO/$CuFe_2O_4$/CPE	Amperometry	H_2O_2	0.52 μM	2.0–200.0 μM	[44]
	DPV		0.064 μM	2.0–100.0 μM	
$CoFe_2O_4$@CdSeQDs/RIF/GCE	Amperometry	H_2O_2	0.38 μM	7.0 μM–1.43 mM	[45]

12.2.2 Application in bioanalysis and pharmaceutical compounds analysis

The detection of biological and pharmaceutical compounds plays an important role in public health. Hence, perceived and enhancing demands exist for simple, cheap, and quick analytical experiments to detect trace concentrations of biologically and pharmaceutically prominent compounds [46]. MNMs-based electrochemical sensors have been considerably investigated to sensitively detect diverse drug and biological compounds.

12.2.2.1 Neurotransmitters

It is widely accepted that neurotransmitters are crucial chemicals to possibly transfer information between neuron cells. In this regard, Otto Loewi was among the first scholars who discovered a kind of neurotransmitter known as acetylcholine in 1921 [47]. It is notable that his investigation considerably influenced human perceptions of how the information is transmitted in animals. Since then other kinds of neurotransmitters have been discovered so that more than 100 distinct kinds have been identified. Nearly every neurotransmitter has been considered to be crucial for human physical and mental health, and thus any abnormality or change in their activities might result in very serious diseases and mental disorders. For this reason, it is of high importance to monitor neurotransmitters for medical treatments and clinical analyses.

A GCE modified with nickel-ferrite MNPs and decorated with MWCNTs ($NiFe_2O_4$–MWCNTs/GCE) as a sensitive and selective electrochemical sensor to determine epinephrine with DPV [48]. Therefore, at the optimized pH 6.0, the electrode response in 0.1 M PBS was proportionate to epinephrine concentration in ranges between 0.9 and 800.0 μM with 0.09 μM LOD. Finally, this modified electrode has reasonable reproducibility to detect epinephrine in pharmaceutical and body fluids.

Zhang and coworkers fabricated a sensitive simplistic dopamine electrochemical sensor based on the magnetism-assisted modification of screen-printed electrode (SPE) with the magnetic multiwalled carbon nanotubes (mMWCNTs) [49]. The synthetic route of mMWCNTs and the modification process for the SPE with redox mechanisms of dopamine is shown in Fig. 12.3A and B, respectively. The linearity was found between 5.0 and 180.0 μM, with a 0.43 μM LOD. Moreover, this modified electrode showed very good selectivity for detecting dopamine with nearly no interference from the ascorbic acid that coexisted with dopamine in several biosamples and that caused common interferences. Ultimately, researchers utilized the new electrode for determining dopamine concentrations in the spiked human blood serum, and reasonable recovery was obtained in ranges between 97.43% and 102.94% with RSDs <2.27%.

Ran's group fabricated a serotonin sensor using a Fe_3O_4-MWCNTs-poly(bromocresol green) [Fe_3O_4-MWCNT-poly(BCG)] nanocomposite film [50]. This sensor exhibited enhanced DPV responses to serotonin in the presence of dopamine. Its peak current (at an oxidation voltage equal to 0.36 V vs Ag/AgCl) was linearly enhanced in the serotonin concentration ranges 0.5–100.0 mM with 80.0 nM LOD. It was employed for determining the serotonin spiked into human serum specimens.

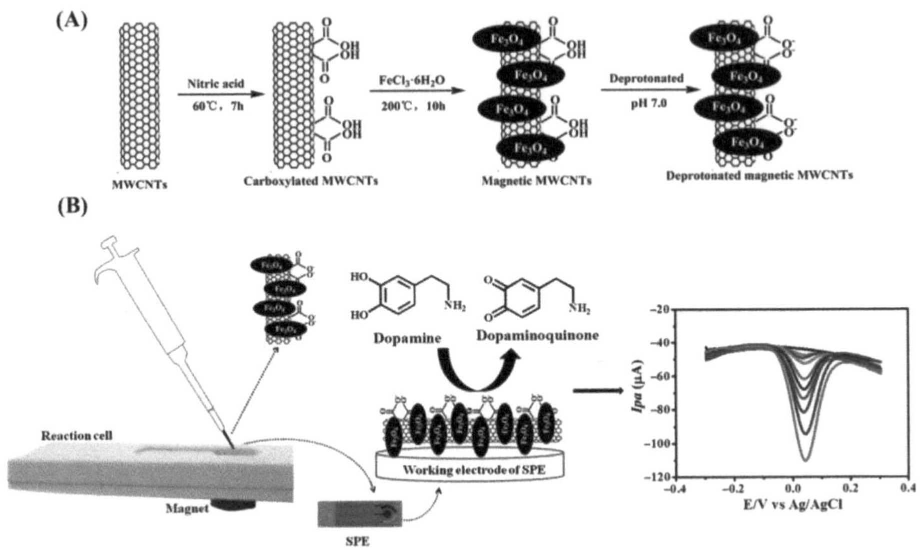

Figure 12.3 (A) Synthetic path of mMWCNTs. (B) Magnetism-assisted modification of SPE with mMWCNTs as well as the electrochemical response of dopamine on the mMWCNTs/SPE.
Source: Reused with permission from Y.M. Zhang, P.L. Xu, Q. Zeng, Y.M. Liu, X. Liao, M. F. Hou, Magnetism-assisted modification of screen printed electrode with magnetic multi-walled carbon nanotubes for electrochemical determination of dopamine, Mater. Sci. Eng. C 74 (2017) 62–69.

Queiroz et al. described the electrochemical procurement of electrocatalytic oxidation/reduction of noradrenaline in modified GCE of cobalt ferrite NPs and MWCNT (MWCNT/FCo_{98}/GCE) [51]. The researchers showed the optimal conditions in an electrode composition of 4 μL of cobalt ferrite and 10 μL of MWCNTs in 0.1 M PBS at a pH equal to 7.0. Therefore the electrode displayed electrochemical behaviors in a broad potential range between −0.4 and 1.0 V versus Ag/AgCl, higher conductivity as well as electrode stability/durability in 0.1 M PBS. Moreover, the catalytic oxidation of noradrenaline was done at an unmodified GCE electrode at +0.60 V versus Ag/AgCl and 0.17 μA current and modified GCE with cobalt ferrite NPs and MWCNTs at +0.54 V versus Ag/AgCl and 0.23 mA current. Considering anodic peak current (I_{pa}) versus the noradrenaline concentration with the amperometric procedure at the modified electrode that was linear in the 0.16−1.91 mM concentration range, LOD equaled 0.76 μM.

12.2.2.2 Amino acids

It is well-known that amino acids have been introduced as a class of significant biomolecules containing the carboxylate and amino groups. An amino acid may be essential, which must be directly obtained from food, or nonessential, which can be produced by the body. They are building blocks of vital proteins and peptides, neurotransmitters, nourishment, poison, antibiotics, and transporters [52]. Therefore analyzing amino acids contributes importantly to the biochemical, drug, and biomedical areas.

Hasanzadeh and coworkers reported the mesoporous silica mobile crystalline material-41 (MCM-41)-Fe_2O_3 as a new magnetic nanosensor to determine the trace amounts of amino acids [53]. The MCM-41-Fe_2O_3 modified GCE (MCM-41-Fe_2O_3/GCE) exhibited potential electrocatalytic activities for the amino acid electrooxidation. Using the amperometry technique, LODs of 56, 145, 94, 53, 84, 112, 85, and 81 nM were obtained for cysteine, glycine, tyrosine, serine, valine, L-arginine, L-phenylalanine, and L-tryptophan, respectively.

Hasanzadeh's group addressed synthesizing and characterizing a new nanocomposite, Fe_3O_4 MNPs/GQDs, to sense some amino acids [54]. Therefore the as-synthesized GQDs and Fe_3O_4 MNPs-GQDs were electrodeposited on the GCE for the first time using the CV in potential ranges between −1.0 and 1.0 V. Moreover, Fe_3O_4 MNP-GQDs was engineered to specifically and efficiently capture and augment electrochemical signals of some amino acids at the physiological pH because of the synergy between GQDs and MNPs, according to the calibration curve for amino acids in PBS at a pH of 7.4. Also, limit of quantifications (LOQs) equaled 0.01 μM for L-cysteine, 0.09 μM for L-tyrosine, 1 μM for L-aspartic acid, and 0.5 μM for L-phenylalanine.

Atacan et al. devised a new electrochemical sensor based on the GCE modified by $CuFe_2O_4$/rGO nanocomposite decorated with gold NPs for the electrochemical detection of L-cysteine [55]. Electrochemical characterizations of the bare and modified GCEs were carried out via EIS and CV. Also, modification of GCE increased the surface area. Besides, $CuFe_2O_4$/rGO−Au demonstrated highly effective electrocatalytic activities to oxidize L-cysteine in 0.1 M PBS (pH = 6.5), and the higher sensitivity equaled 100.01 $\mu A\ mM^{-1}\ cm^{-2}$. Also, the electrocatalytic current was linearly enhanced with L-cysteine concentrations in ranges between 0.05 and 0.2 mM with 0.383 μM LOD.

12.2.2.3 Drugs

Based on studies, drug analysis is performed in the course of diverse stages of drug development for such reasons as analytical examination of the bulk drug substances, medicine formulation, impurity, and degradation products of the medicine materials for controlling quality in the clinical trials, treatment medicine monitoring, as well as cases of the medicine abuse [56−60]. Therefore these quantitative and qualitative analyses require valid and reliable analytical procedures for determining the medicines in real specimens like drug formulations and human body fluids.

Afkhami and coworkers developed an efficient electrochemical sensor for the quick and simultaneous detection of acetaminophen and tramadol based on CPE modified with $NiFe_2O_4$/Gr NPs (Gr/$NiFe_2O_4$) [61]. According to the analysis, the peak current of SWV of acetaminophen and tramadol linearly increased with their concentration ranges between 0.01 and 9 μM. Moreover, LOD equaled 0.0036 and

0.0030 μM, reflecting higher sensitivity and reasonable stability of this sensor for clinical assays of acetaminophen and tramadol.

Afkhami's group also showed simultaneous electrochemical detection of acetaminophen and codeine in drug specimens and biological fluids via the graphene–$CoFe_2O_4$ nanocomposite–modified CPE (Gr/$CoFe_2O_4$/CPE) [62]. The linearity range was observed between 0.03 and 12.0 μM for acetaminophen and codeine. Low LODs of 0.025 μM for acetaminophen and 0.011 μM for codeine were obtained based on three times the standard deviation of the blank over sensitivity. Moreover, this new procedure was free from interference impacts of ascorbic acid, glucose, naproxen, caffeine, phenylalanine, alanine, glycine, etc. Any electrode surface fouling was not seen in the course of the successive scans. Also, its functional utility was reliable and favorable in biological fluid and drug specimen analyses.

Hasanzadeh's group proposed a sensitive, inexpensive, quick, and selective procedure for simultaneously determine amiodarone and atenolol based on electrocatalytic oxidation of these analytes on the magnetic (Fe_2O_3) MCM-41 grafted with 3-aminopropyl groups ($nPrNH_2$) modified CPE (Fe_2O_3-MCM-41-$nPrNH_2$-CPE) [63]. The measurements have been done using DPV in Briton–Robinson buffer (BR-buffer) so that pH equaled 4.00. Based on DPVs, the linear calibration range for detecting analytes have been between 8–205 μM for atenolol and 4.5–710 μM for amiodarone with LOD equal to 1.4 and 2.1 μM, respectively.

Haghshenas et al. made a simplified and novel electrochemical sensor for voltammetric detection of acetaminophen with the use of Au NPs@Fe_3O_4/CPE. The quantitative analysis of acetaminophen was investigated by DPV [64]. The anodic peak current was linear to acetaminophen concentration in ranges between 0.1 and 70.0 μM with the LOD 4.5×10^{-2} μM. Therefore the method was satisfactorily utilized for detecting the concentration of acetaminophen in drug formulations and human sera specimens.

Fernandes's group prepared a GCE modified with the new N-doped CNTs (CTN-N) functionalized with $MnFe_2O_4$ NPs ($MnFe_2O_4$@CNT-N) to electrochemically determine acetaminophen, ascorbic acid, and caffeine [65]. In comparison with the bare CNT-N/GCE and GCE, the $MnFe_2O_4$@CNT-N/GCE may notably ameliorate electrocatalytic activities for oxidizing caffeine, acetaminophen, and ascorbic acid with the increased anodic peak current equal to 52%, 50%, and 55%. Moreover, the SWV anodic peak of the molecules may be differentiated from one another at $MnFe_2O_4$@CNT-N/GCE with the elevated oxidation current. Moreover, LODs for SWV detection of caffeine, acetaminophen, and ascorbic acid equaled 0.83×10^{-6}, 0.83×10^{-6}, and 1.8×10^{-6} mol dm^{-3}, respectively.

Arvand et al. described the development and application of a nanocomposite with GQDs, Fe_3O_4 NPs, and f-MWCNTs for GCE surface modification (Fe_3O_4@GQDs/f-MWCNTs/GCE) as a sensing platform for electrochemically determining levodopa in agricultural products [66]. Using DPV, the prepared sensor was applied to determine levodopa in range 3.0–400.0 μM with 14.3 nM LOD. The amount of levodopa in sunflower seed, pumpkin seed, fava, sesame seed, and bean seed was assessed by this new sensor suggested by researchers.

Madrakian and coworkers reported an electrochemical sensor to detect rizatriptan benzoate with the use of the Fe_3O_4/MWCNTs/GCE [67]. The Fe_3O_4/MWCNTs/GCE showed very high electrocatalytic activities to rizatriptan benzoate oxidation in comparison to the MWCNTs/GCE, Fe_3O_4/GCE, and bare GCE. Moreover, SWV responses of the electrode to rizatriptan benzoate linearly ranged between 0.5 and 100.0 μM with 0.09 μM LOD. The suggested sensor has been satisfactorily utilized for real-life specimens of blood serum and rizatriptan benzoate detection in drugs.

Fallah Shojaei's group fabricated a CPE modified in situ with $ZnFe_2O_4$ MNPs and 1,3-dipropylimidazolium bromide ionic liquid (IL) ($ZnFe_2O_4$/IL/CPE) to detect 5-fluorouracile anticancer drug in medicine specimens [68]. Moreover, electrooxidation peak currents for 5-fluorouracile linearly changed with its concentration in ranges between 0.1 and 1400 μM with SWV. Furthermore, $ZnFe_2O_4$/IL/CPE has been satisfactorily utilized to analyze 5-fluorouracile in drug and urine specimens.

Tammari et al. reported the synthesis and use of a novel magnetic nanocomposite, containing β-alanine (Ala), Pd, and Fe_3O_4 (Fe_3O_4/Ala/Pd) as a strongly sensitive sensor to detect clozapine at the biological pH of 7.2 with the use of GCE [69]. The correlation of the peak current with the concentration of clozapine was linearly ranged between 3 and 70 nM with 1.53 nM LOD via DPV. Therefore this modified electrode showed acceptable sensitivity, reproducibility, and stability. Also, this new sensor was satisfactorily utilized in determining clozapine in clozapine tablets and human sera.

Ensafi and coworkers made a GCE modified with $CoFe_2O_4$/rGO/IL. The modified GCE was utilized for electrochemical detection of methotrexate [70]. The outputs of their examination indicated that the $CoFe_2O_4$/rGO/IL nanocomposite had a synergetic impact on the methotrexate oxidation via diminishing oxidation overpotential and enhanced oxidation peak currents. With the use of this electrochemical sensor, it is possible to measure methotrexate in a range between 0.05 and 7.5 nmol mL^{-1} with 0.01 nmol mL^{-1} as the LOD. Moreover, the RSD for five replicate measurement of 2.0 nmol mL^{-1} methotrexate equaled 4.5%, and the utility and selectivity of this new electrochemical sensor toward methotrexate were examined by analyzing the drug tablets with hopeful outputs.

Afkhami's group applied a nickel-zinc ferrite/Gr nanocomposite ($Ni_{0.5}Zn_{0.5}Fe_2O_4$/Gr) for the determination of omeprazole based on modified GCE [71]. They showed a linear response in a range from 0.03 to 100.0 μM and a low LOD of 0.015 μM. Moreover, via measurement of omeprazole concentrations in the plasma and drug specimens, the functional utilization of the modified electrode was assessed.

Cincotto et al. reported simultaneous electrochemical determination of mood disorder–associated materials like melatonin, tryptophan, and amitriptyline with the use of a new nanomagnetic electrochemical sensor design, entailing Fe_3O_4 NPs decorated with carbon quantum dots (Fe_3O_4/CQDs) [72]. Moreover, amitriptyline, tryptophan, and melatonin were determined by monitoring the oxidation promoted by Fe_3O_4/CQDs/GCE in BR-buffer at a pH of 3.0 that was done according to the organized DPV peaks with LODs of 5.9, 4.4, and 4.2 nM. According to the analyses, considerable interferences were observed from the biological interferents like ascorbic acid, uric acid, estriol, 17β-estradiol, and dopamine.

Amir and coworkers demonstrated the potent utilization of caffeic acid (CA)–functionalized MNPs (CA-Fe_3O_4 NPs) as an efficient modifier to electrochemically oxidize 6-thioguanine drug [73]. The presence of a desirable interaction of the active functional moieties of CA with 6-thioguanine synergized with a larger surface area of MNPs produced a stable electrooxidation signal in the working range between 0.01 and 0.23 μM with sensitivity equal to 0.001 μM. Moreover, the sensor exhibited strong antiinterference potency against conventional coexisting medicine molecules like acetaminophen, benzoic acid, norepinephrine, epinephrine, ascorbic acid, L-cysteine, and glucose. Additionally, 6-thioguanine from the commercial tablets was successfully quantified from the local pharmacy, reflecting the functional capabilities of this new sensor as suggested by the researchers.

Beitollahi's group developed a highly sensitive electrochemical nanostructure sensor based on the GO/Fe_3O_4@SiO_2 nanocomposite modified SPE (GO/Fe_3O_4@SiO_2/SPE) to trace analyze acetaminophen [74]. The electrochemical examination of the modified electrode and its efficiency for simultaneous voltammetric oxidize tryptophan and acetaminophen was investigated. In comparison to the bare SPE, GO/Fe_3O_4@SiO_2/SPE showed very good electrocatalytic activities toward acetaminophen oxidation. Hence, the plot of the catalytic current versus acetaminophen concentrations indicated a linear segment in the concentration ranges between 0.5 and 100.0 μM. Moreover, the LOD of 0.1 μM was found with the use of the calibration plot. Also, GO/Fe_3O_4@SiO_2/SPE showed successful application and validation via the analysis of tryptophan and acetaminophen in the specimens of urine and drug.

Ghapanvari et al. used a novel CPE modified with Fe_3O_4 NPs supported on the MWCNTs and polyacrylonitrile nanofibers (Fe_3O_4@MWCNTs@PANNFs) for the determination of imatinib, as an important anticancer drug [75]. Due to its very good electrocatalytic behavior of Fe_3O_4@MWCNTs@PANNFs/CPE, the electrochemical oxidation of imatinib is facilitated. The anodic peak current of imatinib linearly increased within concentration ranges of 0.0017–0.8500 μM, along with an LOD of 0.4 nM. Finally, the Fe_3O_4@MWCNTs@PANNFs/CPE was employed for electrochemical detection of imatinib in the urine specimens.

Khalilzadeh and coworkers constructed an SPE sensor based on the Fe_3O_4@cellulose nanocrystals/Cu nanocomposite (Fe_3O_4@CNC/Cu) to sensitively detect venlafaxine [76]. Moreover, a Fe_3O_4@CNC/Cu nanocomposite was procured via a green procedure with the use of a plant extract (*Petasites hybridus* leaf) as a stabilizing and reducing agent for copper. Fe_3O_4@CNC/Cu was applied to the modification of a graphite screen-printed electrode (GSPE) to analyze venlafaxine concentrations. According to the tests, their sensor had acceptable functions to electrochemically detect venlafaxine. The results showed linear dynamic ranges for venlafaxine concentration via the modified electrode was 0.05–600.0 μM, with 0.01 μM LOD. Additionally, this technique was exhibited as a simplistic, selective, and accurate sensor for identifying venlafaxine in the specimens of urine, drug, and water formulations.

12.2.2.4 Purine bases in DNA

Studies demonstrated the storage of genetic information through DNA, which is crucial to biosynthesize protein. Moreover, purines have been considered to be the building blocks of DNA with an essential contribution to the storage of genetic information and the cell signaling process. Also, purine nucleobases such as guanine and adenine contribute importantly to biological procedures [77,78]. They extensively play a role in diverse diseases; a majority of them have a very high prominence in the medical and clinical practices, and therefore their presence and content in biological fluids and drugs should be cautiously specified and quantified.

Ensafi's group devised an electrochemical sensor for adenine and guanine in DNA determination with the use of the MWCNTs/$NiFe_2O_4$/GCE [79]. Incorporation of a MWCNTs/$NiFe_2O_4$ nanohybrid over the electrode surface considerably increased oxidation peak currents and reduced the peaks' potential of adenine and guanine. Moreover, researchers utilized the modified electrode to detect purine bases with the help of the linear sweep voltammetry (LSV) as along with hydrodynamic amperometry. Using the hydrodynamic amperometry, they displayed peak current of adenine and guanine as the linear dependence on their concentration ranged between 3.0 and 45.0 μM for guanine and 3.0 and 40.0 μM for adenine. In the case of the use of LSV, peak current linearly depended on guanine in concentration ranges between 0.05 and 3.0 μM and on adenine in concentration ranges between 0.1 and 4.0 μM. Moreover, LOD (S/N = 3) equaled 0.01 and 0.006 μM for adenine and guanine, and LSV was utilized to simultaneously determine guanine and adenine. In the end, this new electrochemical sensor was applied for determining adenine and guanine in single-strand deoxyribonucleic acid (ssDNA) specimens with hopeful outputs.

Yari et al. reported an easy strategy for preparing a MWCNT-Fe_3O_4@Poly dopamine (PDA)-Ag nanocomposite, which may function as a new sensing element of an electrochemical sensor to detect the guanine and adenine content of DNA [80]. Fig. 12.4 depicts the processes in preparing MWCNT-Fe_3O_4@PDA-Ag nanocomposite. The sensor demonstrated a very good electrocatalytic function toward determining adenine and guanine. Results showed the linearity of the oxidation peak currents in a concentration range from 8 to 130 and 10 to 120 μM for guanine and adenine. Moreover, lower LODs of the sensor equaled 5.66 μM for adenine and 1.47 μM for guanine. It was found that MWCNT-Fe_3O_4@PDA-Ag nanocomposite–based sensor was satisfactorily utilized for detecting guanine and adenine in specimens of fish sperm DNA.

Table 12.2 lists applications of MNMs-based electrochemical sensors for bioanalysis and pharmaceutical compounds analysis.

12.2.3 Application in food analysis and environmental monitoring

Food may be polluted by various chemical compounds, resulting in diverse food-borne ailments. According to the World Health Organization (WHO), food-borne and water-borne diarrheal illnesses kill numerous people annually, of whom many are considered to be children [81]. Therefore such an ever increasing public health problem requires developing strongly sensitive technologies to rapidly determine food contaminants. Hence, MNMs-based electrochemical sensors possess a potential for emerging as a key tool for maintenance of the food safety.

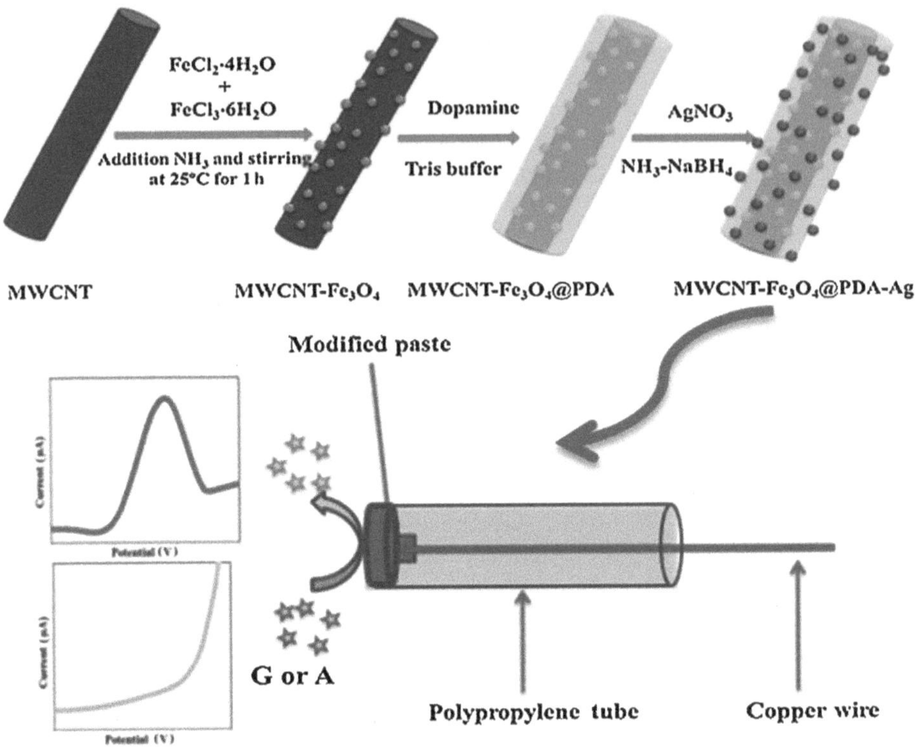

Figure 12.4 Schema to prepare MWCNT-Fe_3O_4@PDA-Ag nanocomposite.
Source: Reused with permission from A. Yari, S. Derki, New MWCNT-Fe_3O_4@PDA-Ag nanocomposite as a novel sensing element of an electrochemical sensor for determination of guanine and adenine contents of DNA, Sens. Actuators B: Chem. 227 (2016) 456–466.

12.2.3.1 Heavy metals

Increased industrialization and urbanization are thought to have caused an increased range of metals in the environment, specifically heavy metals [82]. Heavy metal ions (e.g., those of Ag^+, Cu^{2+}, Pb^{2+}, Zn^{2+}, Cd^{2+}, etc.), and semimetals (e.g., As) are highly toxic and may cause serious damage to health. The level of toxicity depends on the kind of metal, its biological function, and the types of the organisms exposed to it. Hence the initial determination of heavy metal ions is crucial for protecting the environment and avoiding contamination.

Yang and coworkers demonstrated a simplified and selective magnetic electrochemical sensor to detect Ag^+ ions in aqueous solutions using Fe_3O_4@Au NPs [83]. The linear response ranges of Ag^+ ions equaled 0.117–17.7 μM ($R^2 = 0.9909$) with an LOD of 59 nM (S/N = 3). Moreover, some repeatability measurements with 1.0 μM Ag^+ provided reproducible outputs with an RSD equal to 4.5% ($n = 11$). Also, it was possible to eliminate the interference from other metal cations via the addition of coadditive (ethylenediaminetetraacetic acid) for masking metal cations. Recovery ranged between

Table 12.2 Summary of the utilization of MNMs-based electrochemical sensors for bioanalysis and pharmaceutical compounds analysis.

Electrochemical sensor	Method	Analyte	LOD	Linear range	Ref.
$NiFe_2O_4$-MWCNTs/GCE	DPV	Epinephrine	0.09 μM	0.9–800.0 μM	[48]
mMWCNTs/SPE	SWV	Dopamine	0.43 μM	5.0–180.0 μM	[49]
Fe_3O_4-MWCNT-poly(BCG)/GCE	DPV	Serotonin	80.0 nM	0.5–100.0 mM	
MWCNT/FCo_{98}/GCE	Amperometry	Noradrenaline	0.76 μM	0.16–1.91 mM	[50]
MCM-41-Fe_2O_3/GCE	Amperometry	Cysteine	56 nM	100–720 nM	[53]
		Glycine	145 nM	300–1000 nM	
		Tyrosine	94 nM	205–520 nM	
		Serine	53 nM	100–140 nM	
		Valine	84 nM	200–510 nM	
		L-Arginine	112 nM	390–850 nM	
		L-Phenylalanine	85 nM	140–270 nM	
		L-Tryptophan	81 nM	95–210 nM	
Fe_3O_4 MNPs-GQDs-GCE	DPV	L-Cysteine	–	0.01–100 μM	[54]
		L-Tyrosine	–	0.09–230 μM	
		L-Aspartic	–	1–50 μM	
		L-Phenylalanine	–	0.5–650 μM	
$CuFe_2O_4$/rGO–Au-GCE	CV	L-Cysteine	0.383 μM	0.05–0.2 mM	[55]
Gr/$NiFe_2O_4$/CPE	SWV	Tramadol	0.0036 μM	0.01–9 μM	[61]
		Acetaminophen	0.0030 μM	0.01–9 μM	
Gr/$CoFe_2O_4$/CPE	SWV	Acetaminophen	0.025 μM	0.03–12.0 μM	[62]
		Codeine	0.011 μM	0.03–12.0 μM	
Fe_2O_3-MCM-41-nPrNH_2-CPE	DPV	Amiodarone	2.1 μM	4.5–710 μM	[63]
		Atenolol	1.4 μM	8–205 μM	
Au NPs@Fe_3O_4/CPE	DPV	Acetaminophen	4.5×10^{-2} μM	0.1–70.0 μM	[64]
$MnFe_2O_4$@CNT-N/GCE	SWV	Caffeine	0.83×10^{-6} mol dm^{-3}	1–1100 mol dm^{-3}	[65]
		Acetaminophen	0.83×10^{-6} mol dm^{-3}	1–1000 mol dm^{-3}	
		Ascorbic acid	1.8×10^{-6} mol dm^{-3}	2–100 mol dm^{-3}	

(Continued)

Table 12.2 (Continued)

Electrochemical sensor	Method	Analyte	LOD	Linear range	Ref.
Fe_3O_4@GQDs/f-MWCNTs/GCE	DPV	Levodopa	14.3 nM	3.0–400.0 μM	[66]
Fe_3O_4/MWCNTs/GCE	SWV	Rizatriptan benzoate	0.09 μM	0.5–100.0 μM	[67]
$ZnFe_2O_4$/IL/CPE	SWV	5-Fluorouracile	0.07 μM	0.1–1400 μM	[68]
Fe_3O_4/Ala/Pd/GCE	DPV	Clozapine	1.53 nM	3–70 nM	[69]
$CoFe_2O_4$/rGO/IL/GCE	DPV	Methotrexate	0.01 nmol mL^{-1}	0.05–7.5 nmol mL^{-1}	[70]
$Ni_{0.5}Zn_{0.5}Fe_2O_4$/Gr/GCE	DPV	Omeprazole	0.015 μM	0.03–100.0 μM	[71]
Fe_3O_4/CQDs/GCE	DPV	Amitriptyline	5.9 nM	0.05–13.50 μM	[72]
		Melatonin	4.4 nM	0.05–13.50 μM	
		Tryptophan	4.2 nM	0.05–13.50 μM	
CA@Fe_3O_4/GCE	SWV	6-Thioguanine	0.001 μM	0.01–0.23 μM	[73]
GO/Fe_3O_4@SiO_2/SPE	DPV	Acetaminophen	0.1 μM	0.5–100.0 μM	[74]
Fe_3O_4@MWCNTs@PANNFs/CPE	DPV	Imatinib	0.4 nM	0.0017–0.8500 μM	[75]
Fe_3O_4@CNC/Cu/GSPE	DPV	Venlafaxine	0.01 μM	0.05–600.0 μM	[76]
MWCNTs/$NiFe_2O_4$/GCE	Amperometry	Guanine	0.012 μM	3.0–45.0 μM	[79]
	LSV		0.006 μM	0.05–3.0 μM	
	Amperometry	Adenine	0.08 μM	3.0–40.0 μM	
	LSV		0.01 μM	0.1–4.0 μM	
MWCNT-Fe_3O_4@PDA-Ag/CPE	DPV	Guanine	1.47 μM	8–130 μM	[80]
		Adenine	5.66 μM	10–120 μM	

98.6% and 103.99% following the standard addition, demonstrating that the sensor had strong potency in functional utilization.

Bagherzadeh's group reported an electrochemical detection of Cu^{2+} and Pb^{2+} ions with the use of diethylenetriamine pentaacetic acid (DTPA) functionalized dopaminated Fe_3O_4 (Fe_3O_4/DA) MNPs (Fe_3O_4/DA/DTPA) [84]. Analytical studies of the sensor showed two linear ranges from 1.0×10^{-9} to 5.0×10^{-6} M Pb^{2+} and from 1.0×10^{-5} to 1.0×10^{-3} M Pb^{2+} with an LOD of 7.8×10^{-10} M Pb^{2+}, and the linear range between 5.0×10^{-9} and 1.0×10^{-3} M Cu^{2+} with a 9.5×10^{-10} M Cu^{2+} LOD shown in separate determinations of Cu^{2+} and Pb^{2+}. Moreover, two linear ranges from 1.0×10^{-9} to 5.0×10^{-8} M Pb^{2+} and from 5.0×10^{-7} to 1.0×10^{-3} M Pb^{2+} with an LOD of 8.2×10^{-9} M Pb^{2+}, and a linear range between 5.0×10^{-9} and 1.0×10^{-4} M Cu^{2+} with a 2.1×10^{-9} M Cu^{2+} LOD obtained to simultaneously determine Cu^{2+} and Pb^{2+}.

Lee et al. fabricated a new and highly sensitive electrochemical sensor for Zn^{2+}, Cd^{2+}, and Pb^{2+} determination based on Gr supporting a Fe_2O_3 NP composite modified electrode [85]. The Fe_2O_3/Gr nanocomposite was synthesized from the iron–oleate precursor and Gr by a solventless thermal decomposition procedure. They applied differential pulse anodic stripping voltammetry (DPASV) to detect heavy metal ions. The electrode linearly ranged from 1 to 100 $\mu g\ L^{-1}$ for Zn^{2+}, Cd^{2+}, and Pb^{2+}, with LODs of 0.11, 0.08, and 0.07 $\mu g\ L^{-1}$ (S/N = 3). Additionally, this sensor was utilized to determine Zn^{2+}, Cd^{2+}, and Pb^{2+} in tap water specimens, and the output was compatible with the outputs via the inductively coupled plasma mass spectrometry (ICP-MS).

Chimezie and coworkers designed an electrochemical sensor for trace detection of As(III) in water resources using DPASV, on the rGO-Fe_3O_4/SPE [86]. Raman spectroscopy and EIS examination demonstrated that, although As(III) was reduced to As^0 in the course of the deposition step (−0.5 V, 300 seconds), the GO nanosheets were electrochemically reduced to the rGO for providing a more conductive and sensitive substrate. According to the results, anodic peak current was proportional to As(III) concentration in a broad range of 2–300 $\mu g\ L^{-1}$, with an LOD and LOQ of 0.10 and 0.33 $\mu g\ L^{-1}$ (S/N = 3). Their new As(III) electrochemical sensor displayed 3.2% RSD for six replicate analyses of 50 $\mu g\ L^{-1}$ As(III). Moreover, the stability test indicated that this sensor maintained ~93% of its initial signal following the 30 successive measurements as well as ~90% of its initial measurement following 2 weeks of storage at room temperature. Additionally, their sensor was satisfactorily utilized to detect As(III) residue in multiple water specimens, such as lake water, reverse osmosis drinking water, and mineral water. Finally, outputs were consistent with the ICP-MS when compared.

Kong's group devised an electrochemical sensor to quantitatively detect Cd^{2+} and Pb^{2+}. Core–shell Fe_3O_4@PANI NPs were utilized as a sensitive platform for the detection of Pb^{2+} and Cd^{2+} [87]. The fabrication procedure of the electrochemical sensor is shown in Fig. 12.5. The linear ranges for the detection of Pb^{2+} were from 0.1 to 10^4 nM, and the LOD was 0.03 nM. The calculated LOD of Cd^{2+} was 0.3 nM. The procured electrochemical sensor had higher sensitivity and reasonable specificity and very good stability that can be simplified for easier operation.

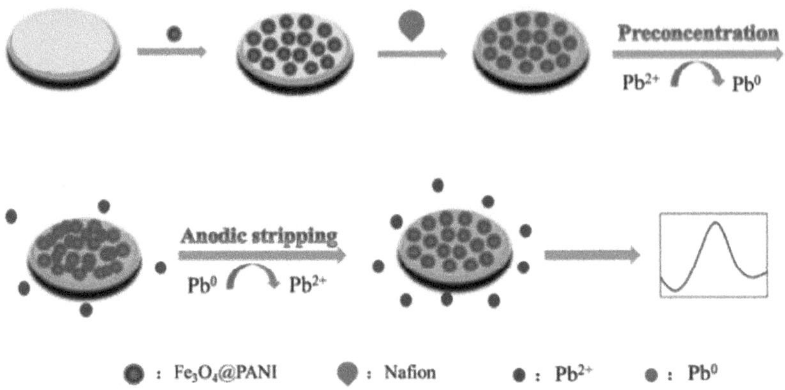

Figure 12.5 Schema to fabricate Fe_3O_4@PANI/GCE.
Source: Reused with permission from Y. Kong, T. Wu, D. Wu, Y. Zhang, Y. Wang, B. Du, et al., An electrochemical sensor based on Fe_3O_4@PANI nanocomposites for sensitive detection of Pb^{2+} and Cd^{2+}, Anal. Methods 10 (2018) 4784–4792.

Baghayeri et al. applied a GCE coated with PAMAM dendrimer – functionalized magnetic GO (GO-Fe_3O_4-PAMAM) to simultaneously detect Pb(II) and Cd(II) in environmental waters [88]. The designed sensor showed linear square wave anodic stripping voltammetry (SWASV) response in a range between 0.4 and 120 $\mu g\ L^{-1}$ for Pb(II), as well as 0.2 and 140 $\mu g\ L^{-1}$ for Cd(II), with the coefficient of determination greater than 0.99. Also, lower LODs of 130 $ng\ L^{-1}$ for Pb(II) and 70 $ng\ L^{-1}$ for Cd(II) were observed. Moreover, probable interferents of some metal ions like Cu(II), Tl(I), Hg(II), Zn(II), Mg(II), Co(II), Mn(II), Ca(II), Fe(III), Ni(II), Cr(III), and In(III) were explored; however, they showed no considerable interference. Ultimately, this assay was utilized to simultaneously detect both cations in the real water specimens, and then outputs were approved by ICP-MS as a confirmatory method.

Mohammadi and coworkers described a technique to detect DPASV of Cd(II) with the use of a CPE modified with the magnetic nanocomposite of Fe_3O_4/eggshell and MWCNT (MWCNT-MES-CPE) [89]. The voltammetric peak current of Cd(II) appeared at a potential of about −0.7 V under optimum conditions. The method was then worked out to detect Cd(II) at lower levels. The response to Cd(II) is linear in a range between 3.0 and 250 $ng\ mL^{-1}$ with 2.4 $ng\ mL^{-1}$ LOD for the Cd(II) ion. The sensor was utilized for determining Cd(II) in the specimens of water, hair, and certified reference substances.

12.2.3.2 Food azo dyes

Azo dyes are the largest group of artificial food dyes, including 70% of the organic dyes generated in the world. Moreover, the commonest azo dyes in the food industry have been considered to be the yellow dyes (sunset yellow and tartrazine) and red dyes (azorubine, ponceau, amaranth, and allura red).

In a majority of cases, they are utilized to color jams, candies, confectionery, ice cream, jellies, alcoholic beverages, soft drinks, and so on. Despite their widespread utilization, some negative impacts on human health have been specified in the output of multiple investigations [90]. So there is enough information stating that the decay of azo dyes gives off products like aromatic amines that have been considered poisonous and carcinogenic. Current investigations have confirmed interactions between human serum albumin and azo dyes [91] and hemoglobin [92]. With regard to some investigations, synthetic food colorants could result in superexcitation and hyperactivity in children, particularly in the case of excessive application [93]. Moreover, they could result in asthmatic and allergic diseases. Consequently, controlling the content of azo dyes in foodstuffs is of high importance.

Han's group reported determining the azo dye amaranth based on the magnetic field–directed self-assembly of the Fe_3O_4@rGO composites as the new magnetic molecularly imprinted electrochemical sensor (MIES) [94]. In molecular imprinting, a complex containing the function monomer of aniline, a template of amaranth, and Fe_3O_4@rGO was preassembled via $\pi-\pi$ stacking and hydrogen bonding interaction and consequently was self-assembled on the GCE surface using magnetic field induction before electropolymerization. The fabrication rout of MIES is shown in Fig. 12.6. The determination of amaranth was observed with the redox probe $K_3[Fe(CN)_6]$ via blockage of the imprinted cavity that prevented interference of oxidation products and amaranth analogs. The linear range for amaranth concentrations by MIES was 0.05–50.0 μM, and the LOD was 50 nM. Also, this MIES showed sufficient selectivity, sensitivity, efficiency, and reproducibility to detect amaranth in fruit drinks. Finally, the average recovery equaled 93.15% to 100.81% with the RSD <3.0%.

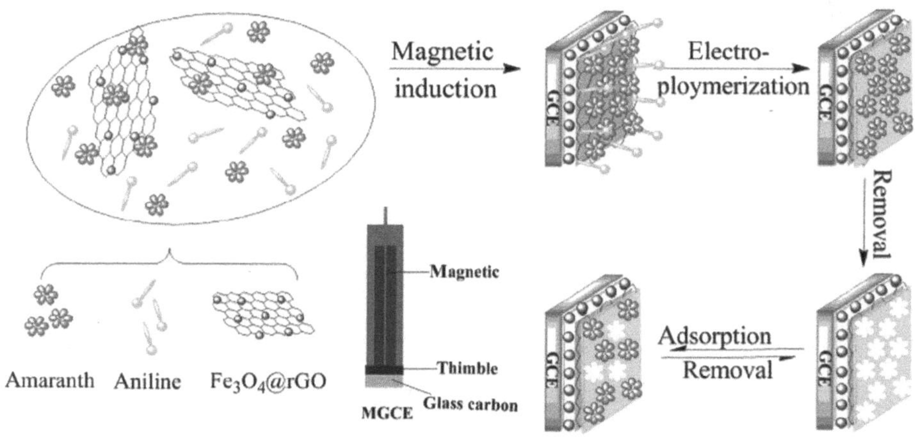

Figure 12.6 Schema to prepare MIES.
Source: Reused with permission from Q. Han, X. Wang, Z. Yang, W. Zhu, X. Zhou, H. Jiang, Fe_3O_4@rGO doped molecularly imprinted polymer membrane based on magnetic field directed self-assembly for the determination of amaranth, Talanta 123 (2014) 101–108.

Li's group synthesized a novel imprinted material based on β-cyclodextrin/IL/gold NPs−functionalized magnetic GO (MGO/β-CD/IL/AuNPs) to modify the GCE surface for constructing the imprinted electrochemical sensor to determine sunset yellow [95]. This new electrochemical sensor exhibited rapid rebinding dynamics that were satisfactorily utilized to detect sunset yellow with a broad linear range between 5.0×10^{-9} and 2.0×10^{-6} M with a 2.0×10^{-9} M LOD. Results indicated the successful application of the electrochemical sensor in detecting sunset yellow in spiked water specimens, Minute Maid, and Mirinda drinks with 97%−105% recovery for the standards.

Arvand and coworkers reported the generation of a new magnetic nanocomposite based on MWCNTs decorated with magnetic core−shell Fe_3O_4@SiO_2 NPs that were utilized for fabricating a modified CPE (Fe_3O_4@SiO_2/MWCNT-CPE). Therefore researchers examined Fe_3O_4@SiO_2/MWCNT-CPE to simultaneously detect the tartrazine and sunset yellow in 0.1 M PBS (pH 6.0) using SWV [96]. The new electrode linearly ranged from 0.5 to 100 μM for sunset yellow and tartrazine with 0.05 and 0.04 μM LODs for sunset yellow and tartrazine. The voltammetric sensor was satisfactorily utilized in the simultaneous determination of tartrazine and sunset yellow in the food products so that it had outputs the same as the ones observed in utilizing a high-performance liquid chromatography (HPLC) at a confidence level of 95%.

Taei's group developed a simplified and sensitive analytical procedure to simultaneously determining sunset yellow and tartrazine based on the $ZnCrFeO_4$−modified CPE [97]. The oxidation response of sunset yellow and tartrazine ameliorated largely at $ZnCrFeO_4$/CPE, which indicated two organized anodic peaks at +0.70 and + 0.99 V versus Ag/AgCl. Moreover, the oxidation reaction was controlled via the diffusion step for tartrazine and via absorption steps for sunset yellow. The simultaneous detection of the dyes demonstrated a wider linear range between 0.07 and 47.5 and 0.05 and 19.0 μM so that the LODs respectively equaled 10.0 and 2.0 nM for tartrazine and sunset yellow. Outputs of the analysis of the real specimens verified that $ZnCrFeO_4$/CPE had notable potential for simultaneously determining tartrazine and sunset yellow in the specimens of soft drinks.

12.2.3.3 Phenolic compounds

Phenolic compounds are utilized in diverse industrial procedures for manufacturing resin, plastic, polymer, plasticizer, detergent, pesticide, and wooden products. As a result, industrial waste products from such phenolic compounds leach into water reservoirs and contaminate nutritional, environmental, and medical matrices [98]. Given that the biologically safe concentration of phenolic compounds in water is 0.001 mg L^{-1} (1 ppm or 10−21 μM) and that concentrations exceeding that level can be very poisonous, monitoring the concentrations of phenolic compounds is crucial for limiting them within the safe range for aquatic life.

Hou et al. designed a new sensitive electrochemical sensor to detect bisphenol A based on carbon black supporting Fe_3O_4 NPs (Fe_3O_4NPs-CB). The Fe_3O_4 NPs-CB nanocomposite made on the GCE exhibited strong electrocatalytic activities toward bisphenol A oxidation [99]. According to the analysis, oxidation overpotentials of bisphenol A considerably declined, and the peak current largely increased in comparison to those of Fe_3O_4 NPs- or CB-modified or bare GCE. DPV was utilized as a sensitive analytical method to detect bisphenol A, and a much broader linear dynamic range of bisphenol A detection ranged from 0.1 nM to 50 μM. The LOD for bisphenol A was found down to 0.031 nM based on S/N = 3. The described sensor was utilized to detect bisphenol A in water specimens with acceptable recovery ranges between 93.2% and 104.1%. Moreover, bisphenol A migration from drinking bottles was assessed via water for simulating the migration into aqueous and acidic foods.

Erogul and coworkers presented the simultaneous detection of hydroquinone and catechol with the use of (3-aminopropyl) triethoxysilane (APTES) as a new electrochemical sensor over AuNPs on combined nanomaterial that was procured with covalently functionalized GO with (3-aminopropyl) triethoxysilane−coated Fe_3O_4 NPs by forming an amide bond (AuNPs/Fe_3O_4-APTES-GO) [100]. Moreover, CV outputs showed the organized oxidation peaks with the anodic peaks at 0.15 V for catechol and 0.05 V for hydroquinone, and the peak-to-peak separation for catechol and hydroquinone was nearly 100 mV. Also, the current response of AuNPs/Fe_3O_4-APTES-GO/GCE was linear for catechol and hydroquinone in the ranges 2−145 and 3−137 μM. LODs for catechol and hydroquinone equaled 0.8 and 1.1 μM. This new sensor was satisfactorily utilized to detect catechol and hydroquinone in the tap water samples with acceptable recoveries.

Zhou's group made a strongly selective and sensitive electrochemical sensor to determine the trace tetra-bromobisphenol-A in the broader specimens with the use of the magnetic molecularly imprinted polymers (MMIPs) [101]. These polymers were procured using the one-pot sol−gel procedure and were consequently utilized as the recognition elements for preparing the MMIPs-modified CPE (MMIPs/CPE). The synergistic impact of the conductive magnetic NPs and selective MIPs ameliorated sensor selectivity and sensitivity. Further, the MMIPs/CPE electrode had very good current responses toward tetrabromobisphenol-A, in comparison to the CPEs consisting of magnetic, nonmolecularly imprinted polymers (MNIPs)/CPE; nonmolecularly imprinted polymers; CPE; and traditional MIPs. Therefore their sensor had acceptable selectivity for tetrabromobisphenol-A in comparison to the compounds with the same structure. Moreover, tetrabromobisphenol-A detection ranged between 5.0 and 2000 nM with 0.77 nM LOD (S/N = 3). Finally, the devised sensor was satisfactorily utilized for the direct detection of tetrabromobisphenol-A in specimens of rain, pool water, and tap water. HPLC−ultraviolet spectrometry was used to validate the outputs.

Karimi-Maleh and coworkers fabricated a voltammetric sensor for the determination of 2-phenylphenol [102]. The sensor was constructed by modifying CPE with a composite of Fe_3O_4-NPs and IL. The presence of Fe_3O_4-NPs and IL in CPE improved the catalytic activity of the electrode toward the oxidation of 2-phenylphenol. The Fe_3O_4-NPs/IL/CPE displayed two well separated oxidation signals in a solution containing

2-phenylphenol and 4-chlorophenol. A peak separation of 380 mV was obtained, which implies that the sensor can be used for the simultaneous determination of the two water contaminants. The Fe_3O_4-NPs/$HMPF_6$/CPE displayed a linear dynamic range of 0.01–520 μM for 2-phenylphenol. The sensor showed good recovery data for the analysis of 2-phenylphenol and 4-chlorophenol in orange rind, lemon rind, and different water samples.

12.2.3.4 Pesticides

It is well-known that pesticides like fungicides, insecticides, and herbicides have a global widespread application, and millions of tons of such pesticides are utilized in industries such as the agricultural and medical fields every year. However, similar compounds can act as nerve poisons and may therefore be employed in the military. Multiple kinds of them are highly poisonous, and in the case of aggregation, they can lead to severe illnesses in living organisms [103]. Hence, to enforce control criteria, it is necessary to provide strongly sensitive, selective, cost-effective analytical procedures for monitoring these compounds.

Li et al. synthesized $Fe_3O_4@ZrO_2$ magnetic core–shell NPs. $Fe_3O_4@ZrO_2$ NPs were modified over the GCE surface to prepare an electrochemical sensor ($Fe_3O_4@ZrO_2$/GCE). The voltammetric behaviors of methyl parathion at the $Fe_3O_4@ZrO_2$/MGCE were studied by CV [104]. The electrochemical sensor was satisfactorily utilized for determining methyl parathion by SWV. The cathodic peak current of methyl parathion linearly increased as the methyl parathion concentrations increased over ranges between 7.60×10^{-8} and 9.12×10^{-5} M with a 1.52×10^{-8} M LOD.

Hunde and coworkers devised an electrochemical sensor by exploiting Fe_3O_4 NPs as modifiers in the graphite paste electrode (GPE). Therefore researchers standardized DPV and CV working variables and utilized them functionally to detect chlorpyrifos pesticide from its aqueous solution [105]. Results showed that a considerable elevation in peak current signals and the magnitude of redox peak potential revealed are an inspiring simplification of the electron transfer procedure by the modifier at the sensor/analyte interface. Moreover, differences in peak current ratio (I_{pc}/I_{pa}), as well as the redox-potential (ΔEp), demonstrated remarkable surface enhancement features of the modifier, which responded to a greater concentration gradient of the pesticide at the interface. Additionally, the lower concentration range of the analyte peak current—1.0–100 μM—directly varied to the pesticide concentration with a 2.8×10^{-6} M LOD. Furthermore, the relative stability of the modified sensor was fine, and the reproducibility of the results equaled 98%, even following a gap of 2 months. Consequently, their new analytical procedure was satisfactory in the case of an application to quantify chlorpyrifos from its aqueous specimens.

Table 12.3 lists applications of MNMs-based electrochemical sensors for food analysis and environmental monitoring.

Table 12.3 Summary of the utilization of MNMs-based electrochemical sensors for food analysis and environmental monitoring.

Electrochemical sensor	Method	Analyte	LOD	Linear range	Ref.
Fe_3O_4@Au/GCE	DPV	Ag^+	59 nM	0.117–17.7 μM	[83]
Fe_3O_4/DA/DTPA/CPE	SWV	Pb^{2+}	7.8×10^{-10} M	1.0×10^{-9}–1.0×10^{-3} M	[84]
		Cu^{2+}	9.5×10^{-10} M	5.0×10^{-9}–1.0×10^{-3} M	
Fe_2O_3/G/GCE	DPASV	Zn^{2+}	0.11 μg L^{-1}	1–100 μg L^{-1}	[85]
		Cd^{2+}	0.08 μg L^{-1}	1–100 μg L^{-1}	
		Pb^{2+}	0.07 μg L^{-1}	1–100 μg L^{-1}	
rGO-Fe_3O_4/SPE	DPASV	As(III)	0.10 μg L^{-1}	2–300 μg L^{-1}	[86]
Fe_3O_4@PANI/GCE	DPV	Pb^{2+}	0.03 nM	0.1–10^4 nM	[87]
		Cd^{2+}	0.3 nM	1.0–0.9×10^4 nM	
GO-Fe_3O_4-PAMAM/GCE	SWASV	Pb(II)	130 ng L^{-1}	0.4–120 μg L^{-1}	[88]
		Cd(II)	70 ng L^{-1}	0.2–140 μg L^{-1}	
MWCNT-MES-CPE	DPASV	Cd(II)	2.4 ng mL^{-1}	3.0–250 ng mL^{-1}	[89]
Fe_3O_4@rGO/GCE	DPV	Amaranth	50 nM	0.05–50.0 μM	[94]
MGO/β-CD/IL/AuNPs/GCE	DPV	Sunset yellow	2.0×10^{-9} M	5.0×10^{-9}–2.0×10^{-6} M	[95]
Fe_3O_4@SiO_2/MWCNT-CPE	SWV	Sunset yellow	0.05 μM	0.5–100 μM	[96]
		Tartrazine	0.04 μM	0.5–100 μM	
ZnCrFeO_4/CPE	DPV	Sunset yellow	2.0 nM	0.07–47.5 μM	[97]
		Tartrazine	10.0 nM	0.05–19.0 μM	
Fe_3O_4NPs-CB/GCE	DPV	Bisphenol A	0.031 nM	0.1 nM–50μM	[99]
AuNPs/Fe_3O_4-APTES-GO/GCE	Amperometry	Catechol	0.8 μM	2–145 μM	[100]
		Hydroquinone	1.1 μM	3–137 μM	
MMIPs/CPE	DPV	Tetrabromobisphenol-A	0.77 nM	5.0–2000 nM	[101]
Fe_3O_4-NPs/IL/CPE	SWV	2-Phenylphenol	0.006 μM	0.01–520 μM	[102]
Fe_3O_4@ZrO_2/GCE	SWV	Methyl parathion	1.52×10^{-8} M	7.60×10^{-8}–9.12×10^{-5} M	[104]
Fe_3O_4/GPE	DPV	Chlorpyrifos	2.8×10^{-6} M	1.0–100 μM	[105]

12.3 Conclusions and perspectives

A combination of the benefits of magnetic particles and nanomaterials, MNMs have the characteristics of small sizes, superparamagnetism, and large surface-atoms/bulk-atoms ratio. Electrochemical sensors are ideal tools to detect various analytes combining higher sensibility, lower costs, and quick responses. MNMs offer specific chances to devise ultrasensitive electrochemical sensors. This chapter has demonstrated the wide potential of the MNMs in electrochemical sensors for clinical diagnosis, biological species and pharmaceutical compounds analysis, and food safety and environmental pollution control. To improve their analytical performance, MNMs have been combined with different nanomaterials (metal and metal oxide NPs, CNTs, Gr, polymers, ILs, and other electrode modifiers).

Although great achievements has been made on MNMs in electrochemical sensors, it is still necessary to develop a powerful modification and synthesis system for preparing MNMs with favorable features such as being very well dispersed, stable, and largely uninformed, as well as having higher electrical conductivity, for increasing the selectivity and sensitivity of electrochemical sensors.

In the end, magnetic NPs—based electrochemical sensors can contribute significantly to electroanalytical science in the coming years. However, for electroanalytical utilizations, additional investigations are necessary to synthesize and devise protocols to synthesize and functionalize MNMs. To provide analytical utilizations, investigations must highlight the enhancement of selectivity and sensitivity of the nanosensors and their lifetimes based on determination methods. Rapid achievements in MNMs-based electrochemical sensors have pointed to their further utilization in a variety of scientific areas.

References

[1] J. Ma, S. Guo, X. Guo, H. Ge, A mild synthetic route to Fe_3O_4@TiO_2-Au composites: preparation, characterization and photocatalytic activity, Appl. Surf. Sci. 353 (2015) 1117–1125.

[2] Y. Wang, S. Wang, H. Niu, Y. Ma, T. Zeng, Y. Cai, et al., Preparation of polydopamine coated Fe_3O_4 nanoparticles and their application for enrichment of polycyclic aromatic hydrocarbons from environmental water samples, J. Chromatogr. A 1283 (2013) 20–26.

[3] M. Baghayeri, Glucose sensing by a glassy carbon electrode modified with glucose oxidase and a magnetic polymeric nanocomposite, RSC Adv. 5 (2015) 18267–18274.

[4] M. Baikousi, A.B. Bourlinos, A. Douvalis, T. Bakas, D.F. Anagnostopoulos, J. Tuček, et al., Synthesis and characterization of γ-Fe_2O_3/carbon hybrids and their application in removal of hexavalent chromium ions from aqueous solutions, Langmuir 28 (2012) 3918–3930.

[5] C. Murray, D.J. Norris, M.G. Bawendi, Synthesis and characterization of nearly monodisperse CdE (E = sulfur, selenium, tellurium) semiconductor nanocrystallites, J. Am. Chem. Soc. 115 (1993) 8706–8715.

[6] J. Zou, W. Zhang, D. Poe, J. Qin, A. Fornara, Y. Zhang, et al., MRI manifestation of novel superparamagnetic iron oxide nanoparticles in the rat inner ear, Nanomedicine 5 (2010) 739–754.

[7] S.A. Majetich, Y. Jin, Magnetization directions of individual nanoparticles, Science 284 (1999) 470–473.
[8] A. Ranzoni, G. Sabatte, L.J. van IJzendoorn, M.W. Prins, One-step homogeneous magnetic nanoparticle immunoassay for biomarker detection directly in blood plasma, ACS Nano 6 (2012) 3134–3141.
[9] L. Zhang, T. Wang, L. Yang, C. Liu, C. Wang, H. Liu, et al., General route to multifunctional uniform yolk/mesoporous silica shell nanocapsules: a platform for simultaneous cancer-targeted imaging and magnetically guided drug delivery, Chem. Eur. J. 18 (2012) 12512–12521.
[10] L. Gloag, M. Mehdipour, D. Chen, R.D. Tilley, J.J. Gooding, Advances in the application of magnetic nanoparticles for sensing, Adv. Mater. 31 (2019) 1904385.
[11] P. Yáñez-Sedeño, S. Campuzano, J.M. Pingarrón, Electrochemical sensors based on magnetic molecularly imprinted polymers: a review, Anal. Chim. Acta 960 (2017) 1–17.
[12] A.C. Power, B. Gorey, S. Chandra, J. Chapman, Carbon nanomaterials and their application to electrochemical sensors: a review, Nanotechnol. Rev. 7 (2018) 19–41.
[13] N.R. Stradiotto, H. Yamanaka, M.V.B. Zanoni, Electrochemical sensors: a powerful tool in analytical chemistry, J. Braz. Chem. Soc. 14 (2003) 159–173.
[14] B. Bansod, T. Kumar, R. Thakur, S. Rana, I. Singh, A review on various electrochemical techniques for heavy metal ions detection with different sensing platforms, Biosens. Bioelectron. 94 (2017) 443–455.
[15] H. Beitollahi, H. Karimi-Maleh, H. Khabazzadeh, Nanomolar and selective determination of epinephrine in the presence of norepinephrine using carbon paste electrode modified with carbon nanotubes and novel 2-(4-oxo-3-phenyl-3, 4-dihydro-quinazolinyl)-N′-phenyl-hydrazinecarbothioamide, Anal. Chem. 80 (2008) 9848–9851.
[16] A.M. Fekry, M. Shehata, S.M. Azab, A. Walcarius, Voltammetric detection of caffeine in pharmacological and beverages samples based on simple nano-Co (II, III) oxide modified carbon paste electrode in aqueous and micellar media, Sens. Actuators B: Chem. 302 (2020) 127172.
[17] H. Beitollahi, M.A. Khalilzadeh, S. Tajik, M. Safaei, K. Zhang, H.W. Jang, et al., Recent advances in applications of voltammetric sensors modified with ferrocene and its derivatives, ACS Omega 5 (2020) 2049–2059.
[18] A. Manal, H.A. Hendawy, G.M. Eldin, Z.A. El-Sherif, Application of nano graphene-modified electrode as an electrochemical sensor for determination of tapentadol in the presence of paracetamol, J. Iran. Chem. Soc. 16 (2019) 1123–1130.
[19] H. Mahmoudi-Moghaddam, S. Tajik, H. Beitollahi, Highly sensitive electrochemical sensor based on La^{3+}-doped Co_3O_4 nanocubes for determination of sudan I content in food samples, Food Chem. 286 (2019) 191–196.
[20] F. Asadi, S.N. Azizi, S. Ghasemi, Preparation of Ag nanoparticles on nano cobalt-based metal organic framework (ZIF-67) as catalyst support for electrochemical determination of hydrazine, J. Mater. Sci. Mater. Electron. 30 (2019) 5410–5420.
[21] T. Dayakar, K.V. Rao, K. Bikshalu, V. Malapati, K.K. Sadasivuni, Non-enzymatic sensing of glucose using screen-printed electrode modified with novel synthesized CeO_2@CuO core shell nanostructure, Biosens. Bioelectron. 111 (2018) 166–173.
[22] S. Lotfi, H. Veisi, Pd nanoparticles decorated poly-methyldopa@GO/Fe_3O_4 nanocomposite modified glassy carbon electrode as a new electrochemical sensor for simultaneous determination of acetaminophen and phenylephrine, Mater. Sci. Eng. C 105 (2019) 110112.
[23] T.W. Chen, S. Chinnapaiyan, S.M. Chen, M.A. Ali, M.S. Elshikh, A.H. Mahmoud, Facile synthesis of copper ferrite nanoparticles with chitosan composite for high-performance electrochemical sensor, Ultrason. Sonochem. 63 (2020) 104902.

[24] R. Monošík, M. Streďanský, E. Šturdík, Application of electrochemical biosensors in clinical diagnosis, J. Clin. Lab. Anal. 26 (2012) 22–34.
[25] S.A. Zaidi, J.H. Shin, Recent developments in nanostructure based electrochemical glucose sensors, Talanta 149 (2016) 30–42.
[26] Z. Shahnavaz, P.M. Woi, Y. Alias, Electrochemical sensing of glucose by reduced graphene oxide-zinc ferrospinels, Appl. Surf. Sci. 379 (2016) 156–162.
[27] P. Vennila, D.J. Yoo, A.R. Kim, Ni-Co/Fe_3O_4 flower-like nanocomposite for the highly sensitive and selective enzyme free glucose sensor applications, J. Alloy Compd. 703 (2017) 633–642.
[28] A. Gevaerd, S.F. Blaskievicz, A.J. Zarbin, E.S. Orth, M.F. Bergamini, L.H. Marcolino-Junior, Nonenzymatic electrochemical sensor based on imidazole-functionalized graphene oxide for progesterone detection, Biosens. Bioelectron. 112 (2018) 108–113.
[29] G. Contreras Jiménez, S. Eissa, A. Ng, H. Alhadrami, M. Zourob, M. Siaj, Aptamer-based label-free impedimetric biosensor for detection of progesterone, Anal. Chem. 87 (2015) 1075–1082.
[30] J.V. Samsonova, V.A. Safronova, A.P. Osipov, Pretreatment-free lateral flow enzyme immunoassay for progesterone detection in whole cows' milk, Talanta 132 (2015) 685–689.
[31] W.M. Allen, Progesterone: how did the name originate? South Med. J. 63 (1970) 1151–1155.
[32] A.H. Hanita, Potential use of single measurement of serum progesterone in detecting early pregnancy failure, Malays. J. Pathol. 34 (2012) 41–46.
[33] M. Arvand, S. Hemmati, Magnetic nanoparticles embedded with graphene quantum dots and multiwalled carbon nanotubes as a sensing platform for electrochemical detection of progesterone, Sens. Actuators B: Chem. 238 (2017) 346–356.
[34] E.P. Randvir, C.E. Banks, Analytical methods for quantifying creatinine within biological media, Sens. Actuators B: Chem. 183 (2013) 239–252.
[35] S.C. Kazmlerczak, F.V. Lente, E.D. Hodges, Diagnostic and prognostic utility of phospholipase: an activity in patients with acute pancreatitis: comparison with amylase and lipase, Clin. Chem. 37 (1991) 356–360.
[36] X. Bo, M. Zhou, L. Guo, Electrochemical sensors and biosensors based on less aggregated graphene, Biosens. Bioelectron. 89 (2017) 167–186.
[37] T. Wen, W. Zhu, C. Xue, J. Wu, Q. Han, X. Wang, et al., Novel electrochemical sensing platform based on magnetic field-induced self-assembly of Fe_3O_4@ Polyaniline nanoparticles for clinical detection of creatinine, Biosens. Bioelectron. 56 (2014) 180–185.
[38] N. Shoaie, M. Daneshpour, M. Azimzadeh, S. Mahshid, S.M. Khoshfetrat, F. Jahanpeyma, et al., Electrochemical sensors and biosensors based on the use of polyaniline and its nanocomposites: a review on recent advances, Microchim. Acta 186 (2019) 465.
[39] R. Suresh, K. Giribabu, R. Manigandan, A. Stephen, V. Narayanan, Fabrication of Ni–Fe_2O_3 magnetic nanorods and application to the detection of uric acid, RSC Adv. 4 (2014) 17146–17155.
[40] M. Arvand, M. Hassannezhad, Magnetic core–shell Fe_3O_4@SiO_2/MWCNT nanocomposite modified carbon paste electrode for amplified electrochemical sensing of uric acid, Mater. Sci. Eng. C 36 (2014) 160–167.
[41] M.R. Majidi, M.H. Pournaghi-Azar, A. Saadatirad, E. Alipour, Simple and rapid amperometric monitoring of hydrogen peroxide at hemoglobin-modified pencil lead electrode

as a novel biosensor: application to the analysis of honey sample, Food Anal. Methods 8 (2015) 1067–1077.
[42] B. Halliwell, M.V. Clement, L.H. Long, Hydrogen peroxide in the human body, FEBS Lett. 486 (2000) 10–13.
[43] X. Yang, L. Wang, G. Zhou, N. Sui, Y. Gu, J. Wan, Electrochemical detection of H_2O_2 based on Fe_3O_4 nanoparticles with graphene oxide and polyamidoamine dendrimer, J. Clust. Sci. 26 (2015) 789–798.
[44] A. Benvidi, M.T. Nafar, S. Jahanbani, M.D. Tezerjani, M. Rezaeinasab, S. Dalirnasab, Developing an electrochemical sensor based on a carbon paste electrode modified with nano-composite of reduced graphene oxide and $CuFe_2O_4$ nanoparticles for determination of hydrogen peroxide, Mater. Sci. Eng. C 75 (2017) 1435–1447.
[45] F. Mollarasouli, S. Kurbanoglu, K. Asadpour-Zeynali, S.A. Ozkan, Non-enzymatic monitoring of hydrogen peroxide using novel nanosensor based on $CoFe_2O_4$@CdSeQD magnetic nanocomposite and rifampicin mediator, Anal. Bioanal. Chem. 412 (2020) 5053–5065.
[46] B.R. Adhikari, M. Govindhan, A. Chen, Carbon nanomaterials based electrochemical sensors/biosensors for the sensitive detection of pharmaceutical and biological compounds, Sensors 15 (2015) 22490–22508.
[47] T. Pradhan, H.S. Jung, J.H. Jang, T.W. Kim, C. Kang, J.S. Kim, Chemical sensing of neurotransmitters, Chem. Soc. Rev. 43 (2014) 4684–4713.
[48] A.A. Ensafi, F. Saeid, B. Rezaei, A.R. Allafchian, $NiFe_2O_4$ nanoparticles decorated with MWCNTs as a selective and sensitive electrochemical sensor for the determination of epinephrine using differential pulse voltammetry, Anal. Methods 6 (2014) 6885–6892.
[49] Y.M. Zhang, P.L. Xu, Q. Zeng, Y.M. Liu, X. Liao, M.F. Hou, Magnetism-assisted modification of screen printed electrode with magnetic multi-walled carbon nanotubes for electrochemical determination of dopamine, Mater. Sci. Eng. C 74 (2017) 62–69.
[50] G. Ran, X. Chen, Y. Xia, Electrochemical detection of serotonin based on a poly (bromocresol green) film and Fe_3O_4 nanoparticles in a chitosan matrix, RSC Adv. 7 (2017) 1847–1851.
[51] D.F.D. Queiroz, T.R.D.L. Dadamos, S.A.S. Machado, M.A.U. Martines, Electrochemical determination of norepinephrine by means of modified glassy carbon electrodes with carbon nanotubes and magnetic nanoparticles of cobalt ferrite, Sensors 18 (2018) 1223.
[52] E.V. Suprun, E.V. Karpova, S.P. Radko, A.A. Karyakin, Advanced electrochemical detection of amino acids and proteins through flow injection analysis and catalytic oxidation on Prussian Blue, Electrochim. Acta 331 (2020) 135289.
[53] M. Hasanzadeh, N. Shadjou, E. Omidinia, Mesoporous silica (MCM-41)-Fe_2O_3 as a novel magnetic nanosensor for determination of trace amounts of amino acids, Colloids Surf. B: Biointerfaces 108 (2013) 52–59.
[54] M. Hasanzadeh, A. Karimzadeh, N. Shadjou, A. Mokhtarzadeh, L. Bageri, S. Sadeghi, et al., Graphene quantum dots decorated with magnetic nanoparticles: synthesis, electrodeposition, characterization and application as an electrochemical sensor towards determination of some amino acids at physiological pH, Mater. Sci. Eng. C 68 (2016) 814–830.
[55] K. Atacan, $CuFe_2O_4$/reduced graphene oxide nanocomposite decorated with gold nanoparticles as a new electrochemical sensor material for L-cysteine detection, J. Alloy Compd. 791 (2019) 391–401.

[56] M.L. Rodríguez, Analytical challenges for forensic screening of drugs of abuse, Med. Lab. Obs. 44 (2012) 8−10.
[57] M.R. Siddiqui, Z.A. AlOthman, N. Rahman, Analytical techniques in pharmaceutical analysis: a review, Arab. J. Chem. 10 (2017) S1409−S1421.
[58] S. Görög, Drug safety, drug quality, drug analysis, J. Pharm. Biomed. Anal. 48 (2008) 247−253.
[59] R. de Vries, L. Vereyken, I. François, L. Dillen, R.J. Vreeken, F. Cuyckens, High sensitivity and selectivity in quantitative analysis of drugs in biological samples using 4-column multidimensional micro-UHPLC-MS enabling enhanced sample loading capacity, Anal. Chim. Acta 989 (2017) 104−111.
[60] W. Misiuk, The role of assay methods in characterizing the quality of bulk pharmaceuticals, J. Pharm. Bioallied Sci. 2 (2010) 88−92.
[61] A. Afkhami, H. Khoshsafar, H. Bagheri, T. Madrakian, Preparation of $NiFe_2O_4$/graphene nanocomposite and its application as a modifier for the fabrication of an electrochemical sensor for the simultaneous determination of tramadol and acetaminophen, Anal. Chim. Acta 831 (2014) 50−59.
[62] A. Afkhami, H. Khoshsafar, H. Bagheri, T. Madrakian, Facile simultaneous electrochemical determination of codeine and acetaminophen in pharmaceutical samples and biological fluids by graphene−$CoFe_2O_4$ nancomposite modified carbon paste electrode, Sens. Actuators B: Chem. 203 (2014) 909−918.
[63] M. Hasanzadeh, M.H. Pournaghi-Azar, N. Shadjou, A. Jouyban, Magnetic nanoparticles incorporated on functionalized mesoporous silica: an advanced electrochemical sensor for simultaneous determination of amiodarone and atenolol, RSC Adv. 4 (2014) 4710−4717.
[64] E. Haghshenas, T. Madrakian, A. Afkhami, A novel electrochemical sensor based on magneto Au nanoparticles/carbon paste electrode for voltammetric determination of acetaminophen in real samples, Mater. Sci. Eng. C 57 (2015) 205−214.
[65] D.M. Fernandes, N. Silva, C. Pereira, C. Moura, J.M. Magalhães, B. Bachiller-Baeza, et al., $MnFe_2O_4$@CNT-N as novel electrochemical nanosensor for determination of caffeine, acetaminophen and ascorbic acid, Sens. Actuators B: Chem. 218 (2015) 128−136.
[66] M. Arvand, S. Abbasnejad, N. Ghodsi, Graphene quantum dots decorated with Fe_3O_4 nanoparticles/functionalized multiwalled carbon nanotubes as a new sensing platform for electrochemical determination of L-DOPA in agricultural products, Anal. Methods 8 (2016) 5861−5868.
[67] T. Madrakian, S. Maleki, M. Heidari, A. Afkhami, An electrochemical sensor for rizatriptan benzoate determination using Fe_3O_4 nanoparticle/multiwall carbon nanotube-modified glassy carbon electrode in real samples, Mater. Sci. Eng. C 63 (2016) 637−643.
[68] A. Fallah Shojaei, K. Tabatabaeian, Sh Shakeri, F. Karimi, A novel 5-fluorouracile anticancer drug sensor based on $ZnFe_2O_4$ magnetic nanoparticles ionic liquids carbon paste electrode, Sens. Actuators B: Chem. 230 (2016) 607−614.
[69] E. Tammari, A. Nezhadali, S. Lotfi, H. Veisi, Fabrication of an electrochemical sensor based on magnetic nanocomposite Fe_3O_4/β-alanine/Pd modified glassy carbon electrode for determination of nanomolar level of clozapine in biological model and pharmaceutical samples, Sens. Actuators B: Chem. 241 (2017) 879−886.
[70] A.A. Ensafi, F. Rezaloo, B. Rezaei, $CoFe_2O_4$/reduced graphene oxide/ionic liquid modified glassy carbon electrode, a selective and sensitive electrochemical sensor for determination of methotrexate, J. Taiwan. Inst. Chem. Eng. 78 (2017) 45−50.

[71] A. Afkhami, A. Bahiraei, T. Madrakian, Application of nickel zinc ferrite/graphene nanocomposite as a modifier for fabrication of a sensitive electrochemical sensor for determination of omeprazole in real samples, J. Colloid Interface Sci. 495 (2017) 1−8.
[72] F.H. Cincotto, D.A. Carvalho, T.C. Canevari, H.E. Toma, O. Fatibello-Filho, F.C. Moraes, A nano-magnetic electrochemical sensor for the determination of mood disorder related substances, RSC Adv. 8 (2018) 14040−14047.
[73] M. Amir, M.M. Tunesi, R.A. Soomro, A. Baykal, N.H. Kalwar, Sensitive determination of 6-thioguanine using caffeic acid-functionalized Fe_3O_4 nanoparticles as an electrochemical sensor, J. Electron. Mater. 47 (2018) 2198−2208.
[74] H. Beitollahi, F. Garkani-Nejad, S. Tajik, M.R. Ganjali, Voltammetric determination of acetaminophen and tryptophan using a graphite screen printed electrode modified with functionalized graphene oxide nanosheets within a Fe_3O_4@SiO_2 nanocomposite, Iran. J. Pharm. Res. 18 (2019) 80.
[75] M. Ghapanvari, T. Madrakian, A. Afkhami, A. Ghoorchian, A modified carbon paste electrode based on Fe_3O_4@multi-walled carbon nanotubes@polyacrylonitrile nanofibers for determination of imatinib anticancer drug, J. Appl. Electrochem. 50 (2020) 281−294.
[76] M.A. Khalilzadeh, S. Tajik, H. Beitollahi, R.A. Venditti, Green synthesis of magnetic nanocomposite with iron oxide deposited on cellulose nanocrystals with copper (Fe_3O_4@CNC/Cu): investigation of catalytic activity for the development of a venlafaxine electrochemical sensor, Ind. Eng. Chem. Res. 59 (2020) 4219−4228.
[77] T. Robak, Novel drugs for chronic lymphoid leukemias: mechanism of action and therapeutic activity, Curr. Med. Chem. 16 (2009) 2212−2234.
[78] T. Robak, A. Korycka, M. Kasznicki, A. Wrzesien-Kus, P. Smolewski, Purine nucleoside analogues for the treatment of hematological malignancies: pharmacology and clinical applications, Curr. Cancer Drug Targets 5 (2005) 421−444.
[79] A.A. Ensafi, M. Jafari-Asl, B. Rezaei, A.R. Allafchian, Simultaneous determination of guanine and adenine in DNA based on $NiFe_2O_4$ magnetic nanoparticles decorated MWCNTs as a novel electrochemical sensor using adsorptive stripping voltammetry, Sens. Actuators B: Chem. 177 (2013) 634−642.
[80] A. Yari, S. Derki, New MWCNT-Fe_3O_4@PDA-Ag nanocomposite as a novel sensing element of an electrochemical sensor for determination of guanine and adenine contents of DNA, Sens. Actuators B: Chem. 227 (2016) 456−466.
[81] A. Sett, S. Das, P. Sharma, U. Bora, Aptasensors in health, environment and food safety monitoring, Open J. Appl. Biosens. 1 (2012) 9−19.
[82] L. Eddaif, A. Shaban, J. Telegdi, Sensitive detection of heavy metals ions based on the calixarene derivatives-modified piezoelectric resonators: a review, Int. J. Environ. Anal. Chem. 99 (2019) 824−853.
[83] H. Yang, X. Liu, R. Fei, Y. Hu, Sensitive and selective detection of Ag^+ in aqueous solutions using Fe_3O_4@Au nanoparticles as smart electrochemical nanosensors, Talanta 116 (2013) 548−553.
[84] M. Bagherzadeh, M. Pirmoradian, F. Riahi, Electrochemical detection of Pb and Cu by using DTPA functionalized magnetic nanoparticles, Electrochim. Acta 115 (2014) 573−580.
[85] S. Lee, J. Oh, D. Kim, Y. Piao, A sensitive electrochemical sensor using an iron oxide/graphene composite for the simultaneous detection of heavy metal ions, Talanta 160 (2016) 528−536.
[86] A.B. Chimezie, R. Hajian, N.A. Yusof, P.M. Woi, N. Shams, Fabrication of reduced graphene oxide-magnetic nanocomposite (rGO-Fe_3O_4) as an electrochemical sensor for

trace determination of As (III) in water resources, J. Electroanal. Chem. 796 (2017) 33–42.
[87] Y. Kong, T. Wu, D. Wu, Y. Zhang, Y. Wang, B. Du, et al., An electrochemical sensor based on Fe_3O_4@PANI nanocomposites for sensitive detection of Pb^{2+} and Cd^{2+}, Anal. Methods 10 (2018) 4784–4792.
[88] M. Baghayeri, H. Alinezhad, M. Fayazi, M. Tarahomi, R. Ghanei-Motlagh, B. Maleki, A novel electrochemical sensor based on a glassy carbon electrode modified with dendrimer functionalized magnetic graphene oxide for simultaneous determination of trace Pb (II) and Cd (II), Electrochim. Acta 312 (2019) 80–88.
[89] S. Mohammadi, M.A. Taher, H. Beitollahi, M. Naghizadeh, Sensitive voltammetric determination of cadmium at a carbon nanotubes/Fe_3O_4/eggshell composites modified carbon paste electrode, Environ. Nanotechnol. Monit. Manag. 12 (2019) 100241.
[90] S.N. Okafor, W. Obonga, M.A. Ezeokonkwo, J. Nurudeen, U. Orovwigho, J. Ahiabuike, Assessment of the health implications of synthetic and natural food colourants–a critical review, UK J. Pharm. Biosci. 4 (2016) 01–11.
[91] D. Masone, C. Chanforan, Study on the interaction of artificial and natural food colorants with human serum albumin: a computational point of view, Comput. Biol. Chem. 56 (2015) 152–158.
[92] A. Basu, G.S. Kumar, Binding of carmoisine, a food colorant, with hemoglobin: spectroscopic and calorimetric studies, Food Res. Int. 72 (2015) 54–61.
[93] L.J. Stevens, T. Kuczek, J.R. Burgess, M.A. Stochelski, L.E. Arnold, L. Galland, Mechanisms of behavioral, atopic, and other reactions to artificial food colors in children, Nutr. Rev. 71 (2013) 268–281.
[94] Q. Han, X. Wang, Z. Yang, W. Zhu, X. Zhou, H. Jiang, Fe_3O_4@rGO doped molecularly imprinted polymer membrane based on magnetic field directed self-assembly for the determination of amaranth, Talanta 123 (2014) 101–108.
[95] J. Li, X. Wang, H. Duan, Y. Wang, Y. Bu, C. Luo, Based on magnetic graphene oxide highly sensitive and selective imprinted sensor for determination of sunset yellow, Talanta 147 (2016) 169–176.
[96] M. Arvand, Y. Parhizi, S.H. Mirfathi, Simultaneous voltammetric determination of synthetic colorants in foods using a magnetic core–shell Fe_3O_4@SiO_2/MWCNTs nanocomposite modified carbon paste electrode, Food Anal. Methods 9 (2016) 863–875.
[97] M. Taei, H. Salavati, M. Fouladgar, E. Abbaszadeha, Simultaneous determination of sunset yellow and tartrazine in soft drinks samples using nanocrystallites of spinel ferrite-modified electrode, Q. J. Iran. Chem. Commun. 8 (2020) 67–79.
[98] I. Gul, M.S. Ahmad, S.S. Naqvi, A. Hussain, R. Wali, A.A. Farooqi, et al., Polyphenol oxidase (PPO) based biosensors for detection of phenolic compounds: a review, J. Appl. Biol. Biotechnol. 5 (2017) 72–85.
[99] C. Hou, W. Tang, C. Zhang, Y. Wang, N. Zhu, A novel and sensitive electrochemical sensor for bisphenol A determination based on carbon black supporting ferroferric oxide nanoparticles, Electrochim. Acta 144 (2014) 324–331.
[100] S. Erogul, S.Z. Bas, M. Ozmen, S. Yildiz, A new electrochemical sensor based on Fe_3O_4 functionalized graphene oxide-gold nanoparticle composite film for simultaneous determination of catechol and hydroquinone, Electrochim. Acta 186 (2015) 302–313.
[101] T. Zhou, Y. Feng, L. Zhou, Y. Tao, D. Luo, T. Jing, et al., Selective and sensitive detection of tetrabromobisphenol-A in water samples by molecularly imprinted electrochemical sensor, Sens. Actuators B: Chem. 236 (2016) 153–162.

[102] H. Karimi-Maleh, C.T. Fakude, N. Mabuba, G.M. Peleyeju, O.A. Arotiba, The determination of 2-phenylphenol in the presence of 4-chlorophenol using nano-Fe_3O_4/ionic liquid paste electrode as an electrochemical sensor, J. Colloid Interface Sci. 554 (2019) 603–610.
[103] M. Tankiewicz, M. Biziuk, Fast, sensitive and reliable multi-residue method for routine determination of 34 pesticides from various chemical groups in water samples by using dispersive liquid-liquid microextraction coupled with gas chromatography mass spectrometry, Anal. Bioanal. Chem. 410 (2018) 1533–1550.
[104] N.N. Li, T.F. Kang, J.J. Zhang, L.P. Lu, S.Y. Cheng, Fe_3O_4@ZrO_2 magnetic nanoparticles as a new electrode material for sensitive determination of organophosphorus agents, Anal, Methods 7 (2015) 5053–5059.
[105] T. Hunde, M. Berhe, A. Tadese, M. Tirfu, A. Woldu, B. Menasbo, et al., Nano Fe_3O_4–graphite paste modified electrochemical sensor for phosphatic pesticide-chlorpyrifos, Momona Ethiop. J. Sci. 9 (2017) 76–89.

Application of magnetic nanomaterials as electrochemical biosensors

13

Hadi Beitollahi[1], *Zahra Dourandish*[1], *Somayeh Tajik*[2] and *Peyman Mohammadzadeh Jahani*[3]
[1]Environment Department, Institute of Science and High Technology and Environmental Sciences, Graduate University of Advanced Technology, Kerman, Iran, [2]Research Center for Tropical and Infectious Diseases, Kerman University of Medical Sciences, Kerman, Iran, [3]Department of Basic Sciences, School of Medicine, Bam University of Medical Sciences, Bam, Iran

13.1 Introduction

Electrochemical biosensors have been considered selective tools with the use of a biological recognition element such as an antibody, enzyme, nucleic acid, etc. that was immobilized on the transducer. Moreover, biosensors can offer quantitative or semiquantitative analytical data that transform the biorecognition reaction of the analyte to a measurable signal. Electrochemical biosensors enjoy diverse benefits due to portability, simplicity, simplified utilization, relatively fast response times, affordability, and, in a majority of situations, disposable. The key utilizations for electrochemical biosensors relate to the control of beverage and food quality, environmental monitoring, security, bioprocessing with widespread application in health care centers [1–3]. Also, two dimensions have presented the greatest problems in developing biosensors: immobilizing the biomolecules in an appropriate matrix and the diffusional limitation of the analyte into this matrix. Thus chemists have largely considered novel immobilization matrices for biomolecules [4,5].

Experts in this field have been largely focused on developing nanotechnology rather than on traditional sensing procedures, leading to the construction of strongly selective biosensors with nanomolar-level capacities of the sensing analytes. A recent development in electrochemical biosensors mostly focuses on using nanomaterials and nanostructures [6,7]. Magnetic nanoparticles (MNPs) are particles of natural or synthetic materials with sizes varying from 5 to 100 nm and having magnetic properties. Examples are crystals of transition metals (Fe, Co, Ni) and their oxides, as well as halides of the iron, chromium, manganese, etc. groups [8,9]. Magnetic nanomaterials (MNMs) have attained significant attraction among the scientific population due to their unique characteristics such as small size, electromagnetic properties, colloidal stability, and biocompatibility [10,11].

Magnetic Nanomaterials in Analytical Chemistry. DOI: https://doi.org/10.1016/B978-0-12-822131-0.00011-X

The application of MNM modified electrodes in manufacturing is extremely attractive. These novel electrodes have been considered essential elements due to their multiple benefits like larger surface areas, lower resistance to electronic transmission, and capability for absorbing chemical (biochemical) analytes, which have resulted in their attractiveness and usefulness in classical electrochemical detection [12,13]. MNMs are involved in minimizing the degradation of the surface of electrodes, enhancing electrocatalytic activities, and simplifying the immobilization procedure of biomolecules like antibodies and enzymes on the electrode surface. Both MNPs and functionalized MNPs have a widespread application in developing electrochemical biosensors for diverse (bio) analytical utilization [14,15].

13.2 MNMs-based electrochemical immunosensors

According to immunoassay, in biochemical determination and measurement techniques, a certain interaction exists for recognizing and binding the antibody and a specific target, called the antigen. Actually, in electrochemical immunoassay, a certain antibody is immobilized at the surface of the electrode, and a particular interaction between antigen and antibody leads to the electrochemical responses [16,17].

13.2.1 *Iron oxide MNPs utilized in electrochemical immunosensors*

Because of this simplistic procurement—the superparamagnetic feature, the same dimension under control as antibodies, and a higher specific surface area for antibody binding—iron oxide MNPs (Fe_2O_3 and Fe_3O_4) have been considered to have widespread utilization in designing immunosensors. Nonetheless, due to magnetic dipolar attraction, as well as a large ratio of the surface area to volume, a majority of unmodified MNPs are readily aggregated into clusters in the case of direct exposure to biological solutions. Therefore to overcome these disadvantages, enhance their biocompatibility, and bring about novel applications for MNPs, experts in the field have synthesized diverse functionalized MNPs [18,19].

Yuan and coworkers prepared a Fe_3O_4@Au-based pseudo-homogeneous electrochemical immunosensor to determine alpha-fetoprotein (AFP), which is one of the popular hepatocellular carcinoma biomarkers. The primary antibody (Ab_1) was immobilized on the Fe_3O_4@Au NPs as a capture probe. Then the secondary antibody (Ab_2) and horseradish peroxidase (HRP) was conjugated on the Au NPs through electrostatic absorption for forming the signal-amplifying labels. After that, a sandwich immunocomplex in the presence of AFP was established by specifically recognize the antigen–antibody in a Fe_3O_4@Au-based pseudo-homogeneous reaction system. Following the capture of the immune complex to the surface of the magnetic glassy carbon electrode (MGCE), the decomposition of H_2O_2 was catalyzed by the labeling HRP and led to a notable current to quantitatively detect AFP. Notably, an amperometric (i–t) procedure was applied for recording the response signal of the immunosensor based on catalyzing the immobilized HRP toward H_2O_2 reduction with hydroquinone (HQ) as a redox mediator. At the optimized condition, the amperometric current response linearly related to the AFP concentration ranges between 20 and 100 ng mL^{-1} with a correlation coefficient equal to 0.9940 and a 0.64 ng mL^{-1} limit of detection (LOD) at a signal/noise ratio [S/N] equal to 3. Additionally, the electrochemical immunosensor had greater antiinterference abilities, reasonable reproducibility, and prolonged stability in detecting AFP [20].

Sun's group developed a strongly sensitive electrochemical immunoassay of chlorpyrifos (CPF) with the use of biocompatible quinone-rich polydopamine nanospheres–modified GCE as the sensor platform and multi-HRP flake-like Fe_3O_4-coated carbon nanotube (CNT) nanocomposites as the signal label. Because of the quinine-rich polydopamine nanospheres, the platform showed very good fixing capacity via a convenient coating of the sticky polydopamine nanospheres and consequent oxidization. In the case of coprecipitation of Fe^{2+} and Fe^{3+} on the polydopamine-modified CNTs by the aiding ethylene glycol (EG), the flake-like Fe_3O_4-coated CNTs (CNTs@f-Fe_3O_4) was synthesized and selected as a carrier of the multienzyme label because of higher loading of HRP Ab_2, as well as the peroxidase mimic activities of Fe_3O_4. At the optimized condition, the immunosensor may determine CPF in a broad range with 6.3 pg mL^{-1} LOD. In addition, higher reproducibility, stability, and specificity of the immunosensor were demonstrated. Finally, primary utilization in real samples indicated acceptable recovery that revealed a promising future in rapidly analyzing CPF in aquatic environments [21].

Feng and coworkers developed a sandwich-type electrochemical immunosensor to sensitively detect carcinoembryonic antigen (CEA) with the use of ferroferric oxide@silica–amino groups (Fe_3O_4@SiO_2–NH_2) as the carriers and Au NPs–graphene oxide (GO) as the platform. Researchers utilized the surface of Fe_3O_4@SiO_2–NH_2 as the linked reagents to co-immobilize ferrocene carboxylic acid (Fc-COOH) and the secondary anti-CEA (Ab_2) for preparing the signal probe. Thus it could accelerate decomposing the H_2O_2 for amplifying the signals. Fig. 13.1A and B illustrates preparing the Fe_3O_4@SiO_2-Fc-Ab_2/HRP bioconjugates and modification procedure for electrochemical and electrode determination of CEA. Also, differential pulse voltammetry (DPV) was satisfactorily utilized for the CEA quantification. At the optimal condition, their new immunosensor showed a very good analytical function in the broad dynamic response ranges of CEA concentration between 0.001 and 80 ng mL^{-1} with low LOD equal to 0.0002 ng mL^{-1} (S/N = 3), as well as high specificity and acceptable reproducibility. This new immunosensor was substantially utilized for determining CEA in the spiked human serum specimens [22].

In their research, Zhang and coworkers, prepared a hybrid nanostructure of the Fe_3O_4 nanospheres as well as Ag@Au nanorods via polydopamine coating as the nanoelectrocatalyst for constructing a new sandwich-type electrochemical immunosensor. A Ag@Au-Fe_3O_4 nanohybrid modified electrode had highly electrocatalytic activities for reducing H_2O_2 in comparison to Fe_3O_4 nanospheres or Ag@Au nanorods as a result of the synergetic catalytic impacts. Therefore the immunosensor was procured via immobilization of the captured antibodies on an amine-terminated

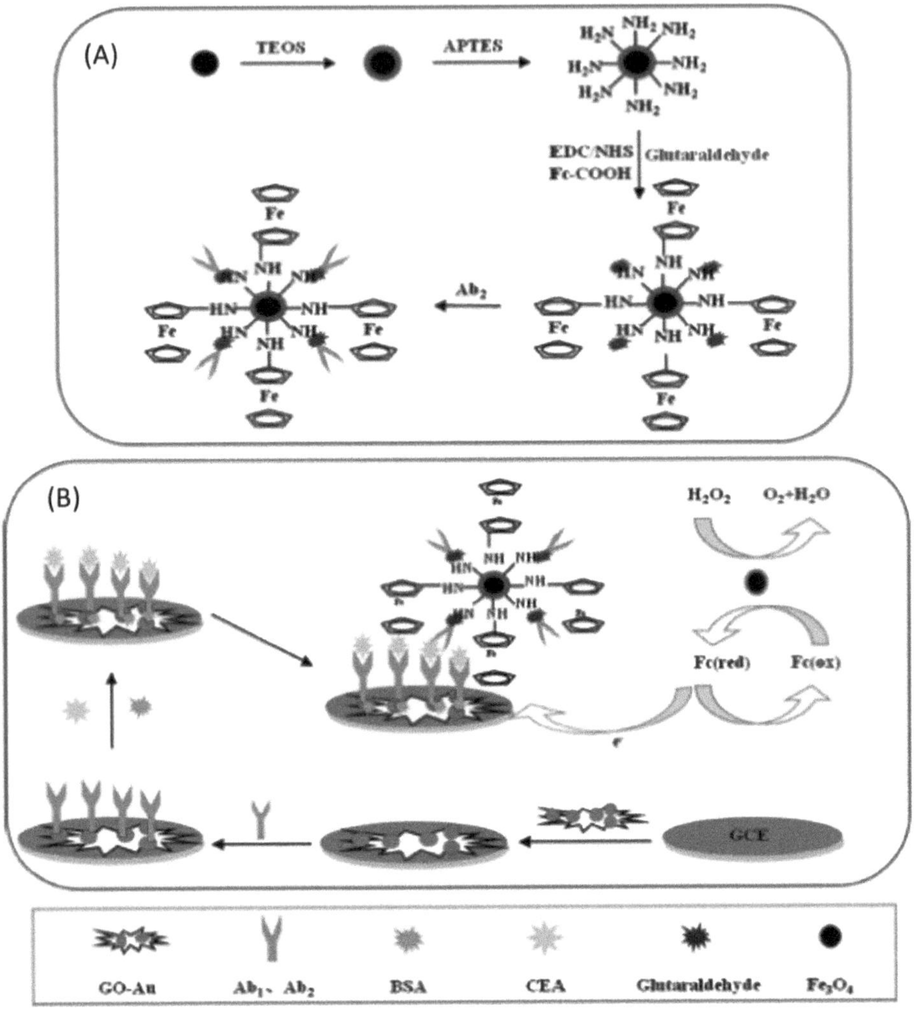

Figure 13.1 (A) Process to prepare $Fe_3O_4@SiO_2$-Fc-Ab_2/HRP bioconjugates; (B) preparation of the functionalized electrode as well as CEA electrochemical determination. *Source*: Reused with permission from T. Feng, X. Qiao, H. Wang, Z. Sun, C. Hong, A sandwich-type electrochemical immunosensor for carcinoembryonic antigen based on signal amplification strategy of optimized ferrocene functionalized $Fe_3O_4@$ SiO_2 as labels, Biosens. Bioelectron. 79 (2016) 48–54.

nanocomposite of carbon nanofibers–chitosan, whereas the trace was procured by loading determination antibodies on a Ag@Au-Fe_3O_4 nanocomposite. Upon the parameter optimization, the amperometric signal enhanced linearly with the human IgG concentration in a wide range equal to 0.1 pg mL^{-1} to 5 μg mL^{-1} with a 50 fg mL^{-1} LOD. Also, the enzyme-free catalyst–based immunosensor exhibited reasonable stability, selectivity, and reproducibility [23].

Martín's group devised a novel amperometric magneto immunoassay based on the utilization of the core–shell MNPs and screen-printed carbon electrodes (SPCEs) to selectively detect *Legionella pneumophila* SG1. A particular captured antibody (Ab) was linked to poly(dopamine)-modified MNPs (MNPs@pDA-Ab) and then incubated with bacteria. These captured bacteria were sandwiched with the use of the antibody labeled with HRP (Ab-HRP), and thus the resultant MNPs@pDA-Ab-*Legionella pneumophila*-Ab-HRP were captured by a magnetic field on the surface of the electrode. Also, amperometric response gauged at −0.15 V versus Ag pseudo-reference electrode of SPCE following H_2O_2 addition in the presence of HQ was utilized as the transduction signal. According to the analyses, the observed LOD, without preconcentration or preenrichment steps equaled 104 colony forming units (CFUs) mL^{-1}. This procedure demonstrated an acceptable selectivity, and the MNPs@pDA-Ab showed reasonable stability over 30 days. Possible detection of *L. pneumophila* at the 10 CFU mL^{-1} level in less than 3 hours following the membrane-based preconcentration stage was also verified in their study [24].

In another work, Wang's group developed an ultrasensitive electrochemical immunosensor for detecting CEA based on Ag/MoS_2@Fe_3O_4 and an analogous enzyme-linked immunosorbent assay (ELISA) procedure. Therefore researchers utilized a 96-well microplate, and a determination procedure called analogous ELISA was implemented on the MGCE. Then total internal reflection fluorescence was applied for verifying the feasibility of their new technique. One of the sandwich-type strategies was utilized to fabricate the immunosensor, and Ag/MoS_2@Fe_3O_4 was applied as the label. Ag NPs were employed to modify MGCE as the sensing platform because of their very good electroconductivity. Moreover, Ag/MoS_2@Fe_3O_4 can strongly attach to the MGCE surface as a result of the magnetic feature of Fe_3O_4. Furthermore, DPV was utilized to detect CEA, as well as electrochemical impedance spectroscopy (EIS), which verified the satisfactory construction of the immunosensor. Based on the optimized experimental condition, a linear response was shown in ranges between 0.0001 and 20 ng mL^{-1} with a 0.03 pg mL^{-1} LOD [25].

Omidfar and coworkers developed one of the new chip format sandwich electrochemical immunoassays for the ultrasensitive detection of the epidermal growth factor receptor (EGFR) with the use of the Fe_3O_4/*N*-trimethyl chitosan/gold (Fe_3O_4/TMC/Au) nanocomposite as a tracing tag for labeling the anti-EGFR VHH antibody (or nanobody) and polythiophene (PT) as the immobilization platform. This EGFR was regarded as one of the prominent biomarkers because its overexpression was generally obtained in several kinds of aggressive cancers. Moreover, nanoimmunoassay (the combined nanocomposite and nanobody) was accomplished on a commercial SP platinum electrode and showed sensitivity in the femtogram/milliliter level of soluble EGFR in 25-μL specimens. Also, the prolonged stability showed reasonable efficacy of this nanoimmunosensor for more than 8 weeks. Regardless of the utilization of nanobodies for nanoelectrochemical-based immunoassays, very low and clinically respective LODs of EGFR demonstrated the direct utility of the nanoimmunosensor for quick sensitive detection of biomarkers in the clinical diagnostic centers [26].

Peng's group developed an electrochemical immunosensor with regard to a signal amplification approach to sensitively determine the CEA. GO/chitosan-ferrocene (GO/CS-Fc) was utilized as a matrix for immobilizing the primary antibodies (Ab_1). Due to the greater surface areas, much content of Ab_1 was captured on the GO/CS-Fc sensor platform, which amplified the detection signal. In addition, the functionalization of Fe_3O_4/Au NPs was done as a label for conjugating with the Ab_2. Due to the homogeneous and narrow size distribution of the Fe_3O_4/Au NPs, multiple Ab_2s were bonded over each nanoparticle, which augmented sensitivity by introducing Fe_3O_4/Au NPs/Ab_2 into the electrode surface using sandwich immunoreactions. According to the of *p*-aminophenol (AP) redox cycling amplification approach, their new immunosensor exhibited a 10-fold enhancement in determining signals in comparison with the immunosensor with any Fe_3O_4/Au NPs labeling. This new immunosensor sensitively determined CEA in ranges between 0.001 and 30 ng mL^{-1} with 0.39 pg mL^{-1} LOD [27].

Fan et al. synthesized the trimetallic AgPtCo nanodendrites (NDs) as a robust redox probe and strongly effective signal amplifier for electrochemical signal readout. One of the powerful and ultrasensitive electrochemical immunosensors based on the AgPtCo NDs as well as magnetic nanocomposites (Fe_3O_4@SiO_2@Au MNCs) were satisfactorily devised for detecting the serum human epididymis protein 4 (HE4). Therefore this procedure showed electrochemical activities of the AgPtCo NDs and enrichment, the separation impact of MNCs, the broad linear ranges between 0.001 and 50 ng mL^{-1}, as well as low LOD equal to 0.487 pg mL^{-1}. Moreover, this new sensor was utilized for analyzing HE4 concentration in real specimens, and outputs were compatible with the ones observed in the clinical procedures [28].

Li and coworkers fabricated a sensitive sandwich-type electrochemical immunosensor to quantitatively detect the CEA. For constructing the immunosensor base, Au NPs were immobilized on the bare glassy carbon electrode (GCE) surface via the electrochemical reduction of $HAuCl_4$ solution. It was found that electrodeposited Au NPs both effectively improved the immobilization of the Ab_1 and accelerated electron transfer on the electrode interface due to higher specific surface areas, acceptable biocompatibility, and more reasonable electrical conductivity. Moreover, amino-functionalized magnetic GO–loaded gold and silver core–shell NPs for adsorbing the nickel ion (GO-Fe_3O_4/Au@Ag/Ni^{2+}) were utilized as a new label for loading Ab_2. It is notable that the signal amplification approach used the synergetic impact in Au@Ag/Fe_3O_4-GO/Ni^{2+} for improving the immobilization of Ab_2, increasing the reduction abilities of nanocomposites toward H_2O_2, and ameliorating the immunosensor sensitivity. Moreover, at the optimized condition, the current signals and CEA concentrations were linearly related in ranges between 0.1 pg mL^{-1} and 100 ng mL^{-1}, and the LOD CEA equaled 0.0697 pg mL^{-1}. Also, the as-suggested immunosensor demonstrated very good functions for determining human serum specimens [29].

13.2.2 Other MNPs used in electrochemical immunosensors

Shore's group described the procurement and features of mixed-valence manganite NPs of $La_{0.67}Sr_{0.33}MnO_3$ to be utilized in electrochemical immunoassays. These NPs were synthesized with the use of the reverse micelle technique, and their surface was functionalized with silica NPs. Further, silica NPs immobilized on $La_{0.67}Sr_{0.33}MnO_3$ NPs had a ~20 nm diameter. Moreover, $La_{0.67}Sr_{0.33}MnO_3@SiO_2$ NPs were employed for developing an electrochemical immunoassay for human IgG. Furthermore, antibody (antihuman IgG) was immobilized on silica NPs over the magnetic core, and the procedure was monitored by cyclic voltammetry (CV) and square wave voltammetry (SWV). Therefore forming immunocomplex (Ab/Ag) on the electrode surface was monitored with the use of the hexacyanoferrate redox system as a redox marker. According to the results, the assay possessed a linear working range up to an IgG concentration of 5 ng mL^{-1} with a 0.6 ng mL^{-1} LOD [30].

Wu et al. designed a signal-amplified immunosensor with the use of a microwave-assisted self-assembly procedure for synthesizing the $ZnFe_2O_4$-Ag/reduced graphene oxide (rGO) nanocomposites. Outputs indicated improvement in the conductivity of $ZnFe_2O_4$-rGO nanocomposites because of the efficient suppression of rGO accumulation via inserting $ZnFe_2O_4$ and Ag NPs into the graphene (Gr) sheets. Very good sensitivity and reproducibility were observed by the microwave-assisted preparation of $ZnFe_2O_4$-Ag/rGO nanocomposites as the substrate so that Ag NPs enhanced the signal due to the effective conductive matrix. Moreover, the layer assembly procedure of the immune-sensor was approved by CV and EIS. Based on the optimized condition, this new immunosensor had acceptable linearity in broad concentration ranges between 1 pg mL^{-1} and 200 ng mL^{-1} with a 0.98 pg mL^{-1} LOD. These features may lend themselves to the satisfactory utilization of a label-free immunosensor in detecting AFP in human sera [31].

Li and coworkers designed a magneto-controlled microfluidic instrument for direct electrochemical detection of squamous cell carcinoma antigen (SCC-Ag) in serum through the use of the anti-SCC antibody (SCC-Ab)-functionalized magnetic mesoporous nanogold/thionine/$NiCo_2O_4$ hybrid nanostructures as the immune sensing probes (P_1-Ab) and HRP-SCC-Ab conjugates—labeled nanogold/Gr nanosheets as the signal tags (P_2-Ab). The sandwich immunocomplex was formed in the presence of the analyte SCC-Ag between immunosensing probes as well as signal tags. Using the external magnet, the established immunocomplex was bound to the microfluidic tool. This assay was run in the newborn calf serum consisting of 2.5 mM H_2O_2 on the labeled peroxidase on P_2-Ab toward the catalytic reduction of H_2O_2. Based on the optimized condition, an enhancement in the current was proportionate with SCC-Ag concentration from 2.5 pg mL^{-1} to 15 ng mL^{-1} and LOD equaled 1.0 pg mL^{-1} SCC-Ag [32].

Other research conducted by Li and coworkers demonstrated a simplified and possible approach for developing advanced electrochemical immunosensors based on chemically functionalized magnetic mesoporous organic or inorganic hybrid nanomaterials and procurement, description, and measurement of the respective features of the immunosensor to detect CEA in the clinical immunoassays. The as-prepared nanomaterials containing a magnetic mesoporous $NiCo_2O_4$ nanosheet, an interlayer of Nafion/thionine organic molecules, and a nanogold layer showed acceptable absorption features to attach HRP-labeled secondary anti-CEA antibody (HRP-anti-CEA). Using a sandwich-type immunoassay format, functional bionanomaterials presented reasonable analytical features for facilitating and modulating the way it was integrated into electrochemical immunosensors and allowed the detection of CEA at 0.5 pg mL^{-1} concentration [33].

Devi's group fabricated an electrochemical immunosensor to determine amyloid beta-peptide (βA_{1-42}) based on Au NP/nickel ferrite–decorated GO-CS nanocomposite (Au/$NiFe_2O_4$@GO-CS)–modified GCE as one of the efficient sensing platforms. Experts in the field analyzed βA_{1-42} as one of the potent biomarkers for use in monitoring Alzheimer's disease. A combined strongly conducting Au and $NiFe_2O_4$ NPs on the two-dimensional (2-D) GO nanosheets provided a very good platform for sensitively and selectively sensing utilizations. A miniaturized Au/$NiFe_2O_4$@GO-CS/GCE immunosensor was procured via immobilizing βA antibody over Au//$NiFe_2O_4$@GO-CS/GCE through carbodiimide coupling. According to the DPV outputs, the immunosensor selectively determined βA_{1-42} and verified a broad linear range between 1 pg mL^{-1} to 1 ng mL^{-1} and a 3.0 pg mL^{-1} LOD. Consequently, the immunosensor demonstrated its clinical viability via the detection of βA_{1-42} in cerebrospinal fluid [34].

13.3 MNMs-based electrochemical aptasensors

According to research reports, aptamers were introduced as short oligonucleotides of variable lengths that are generated in vitro by a procedure known as SELEX (Systematic Evolution of the Ligands by Exponential Enrichment), which can attach with higher affinity and specificity to diverse target molecules. Aptamers can be chemically labeled so as to facilitate their incorporation into the biosensing tools to be employed in diverse utilizations. One of the usual electrochemical aptamer-based sensors (aptasensors) utilizes the electrode surface as a platform of immobilization over which the analyte binding event is monitored for the electrochemical variations observed in the redox indicators [35,36].

13.3.1 *Iron oxide MNPs used in electrochemical aptasensors*

Jahanbani and Benvidi ameliorated two aptasensors based on a modified carbon paste electrode (CPE) with oleic acid (OA) and a magnetic bar CPE (MBCPE) with the Fe_3O_4 magnetic NPs and OA. Upon the completion of the immobilization procedure of anti-TET at the surface of the electrode, aptasensors were called CPE/OA/anti-TET and MBCPE/Fe_3O_4NPs/OA/anti-TET. These researchers made a comparison between tetracycline determine with the use of the CPE/OA/anti-TET and MBCPE/Fe_3O_4NPs/OA/anti-TET aptasensors. The linear range, as well as LOD for TET with CPE/OA/anti-TET aptasensor, equaled 1.0×10^{-12} to 1.0×10^{-7} and 3.0×10^{-13} M using EIS. A linear range and LOD for TET with CPE/OA/anti-TET aptasensor equaled 1.0×10^{-10} to 1.0×10^{-7} M with 2.9×10^{-11} M LOD with the use of DPV. Therefore MBCPE/Fe_3O_4NPs/OA/anti-TET aptasensor was utilized to detect TET, and 1.0×10^{-14} to 1.0×10^{-6} M linear range 3.8×10^{-15} M LOD was shown with EIS. Furthermore, 1.0×10^{-12} to 1.0×10^{-6} M linear range and

3.1×10^{-13} M LOD were demonstrated for MBCPE/Fe_3O_4NPs/OA/anti-TET aptasensor through the DPV method. Their new aptasensors were employed to detect tetracycline in real specimens like honey and blood serum [37].

Zhu's group applied their new magnetic biocomposite (Fe_3O_4@Au-S1/S2) for analyzing thrombin. Fe_3O_4@Au-S1/S2 contained Fe_3O_4@Au as the carriers for magnetic separation, thiolated complementary strand (S1) anchored based on an Au-S bond, thrombin binding aptamer (S2) as a recognition element, and magnetic field–induced self-assembly. Methylene blue (MB), as a redox indicator, may be absorbed into the DNA anchored over the Fe_3O_4@Au NPs surface via electrostatic interactions. Moreover, MB in the absence of thrombin was absorbed onto double-stranded DNA (S1/S2) that was anchored on Fe_3O_4@Au NPs so that a higher electrochemical signal of MB was registered via DPV. Oppositely, the complementary strand (S1) was exposed after thrombin was bonded with an aptamer. Moreover, introducing Pb^{2+}-dependent DNAzyme (S3) split S1 at the particular rA site, which leads to the significantly decreased adsorption capacity of MB. Hence, thrombin determination can be registered by monitoring the electrochemical signal reduction of MB by incubating thrombin with S3. Their new procedure showed higher sensitivity toward thrombin with wide linear ranges between 5 pmol L^{-1} and 5 nmol L^{-1} and a 1.8 pmol L^{-1} LOD [38].

Yin et al. employed a simplified electrochemical aptasensor as a novel signal amplification approach for sensitively and selectively detecting streptomycin. An aptasensor was procured via a sequential drip of the NP-functionalized magnetic multiwalled carbon nanotube (MWCNT) (Au@MWCNTs–Fe_3O_4) composite and nanoporous PtTi (NP-PtTi) alloy over the surface of GCE. According to the output, the Au@MWCNTs–Fe_3O_4/NP-PtTi composite revealed a notable synergistic impact toward the aptasensor and presented a helpful immobilization ground for the streptomycin aptamer carrier. Finally, at the optimized condition, 0.05–100 ng mL^{-1} linear range and a 7.8 pg mL^{-1} LOD were observed [39].

Hashkavayi and coworkers describe a label-free electrochemical aptasensor to selectively detect epirubicin. Therefore researchers self-assembled 5′-thiol-terminated aptamer on the carbon screen–printed electrode that was modified with the electrodeposited Au NPs on the magnetic double-charged diazoniabicyclo [2.2.2] octane dichloride silica hybrid (Fe_3O_4@SiO_2/DABCO) through the Au-S bonds. CV, linear sweep voltammetry, and EIS were applied to examine the interaction between epirubicin and the aptamer on AuNPs/Fe_3O_4@SiO_2/DABCO/SPE. Based on the optimum condition, the peak current of epirubicin experienced a linear enhancement as epirubicin concentration increased due to the switching in the aptamer conformation as well as the formation of the aptamer–epirubicin complex rather than aptamer on the surface of the modified electrode. Outputs revealed sensitivity and selectivity of the Apt/AuNPs/Fe_3O_4@SiO_2/DABCO/SPE with two linear ranges between 0.07 and 1.0 μM and 1.0 and 21.0 μM with a 0.04 μM LOD [40].

Luo et al. constructed an ultrasensitive voltammetric aptasensor for analyzing the cardiac troponin I (cTnI). The sensor was based on the DNA nanotetrahedron (NTH)–linked dual aptamer (dAPT) and the magnetic metal-organic frameworks (mMOFs) of the type Fe_3O_4@UiO-66. DNA NTH–linked dAPT (Tro4 and Tro6) was immobilized on the Au electrode to ameliorate cTnI capture efficiency. This new mMOFs Fe_3O_4@UiO-66 was decorated by the Au@Pt NPs (Au@PtNPs), G-quadruplex/hemin (GQH) DNAzyme as well as HRP and then by two kinds of aptamers for forming the signaling nanoprobes. On the other hand, an aptamer-protein-nanoprobe sandwich-type structure was created in the presence of cTnI. Then nanoprobes such as an enzyme, Fe_3O_4@UiO-66/Au@PtNP, and GQH DNAzyme were employed to catalyze HQ oxidation via H_2O_2 for electrochemical signal amplification usually at the working potential equal to -0.1 V versus Ag/AgCl. Finally, the voltammetric signal linearly increased in the concentration range between 0.01 and 100 ng mL^{-1} cTnI with a5.7 pg mL^{-1} LOD [41].

Zheng et al. developed an ultrasensitive electrochemical sensor to quantitatively determine CEA. Therefore the first copper-silicate integrated with the nitrogen-doped magnetic carbon microtubes (NCMTs@Fe_3O_4@Cusilicate) was synthesized and utilized for enriching concanavalin A (ConA), which can be applied as one of the affinity probes for selective capturing CEA. The chemical chelation between a copper ion and Con A occurred, the combined Cu silicate/Con A largely increased, and Au nanoclusters (Au NCs) immobilized with aptamers were created for signal amplification. Moreover, in the presence of the targeted protein, CEA was formed in a sandwich manner, and Au NCs were collected on the electrode via a magnetic field. Electrochemical oxidation of Au NCs to $AuCl_4^-$ was conducted, and DPV was directly implemented, which led to electrochemical responses due to $AuCl_4^-$ reduction. This new sensing system created an ultrasensitive determination scheme for CEA with the linear ranges between 0.03 and 6.00 ng mL^{-1} and 5.38 pg mL^{-1} LOD [42].

Zhang's group employed a basic atomic layer deposition (ALD) method for the controllable procurement of a smooth Fe_2O_3-coated Gr nanostructure (Fe_2O_3@Gr). It was thus possible for the researchers to adjust the Fe_2O_3 coating thickness of the Fe_2O_3@Gr nanostructure via changing the cycle number of Fe_2O_3 ALD. It was found that the generated Fe_2O_3@Gr composites were described by X-ray photoelectron spectroscopy, transmission electron microscopy, CV, and Raman spectroscopy. Fe_2O_3 was efficiently deposited on the Gr surface. In combination with the higher affinity and specificity of an aptamer, a simplistic, fast, and sensitive electrochemical aptasensor was satisfactorily built to detect thrombin. Finally, the resultant electrochemical aptasensor displayed a linear response to thrombin in concentration ranges between 1.0×10^{-11} and 4.0×10^{-9} M with a 1.0×10^{-12} M LOD (S/N $= 3$) [43].

Yin's group built a strongly sensitive electrochemical aptasensor to sensitively detect streptomycin antibiotics. Specifically, their aptasensor was made with porous carbon nanorods formed by porous carbon nanospheres and multifunctional Gr composite (Gr-Fe_3O_4-AuNPs) as the biosensing substrate (as shown in Fig. 13.2). Based on the optimum condition, this aptasensor showed higher sensitivity as well as broader linearity to streptomycin in ranges between 0.05 and 200 ng mL^{-1} with a 0.028 ng mL^{-1} LOD. The as-prepared aptasensor was satisfactorily utilized for the determination of streptomycin in real specimens [44].

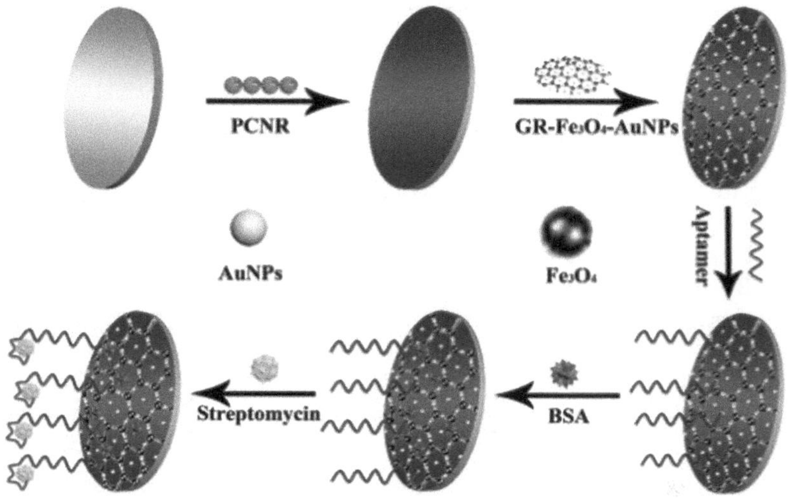

Figure 13.2 Schematic diagram of the streptomycin aptasensor.
Source: Reused with permission from J. Yin, W. Guo, X. Qin, J. Zhao, M. Pei, F. Ding, A sensitive electrochemical aptasensor for highly specific detection of streptomycin based on the porous carbon nanorods and multifunctional graphene nanocomposites for signal amplification, Sens. Actuators B Chem. 241 (2017) 151–159.

13.3.2 Other MNPs used in electrochemical aptasensors

Moreover, Baghayeri and coworkers built an electrochemical aptasensor for the label-free determination of bisphenol A (BPA) with the use of Au NPs immobilized on the functional cupper magnetic NPs ($CuFe_2O_4$-SH) and MWCNTs modified with aptamer and 6-mercapto-1-hexanol (MCH). According to the analysis, this aptasensor possessed a reasonable linearity index (0.05–9 nM) with an ultra-low LOD equal to 25.2 pM) for determining BPA. Electrochemical tests were performed with a $[Fe(CN)_6]^{3-}$[Fe $(CN)_6]^{4-}$ redox system. Based on electrochemical outputs, the presence of Au NPs, as well as MNPs and MWCNTs in the nanocomposite, resulted in greater synergism over the modified electrode surface, thus simplifying the BPA effective sensing [45].

13.4 MNMs-based electrochemical enzymatic biosensors

As shown by different studies, an ever growing attraction has emerged for designing and developing MNPs for enzymatic biosensing. MNPs have been considered to be one of the attractive materials for immobilizing favorable enzymes due to a number of desirable features like electroconductivity, biocompatibility, and simplified synthesis. Therefore enzyme-immobilized MNPs will probably exhibit certain features like larger surface areas, higher bioactivity, and very good stability. Specifically, enzyme immobilization of these NPs surfaces have offered multiple benefits like higher enzymatic activities and lower mass transfer procedures related to the substrates' detection via the enzyme [46–48].

13.4.1 Iron oxide MNPs used in electrochemical enzymatic biosensors

Qiu and coworkers developed an amperometric glucose biosensor via entrapment of glucose oxidase (GOx) in the CS composite doped with the ferrocene monocarboxylic acid–modified magnetic core–shell $Fe_3O_4@SiO_2$ NPs (FMC-AFSNPs). According to the analyses, these magnetic bio-NPs bound to the CPE surface by a permanent magnet exhibited very good electrochemical features and simultaneously functioned as a mediator for transferring the electrons between electrode and enzymes. Based on the optimized condition, the new biosensor could determine glucose in linear ranges between 1.0×10^{-5} and 4.0×10^{-3} M with 3.2 μM (S/N = 3) LOD [49].

In another study, Martin et al. addressed the preparation of the new core–shell Fe_3O_4@poly(dopamine) magnetic NPs via an in situ self-polymerization. Core–shell NPs were utilized as the solid supports to covalently immobilize the HRP, and final biofunctionalized magnetic NPs were applied for constructing an amperometric biosensor for H_2O_2. Moreover, the enzyme biosensor exhibited higher sensitivity equal to 442.14 mA M^{-1} cm^{-2}, low LOD equal to 182 nM, wider linear ranges between 6.0×10^{-7} and 8.0×10^{-4} M with higher stability for 1 month [50].

Gan et al. designed a disposable organophosphorus pesticides (OPs) enzyme biosensor based on the magnetic composite NP-modified SPCE. At first, acetylcholinesterase (AChE)-coated Fe_3O_4/Au (GMP) magnetic NP (GMP-AChE) was synthesized, and GMP-AChE was adsorbed on the SPCE surface modified by CNTs/nano-ZrO_2/Prussian blue (PB)/Nafion (Nf) composite membrane via an external magnetic field. After that, the biosensor (SPCE|CNTs/ZrO_2/PB/Nf|GMP-AChE) for OPs was made, and the inhibition degree (A%) of the AChE by OPs was determined via measurement of reduction current of PB produced by the AChE-catalyzed hydrolysis of acetyl-thiocholine (ATCh). However, in pH = 7.5 KNO_3 solution, a linearly associated dimethoate concentration ranged between 1.0×10^{-3} and 10 ng mL^{-1} with a 5.6×10^{-4} ng mL^{-1} LOD. Recovery rates in Chinese cabbage ranged between 88% and 105% [51].

Onay's group developed an amperometric determination procedure for the levels of blood glucose from GOx enzyme immobilization on the NP-modified graphite rod (GR) electrode. Moreover, Fe_3O_4-CS-Au magnetic NPs were synthesized and utilized to modify the GR electrode. The modified electrode was employed to detect the level of glucose amperometrically at 0.6 V. In addition, the calibration graph was linear in ranges between 5 and 30 mM glucose concentration with (R^2) 0.9971 coefficient of determination. Furthermore, LOD as well as LOQ equaled 0.55 and 1.83 mM. The utility of the devised procedure was investigated in real blood specimens via a comparison of the outputs of the commercial glucose sensor [52].

Hou and coworkers prepared the MOFs (ZIF-8)-coated Fe_3O_4/polypyrrole magnetic nanocomposite (Fe_3O_4/PPy@ZIF-8) with larger surface areas and higher conductivity via polymerization of pyrrole on the Fe_3O_4 NPs and self-assembling ZIF-8 on the composites. Because of the considerable specific areas as well as pore volume of MOFs on the resultant nanocomposite, GOx was efficiently immobilized on the Fe_3O_4/PPy@ZIF-8 for fabricating the new amperometric glucose biosensor. Notably, the as-procured GOx/Fe_3O_4/PPy@ZIF-8/GCE biosensor demonstrated extraordinary electrode termination function for glucose with wider linear ranges between 1 μM and 2 mM and a 0.333 μM (S/N = 3) LOD [53].

Borisova et al. explored the assembly of their new layer-by-layer biosensor architecture with the use of hybrid nanomaterials to build amperometric enzyme biosensors. This nanostructured sensing interface was procured with poly(dopamine)-modified magnetic NPs that were covalently coated with the four-generation ethylene-diamine core polyamidoamine G-4 dendrimers and additionally decorated with platinum NPs. The researchers completely described the nanohybrid and layered-on GCEs coated with a Gr oxide-carboxy-methylcellulose hybrid nanomaterial via electrostatic interactions. Then the nanostructured surface was utilized as the scaffold for covalently immobilizing the enzyme xanthine oxidase via a glutaraldehyde-mediated cross-linking. The fabrication processes of the enzyme electrode are indicated in Fig. 13.3. The enzyme electrode enabled the amperometric determination of xanthine in ranges between 50 nM and 12 μM with a higher sensitivity equal to 140 mA M^{-1} cm^{-2} and a lower LOD equal to 13 nM. Finally, the biosensor showed higher repeatability and reproducibility and satisfactorily quantified xanthine in fish specimens experimentally [54].

A highly sensitive acetylcholinesterase electrochemical biosensor to detect acetylcholine (Ach) was developed by da Silva and Brett. The researchers immobilized AChE on a poly(neutral red) film established by potential cycling electropolymerization on the Fe_2O_3 magnetic NP–modified electrodes in the acid-doped ethaline deep eutectic

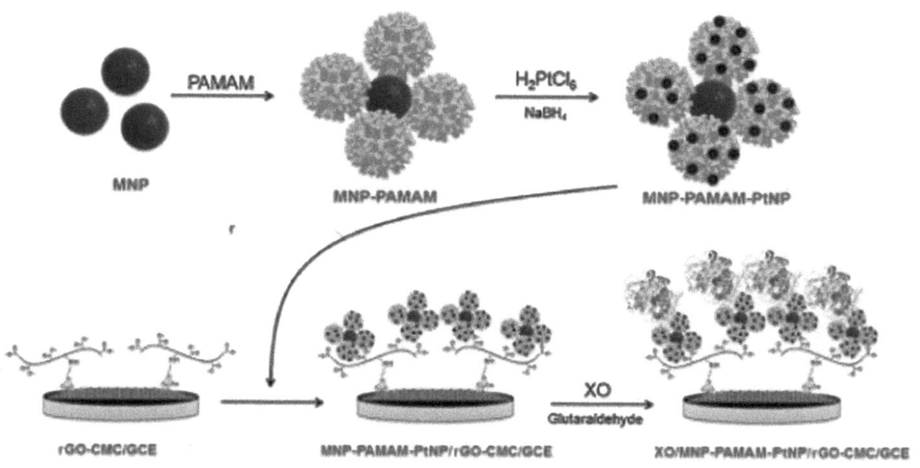

Figure 13.3 Procurement of the MNP-PAMAM-PtNP, as well as XO/MNP-PAMAM-PtNP/rGO-CMC/GCE enzyme electrode.

Source: Reused with permission from B. Borisova, A. Sánchez, S. Jiménez-Falcao, M. Martín, P. Salazar, C. Parrado, et al., Reduced graphene oxide-carboxymethylcellulose layered with platinum nanoparticles/PAMAM dendrimer/magnetic nanoparticles hybrids. Application to the preparation of enzyme electrochemical biosensors, Sens. Actuators B Chem. 232 (2016) 84–90.

solvent. Then they examined the influences of diverse acid dopants with NO_3^-, SO_4^{2-}, Cl^-, and ClO_4 anions on the growth rate as well as the electrochemical features of the polymer films and compared them with PNR growth in the aqueous solution. The modified electrode was described by CV, EIS, and scanning electron microscopy (SEM). AChE was immobilized with the use of glutaraldehyde cross-linking on the best polymer film Fe_2O_3 NP support, and the polymer formed in nitric acid–doped ethaline. Finally, this new biosensor had 1.04 μM LOD for Ach and was satisfactorily utilized for determining ACh in synthetic urine with very good recovery [55].

Sarika's group devised an amperometric catechol biosensor on the basis of α-Fe_2O_3 nanocrystals (NCs) incorporated into the CPE. Therefore laccase enzyme was assembled in the surface of the modified electrode to form a nanobiocomposite that enhanced the electron transfer reaction at the enzyme's active metal centers to oxidize catechol. Moreover, the biosensor provided acceptable sensitivity with the linear determination response in ranges between 8 and 800 μM with a 4.28 μM LOD. The NCs enjoy much potency in constructing biosensors to sensitively detect phenol derivatives [56].

Umar's group report fabricating the strongly selective and sensitive voltammetric or amperometric cholesterol biosensor based on cholesterol oxidase co-immobilized with α-Fe_2O_3 micro pine–shaped hierarchical structures. α-Fe_2O_3 micro pine–shaped hierarchical structures were synthesized via a simplistic hydrothermal procedure in larger quantities and described comprehensively with diverse procedures. These new cholesterol biosensors displayed a very high and reproducible sensitivity equal to 78.56 $\mu A\ mM^{-1}\ cm^{-2}$, and LOD equaled 0.018 mM. Moreover, the biosensor demonstrated a linear dynamic range between 0.1 and 8.0 mM, and the correlation coefficient was equal to $R = 0.9951$. Notably, the lower values of the obvious Michaelis–Menten constant (K_m^{app}) of 0.007 mM displayed a higher affinity between cholesterol and cholesterol oxidase (CHOx) immobilized on the α-Fe_2O_3 micro pine–shaped hierarchical structures [57].

Sethuraman et al. investigated the fabrication and analytical function for a catechol biosensor (PEDOT-rGO-Fe_2O_3-PPO composite modified GCE). The GO-doped conducting polymer poly(3,4-ethylene dioxythiophene) (PEDOT) was procured via electrochemical polymerization through potential cycling. Moreover, PEDOT-GO was reduced by the amperometric procedure. Then the synthesis of the Fe_2O_3 NPs was done in ethanol via the hydrothermal procedure. A mixture of PPO, glutaraldehyde, and Fe_2O_3 was cast on the PEDOT-rGO electrode. After that, catechol was detected successfully via DPV. Thus this new biosensor showed the greatest current response at a pH of 6.5. Consequently, the catechol biosensor displayed wider sensing linear ranges between 4×10^{-8} and 6.20×10^{-5} M and a decreased LOD equal to 7×10^{-9} M. Finally, the activation energy (E_a) of the enzyme electrode equaled 35.93 $kJ\ mol^{-1}$ at a temperature of 50°C, and the PEDOT-rGO-Fe_2O_3-PPO biosensor has stability for at least 75 days in the case of storage in a buffer at nearly 4°C [58].

13.4.2 Other MNPs used in electrochemical enzymatic biosensors

Singh and coworkers reported outputs of examinations associated with in situ syntheses of a new ternary $NiFe_2O_4$/CuO/FeO-CS nanocomposite that may be applicable as a cholesterol biosensor. Cholesterol was quantified via immobilization of the CHOx over a CS-$NiFe_2O_4$/CuO/FeO nanocomposite ($NiFe_2O_4$/CuO/FeO-CS NC) that was deposited on the indium-tin-oxide (ITO) glass substrate through the sol–gel procedure. Electrochemical outputs of the biocompatible CHOx/$NiFe_2O_4$/CuO/FeO-CS/ITO electrode revealed acceptable linearity (50–5000 mg L^{-1}), a lower LOD (313 mg L^{-1}), higher sensitivity (0.043 A (mg L^{-1} cm^{-2})$^{-1}$), faster response time (10 seconds), as well as a shelf life equal to 3 months. Moreover, the lower Michaelis–Menten constant (K_m) of 80 mg L^{-1} (0.21 mM) indicated a higher affinity of CHOx to the analytes. The bioelectrode could be utilized in clinics for estimating the level of cholesterol with intangible interference (2%) from the analytes found in human sera specimens [59].

Yardımcı and coworkers developed a novel H_2O_2 biosensor based on HRP that was immobilized in the $CoFe_2O_4$–CS nanocomposite to determine H_2O_2. HRP was entrapped in $CoFe_2O_4$–CS nanocomposite film, and the immobilized enzyme was able to maintain its bioactivity. This biosensor revealed a quick amperometric response to H_2O_2, and the linear range for detection of H_2O_2 ranged between 3×10^{-2} and 8 mM, with a 2×10^{-3} mM LOD [60].

Singh et al. explored a novel hybrid nanocomposite by means of nanostructured CS and $NiFe_2O_4$ (n-$NiFe_2O_4$) to immobilize bioenzyme (cholesterol esterase (CHEt) and CHOx) as the total cholesterol biosensor. This CHEt–CHOx/n-$NiFe_2O_4$–CS nanocomposite film showed larger specific areas, higher conductivity, an acceptable biocompatibility, faster redox features, as well as greater antimicrobial activities. Moreover, the constructed CHEt–CHOx/n-$NiFe_2O_4$–CS/ITO bioelectrode exhibited great improvement in the amperometric biosensing function: acceptable linearity (5–400 mg dL^{-1}), lower LOD of 24.46 mg dL^{-1} cm^{-2}, higher sensitivity equal to 1.73 μA (mg dL^{-1} cm^{-2})$^{-1}$, reproducibility of more than 15 times, a faster response time of 15 seconds, a shelf life of nearly 90 days, as well as a lower Michaelis–Menten constant (K_m) value equal to 7.05 mg dL^{-1} (0.1825 mM). Finally, their modified bioelectrode could be employed to approximate the total cholesterol in human sera specimens [61].

Li's group prepared a self-assembled $NiFe_2O_4$/CNTs sponge via the ice-templating technique. This tool synergized the beneficial properties of both three-dimensional (3D) porous nanostructure and the catalytic features of CNTs with $NiFe_2O_4$ and GOx NPs. Constructing the porous network from $NiFe_2O_4$/CNTs sheets provided wider specific surfaces to immobilize GOx and presented opportunities to simplify electron transport and reactants diffusion. Moreover, using abnormal-valence elements of Ni and Fe, double catalysis occurred, and researchers observed greater glucose biosensing functions. Finally, the constructed glucose biosensor showed two larger linear ranges (0–3.0 and 3.2–12.4 mM) and a distinctive sensitivity of 84.1 and 24.6 A mM^{-1} cm^{-2} [62].

13.5 MNMs-based electrochemical DNA biosensors

It is widely accepted that DNA was introduced as a carrier of genetic information as well as the foundation material of biological heredity. Moreover, sensing specific DNA sequences contributes importantly to clinical diagnostics, gene therapy, the safety of foodstuffs, environmental analyses, as well as biodefense utilizations [63]. It is notable that since Watson and Crick suggested the double helix structure of DNA in 1953 [64,65], numerous scholars were attracted by DNA sensors. During recent years, in order to meet ever increasing demands for ultrasensitive determination, scholars have devised many procedures for enhancing DNA response based on electrochemical sensing via the modification of DNA with diverse functional substances. Hence, as nanotechnology is continuously developed, researchers have largely considered diverse metal NPs [66,67]. Due to electroconductivity, biocompatibility, and simplistic synthesis, the MNPs ameliorated analytical functions for obtaining amplified determination signals and the stabilized recognition probes or biosensing interfaces [68].

13.5.1 Iron oxide MNPs used in electrochemical DNA biosensor

Li's group developed a Fe_2O_3@Au core/shell NP-based electrochemical DNA biosensor to amperometrically detect *Escherichia coli*. Moreover, the magnetic Fe_2O_3@Au NPs were procured via the reduction of $HAuCl_4$ over the Fe_2O_3 NPs. The DNA biosensor was established on the sandwich determination approach that included immobilizing the capture probe on the MNPs, target, and reporter probe labeled with the HRP. In the case of the addition of the magnetic field, such sandwich complexes were magnetically isolated, and HRP limited at the MNPs surface is able to catalyze the enzyme substrate and produce electrochemical signals. Notably, the biosensor can determine concentrations above 0.01 pM DNA target and 500 CFU mL^{-1} of *E. coli* with no nucleic acid amplification steps. Finally, it is possible to lessen the LOD to 5 CFU mL^{-1} of *E. coli* following 4.0 hours of incubation [69].

Benvidi and Jahanbani designed a simplistic novel electrochemical biosensor to detect DNA based on a MBCPE modified with iron oxide as well as silver NPs (Fe_3O_4@Ag). The researchers utilized the MBCPE/Fe_3O_4@Ag composite in connection with a self-assembly monolayer (SAM) of thiolated single-stranded DNA (SH-DNA). Moreover, electrochemical procedures were utilized to describe the devised biosensor. Notably, experimental conditions—that is, immobilization and hybridization procedures—were optimized CV and EIS. In addition, MBCPE/Fe_3O_4@Ag/ssDNA biosensor demonstrated a broad linear range and lower LOD equal to 0.03 fM to detect *BRCA1* 5382 mutation by EIS. Based on the optimized condition, this new biosensor showed higher selectivity to discriminate complementary sequences from the noncomplementary ones [70].

Miao et al. demonstrated an electrochemical biosensor on the basis of DNA-modified Fe_3O_4@Au magnetic NPs. Three DNA probes were devised for containing specific mismatched base pairs: One was thiolated and modified on Fe_3O_4@Au NPs surface (DNA 1). The other two probes (DNA 2 and 3) were labeled with two independent electrochemical specimens. The stable structure of the cytosine-Ag^+-cytosine and thymine-Hg^{2+}-thymine established in the presence of Hg^{2+} and Ag^+ may help to hybridize DNA 1/DNA 2 and DNA 1/DNA 3, which located the corresponding electrochemical specimens over the magnetic NPs surface. Consequently, these nanocomposites were utilized as the selective electrochemical probes to determine the heavy metal ions via registering SWV signals. According to the analysis, the simultaneous determination of Hg^{2+} and Ag^{+} revealed without considerable interferences, and their great sensitivity was maintained, meeting the U.S. Environmental Protection Agency (USEPA) criteria. Both of the LODs for Ag^{+} and Hg^{2+} were found to be 3.4 and 1.7 nM, respectively. Outputs confirmed that the DNA-modified magnetic NPs—based sensing procedure can enjoy potent utilizations to monitor heavy metal ions in the real sample analyses [71].

Arvand and coworkers designed a simplistic and efficient electrochemical DNA biosensor (CPE modified with ds-DNA/poly(L-cysteine)/Fe_3O_4 NPs-GO (ds-DNA/p (L-Cys)/Fe_3O_4 NPs-GO/CPE)) for sensitively detect guanine (G) and adenine (A). They examined electrocatalytic oxidation of G and A on the electrode using DPV and CV. Their new sensor showed isolated as well as organized peaks for G and A, through which it is possible to detect such biological bases simultaneously or individually. Moreover, ds-DNA/p(L-Cys)/Fe_3O_4 NPs-GO/CPE showed the increased peak current, as well as electron transfer kinetics and decreased overpotential to the oxidation reaction of G and A. Based on the optimized condition, a linear correlation was shown between peak current and the analytes' concentration in ranges between 0.01 and 30.0 μM and 0.01 and 25.0 μM to simultaneously determine G and A with 3.48 and a 1.59 nM LOD. Moreover, individual detection led to two linear concentration ranges between 0.01 and 30.0 μM for A and 0.01–25.0 μM for G with a 3.90 and 1.58 nM LOD for A and G. Consequently, thermal denatured single-stranded DNA was measured and the (G + C)/(A + T) value of DNA was computed to be nearly 0.77 for diverse DNA specimens [72].

Gu et al. constructed a DNA/CS-Fe_3O_4 MNP biocomplex film to immobilize the HRP on GCE. HRP was mixed with CS, DNA, and Fe_3O_4 NPs and utilized at the surface of the electrode for forming an enzyme-incorporated polyion complex film. Further, SEM was utilized for examining the surface characteristics of the DNA/CS/Fe_3O_4/HRP layer. Outputs of the EIS showed that Fe_3O_4 and enzyme were satisfactorily immobilized over the electrode surface with DNA/CS biopolyion complex membranes. Moreover, direct electron transfer (DET) as well as the bioelectrocatalysis of HRP in the DNA/CS/Fe_3O_4 film were examined by CV and fixed potential amperometry. According to the analysis, the HRP-immobilized electrode underwent DET and showed a rapid electron transfer rate constant equal to 3.7 $second^{-1}$. Furthermore, CV outputs reflected that the modified electrode results in the organized peaks in PBS that corresponded to the electrochemical redox reactions between HRP (Fe(III)) and HRP(Fe(II)). Consequently, this electrode showed an electrocatalytic reduction behavior toward H_2O_2, and the final DNA/CS/Fe_3O_4/HRP/GCE showed higher sensitivity (20.8 A cm^{-2} M^{-1}) toward H_2O_2. Finally, a linear response to H_2O_2 measurements was observed in ranges between 2 μM and 100 μM ($R^2 = 0.99$) and an amperometric LOD equal to 1 μM (S/N = 3) [73].

Yang's group decorated 3D-GO with Fe_3O_4 for the formation of Fe_3O_4@3D-GO nanocomposites, which were additionally utilized as the electrode substances in an electrochemical DNA sensor to determine Ag^{+}. Fig. 13.4 outlines the procurement process of the DNA electrochemical biosensor based on the 3D-GO@Fe_3O_4 nanocomposite to determine the Ag^{+} ions. As a result of the chemical functions as well

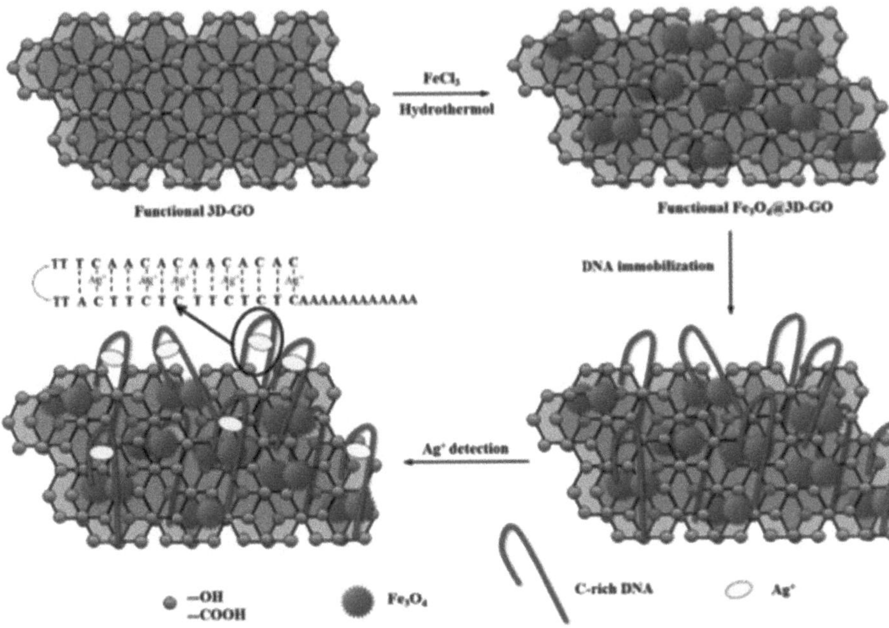

Figure 13.4 Schema of the electrochemical sensor based on functional Fe_3O_4@3D-GO nanocomposite to detect Ag^+ in the aqueous solution.
Source: Reused with permission from Y. Yang, M. Kang, S. Fang, M. Wang, L. He, X. Feng, et al., A feasible C-rich DNA electrochemical biosensor based on Fe_3O_4@3D-GO for sensitive and selective detection of Ag^+, J. Alloy Compd. 652 (2015) 225–233.

as porous structures of these new nanocomposites, single-stranded and cytosine-rich oligonucleotides can be anchored over their surfaces or interior accompanied by a selective determination of Ag^+ via forming the C-Ag^+-C complex forms. Outputs of the electrochemical measurement indicated that Fe_3O_4@3D-GO nanocomposites exhibited higher sensitivities to determine Ag^+ with 2.0 pM LOD in ranges between 0.01 and 100 nM Ag^+ (S/N = 3). Thus the devised electrochemical biosensor exhibited greater selectivity and acceptable repeatability and stability in determining Ag^+ ions [74].

13.5.2 Other MNPs used in electrochemical DNA biosensing

Sadat Vajedi and Dehghani fabricated a new highly sensitive electrochemical DNA biosensor for the electroanalytical study of etoposide using a new ZnAl/layered-double hydroxide–modified cobalt ferrite/GO nanocomposite (GO/$CoFe_2O_4$/ZnAl-LDH) that was electrophoretically deposited on fluorine tin oxide (FTO) substrate. This DNA biosensor was prepared via electrostatic adsorption of DNA onto the GO/$CoFe_2O_4$/ZnAl-LDH/FTO electrode. The electrochemical behavior of electrodes was described with CV and EIS. EIS plot demonstrated a rapid electron transport at low frequencies for

$GO/CoFe_2O_4/ZnAl$-LDH. DPV and square wave anodic stripping voltammetry were employed for the electrochemical determination of ETO. The results revealed that $DNA/GO/CoFe_2O_4/ZnAl$-LDH/FTO bioelectrode had ultrahigh sensitivity to the ETO with a 0.0010 μM LOD in a linear range between 0.2 and 10 μM. The developed biosensor revealed precise reproducibility and excellent stability of about 95% of the initial activity after 6–7 weeks. On the other hand, the present bioelectrode was also capable of distinguishing different interferences and was also used to detect etoposide in real specimens like human blood plasma, sera, and urine with reasonable recoveries, ranging from 97.0% to 104.0% [75].

13.6 Conclusions and perspectives

Nanotechnology provided specific chances to devise ultrasensitive biosensors and bioassays. In biosensing devices, the MNPs could be utilized in the surface of the sensor or as the labels. Moreover, magnetic labeling of the biomolecules was considered to be one of the interesting propositions due to the lack of a magnetic background in nearly every biological specimen. Nonetheless, implementing the magnetic labels demands biocompatibility, monodispersion, and sufficient functionalization for reducing nonspecific binding. Also, the functionalized MNPs with appropriate functional groups, as well as the surface immobilization procedure, may contribute vitally to the considerable improvements in the sensitivity of biosensing tools. Hence, investigations concentrated on synthesizing and characterizing MNP composites and the respective behaviors in biosensing tools will be highly crucial. Consequently, additional studies on properly functionalized MNMs are suggested, which would be appropriate for multianalyte determination systems in the coming years. Thus it is expected that the iron and iron oxide MNPs–based electrochemical biosensors contribute increasingly to electroanalytical science in the coming years. To have electroanalytical utilizations, it is necessary to do investigations of the advancement in the protocols for synthesizing and functionalizing iron and iron oxide MNPs. Therefore, for analytical and bioanalytical applications, researchers must consider the greater sensitivity and selectivity of nanosensors as well as lifetime-based determination procedures. Rapid advancements in the MNP-based electrochemical biosensors have highlighted their further utilization in different scientific areas. However, key influences may be related to the initial determination of diseases, biotargets, and genetic mutations. Moreover, electrochemical biosensors have provided quick, simplified, and sensitive determinations systems for cancer, which could offer powerful instruments for anticancer biosensor investigations.

References

[1] T.A. Silva, F.C. Moraes, B.C. Janegitz, O. Fatibello-Filho, Electrochemical biosensors based on nanostructured carbon black: a review, J. Nanomater. 2017 (2017). Article ID 4571614, 14 pages.

[2] A. Khanmohammadi, A. Aghaie, E. Vahedi, A. Qazvini, M. Ghanei, A. Afkhami, et al., Electrochemical biosensors for the detection of lung cancer biomarkers: a review, Talanta 206 (2020) 120251.

[3] S. Gupta, C.N. Murthy, C.R. Prabha, Recent advances in carbon nanotube based electrochemical biosensors, Int. J. Biol. Macromol. 108 (2018) 687–703.
[4] C. Mousty, Biosensing applications of clay-modified electrodes: a review, Anal. Bioanal. Chem. 396 (2010) 315–325.
[5] W. Putzbach, N.J. Ronkainen, Immobilization techniques in the fabrication of nanomaterial-based electrochemical biosensors: a review, Sensors 13 (2013) 4811–4840.
[6] S. Barua, H.S. Dutta, S. Gogoi, R. Devi, R. Khan, Nanostructured MoS_2-based advanced biosensors: a review, ACS Appl. Nano Mater. 1 (2017) 2–25.
[7] T. Kuila, S. Bose, P. Khanra, A.K. Mishra, N.H. Kim, J.H. Lee, Recent advances in graphene-based biosensors, Biosens. Bioelectron. 26 (2011) 4637–4648.
[8] A.E. Urusov, A.V. Petrakova, A.V. Zherdev, B.B. Dzantiev, Application of magnetic nanoparticles in immunoassay, Nanotechnol. Russ. 12 (2017) 471–479.
[9] J. Llandro, J.J. Palfreyman, A. Ionescu, C.H.W. Barnes, Magnetic biosensor technologies for medical applications: a review, Med. Boil. Eng. Comput. 48 (2010) 977–998.
[10] N.V. Vallabani, S. Singh, A.S. Karakoti, Magnetic nanoparticles: current trends and future aspects in diagnostics and nanomedicine, Curr. Drug Metab. 20 (2019) 457–472.
[11] T.A. Rocha-Santos, Sensors and biosensors based on magnetic nanoparticles, TrAC Trends Anal. Chem. 62 (2014) 28–36.
[12] A. Rios, M. Zougagh, Recent advances in magnetic nanomaterials for improving analytical processes, TrAC Trends Anal. Chem. 84 (2016) 72–83.
[13] M. Duan, J.G. Shapter, W. Qi, S. Yang, G. Gao, Recent progress in magnetic nanoparticles: synthesis, properties, and applications, Nanotechnology 29 (2018) 452001.
[14] B. Jiang, L. Lian, Y. Xing, N. Zhang, Y. Chen, P. Lu, et al., Advances of magnetic nanoparticles in environmental application: environmental remediation and (bio) sensors as case studies, Environ. Sci. Pollut. Res. 25 (2018) 30863–30879.
[15] Y. Xu, E. Wang, Electrochemical biosensors based on magnetic micro/nano particles, Electrochim. Acta 84 (2012) 62–73.
[16] Y. Chen, Y. Li, D. Deng, H. He, X. Yan, Z. Wang, et al., Effective immobilization of Au nanoparticles on TiO_2 loaded graphene for a novel sandwich-type immunosensor, Biosens. Bioelectron. 102 (2018) 301–306.
[17] W.T. Li, Y. Wang, F.F. Deng, L.L. Liu, H.J. Nan, H. Li, Electrochemical immunosensor for carcinoembryonic antigen detection based on Mo–Mn_3O_4/MWCNTs/chits nanocomposite modified ITO electrode, Nano 10 (2015) 1550111.
[18] I.M. Hsing, Y. Xu, W. Zhao, Micro-and nano-magnetic particles for applications in biosensing, Electroanalysis 19 (2007) 755–768.
[19] Y. Zhuo, P.X. Yuan, R. Yuan, Y.Q. Chai, C.L. Hong, Bienzyme functionalized three-layer composite magnetic nanoparticles for electrochemical immunosensors, Biomaterials 30 (2009) 2284–2290.
[20] Y. Yuan, S. Li, Y. Xue, J. Liang, L. Cui, Q. Li, et al., A Fe_3O_4@ Au-basedpseudo-homogeneous electrochemical immunosensor for AFP measurement using AFP antibody-GNPs-HRP as detection probe, Anal. Biochem. 534 (2017) 56–63.
[21] Z. Sun, W. Wang, H. Wen, C. Gan, H. Lei, Y. Liu, Sensitive electrochemical immunoassay for chlorpyrifos by using flake-like Fe_3O_4 modified carbon nanotubes as the enhanced multienzyme label, Anal. Chim. Acta 899 (2015) 91–99.
[22] T. Feng, X. Qiao, H. Wang, Z. Sun, C. Hong, A sandwich-type electrochemical immunosensor for carcinoembryonic antigen based on signal amplification strategy of optimized ferrocene functionalized Fe_3O_4@ SiO_2 as labels, Biosens. Bioelectron. 79 (2016) 48–54.

[23] H. Zhang, L. Ma, P. Li, J. Zheng, A novel electrochemical immunosensor based on nonenzymatic Ag@ Au-Fe_3O_4 nanoelectrocatalyst for protein biomarker detection, Biosens. Bioelectron. 85 (2016) 343–350.

[24] M. Martín, P. Salazar, C. Jiménez, M. Lecuona, M.J. Ramos, J. Ode, et al., Rapid *Legionella pneumophila* determination based on a disposable core–shell Fe_3O_4@ poly (dopamine) magnetic nanoparticles immunoplatform, Anal. Chim. Acta 887 (2015) 51–58.

[25] Y. Wang, G. Zhao, Y. Zhang, X. Pang, W. Cao, B. Du, et al., Sandwich-type electrochemical immunosensor for CEA detection based on Ag/MoS_2@ Fe_3O_4 and an analogous ELISA method with total internal reflection microscopy, Sens. Actuators B Chem. 266 (2018) 561–569.

[26] K. Omidfar, M. Darzianiazizi, A. Ahmadi, M. Daneshpour, H. Shirazi, A high sensitive electrochemical nanoimmunosensor based on Fe_3O_4/TMC/Au nanocomposite and PT-modified electrode for the detection of cancer biomarker epidermal growth factor receptor, Sens. Actuators B Chem. 220 (2015) 1311–1319.

[27] D. Peng, R.P. Liang, H. Huang, J.D. Qiu, Electrochemical immunosensor for carcinoembryonic antigen based on signal amplification strategy of graphene and Fe_3O_4/Au NPs, J. Electroanal. Chem. 761 (2016) 112–117.

[28] L. Fan, Y. Yan, B. Guo, M. Zhao, J. Li, X. Bian, et al., Trimetallic hybrid nanodendrites and magnetic nanocomposites-based electrochemical immunosensor for ultrasensitive detection of serum human epididymis protein 4, Sens. Actuators B Chem. 296 (2019) 126697.

[29] Y. Li, Y. Zhang, F. Li, M. Li, L. Chen, Y. Dong, et al., Sandwich-type amperometric immunosensor using functionalized magnetic graphene loaded gold and silver core-shell nanocomposites for the detection of Carcinoembryonic antigen, J. Electroanal. Chem. 795 (2017) 1–9.

[30] A. Shore, Z. Mazzochette, A. Mugweru, Mixed valence Mn, La, Sr-oxide based magnetic nanoparticles coated with silica nanoparticles for use in an electrochemical immunosensor for IgG, Microchim. Acta 183 (2016) 475–483.

[31] M. Wu, Y. Yang, K. Cao, C. Zhao, X. Qiao, C. Hong, Microwave-assisted preparation of $ZnFe_2O_4$-Ag/rGO nanocomposites for amplification signal detection of alpha-fetoprotein, Bioelectrochemistry 132 (2020) 107434.

[32] Q. Li, D. Tang, J. Tang, B. Su, G. Chen, M. Wei, Magneto-controlled electrochemical immunosensor for direct detection of squamous cell carcinoma antigen by using serum as supporting electrolyte, Biosens. Bioelectron. 27 (2011) 153–159.

[33] Q. Li, L. Zeng, J. Wang, D. Tang, B. Liu, G. Chen, et al., Magnetic mesoporous organic – inorganic $NiCo_2O_4$ hybrid nanomaterials for electrochemical immunosensors, ACS Appl. Mater. Interfaces 3 (2011) 1366–1373.

[34] R. Devi, S. Gogoi, H.S. Dutta, M. Bordoloi, S.K. Sanghi, R. Khan, Au/$NiFe_2O_4$ nanoparticle-decorated graphene oxide nanosheets for electrochemical immunosensing of amyloid beta peptide, Nanoscale Adv. 2 (2020) 239–248.

[35] G. Castillo, K. Spinella, A. Poturnayová, M. Šnejdárková, L. Mosiello, T. Hianik, Detection of aflatoxin B1 by aptamer-based biosensor using PAMAM dendrimers as immobilization platform, Food Control. 52 (2015) 9–18.

[36] S. Forouzanfar, F. Alam, N. Pala, C. Wang, A review of electrochemical aptasensors for label-free cancer diagnosis, J. Electrochem. Soc. 167 (2020) 067511.

[37] S. Jahanbani, A. Benvidi, Comparison of two fabricated aptasensors based on modified carbon paste/oleic acid and magnetic bar carbon paste/Fe_3O_4@ oleic acid nanoparticle electrodes for tetracycline detection, Biosens. Bioelectron. 85 (2016) 553–562.

[38] C. Zhu, W. Zhu, L. Xu, X. Zhou, A label-free electrochemical aptasensor based on magnetic biocomposites with Pb^{2+}-dependent DNAzyme for the detection of thrombin, Anal. Chim. Acta 1047 (2019) 21–27.

[39] Y. Yin, X. Qin, Q. Wang, Y. Yin, A novel electrochemical aptasensor for sensitive detection of streptomycin based on gold nanoparticle-functionalized magnetic multi-walled carbon nanotubes and nanoporous PtTi alloy, RSC Adv. 6 (2016) 39401–39408.

[40] A.B. Hashkavayi, J.B. Raoof, Design an aptasensor based on structure-switching aptamer on dendritic gold nanostructures/Fe_3O_4@ SiO_2/DABCO modified screen printed electrode for highly selective detection of epirubicin, Biosens. Bioelectron. 91 (2017) 650–657.

[41] Z. Luo, D. Sun, Y. Tong, Y. Zhong, Z. Chen, DNA nanotetrahedron linked dual-aptamer based voltammetric aptasensor for cardiac troponin I using a magnetic metal-organic framework as a label, Microchim. Acta 186 (2019) 374.

[42] J. Zheng, J. Wang, D. Song, J. Xu, M. Zhang, Electrochemical aptasensor of carcinoembryonic antigen based on concanavalin A-functionalized magnetic copper silicate carbon microtubes and gold-nanocluster-assisted signal amplification, ACS Appl. Nano Mater. 3 (2020) 3449–3458.

[43] H. Zhang, B. Zhang, A. Chen, Y. Qin, Controllable n-Fe_2O_3@ graphene nanomaterials by ALD applied in an aptasensor with enhanced electrochemical performance for thrombin detection, Dalton Trans. 46 (2017) 7434–7440.

[44] J. Yin, W. Guo, X. Qin, J. Zhao, M. Pei, F. Ding, A sensitive electrochemical aptasensor for highly specific detection of streptomycin based on the porous carbon nanorods and multifunctional graphene nanocomposites for signal amplification, Sens. Actuators B Chem. 241 (2017) 151–159.

[45] M. Baghayeri, R. Ansari, M. Nodehi, I. Razavipanah, H. Veisi, Label-free electrochemical bisphenol A aptasensor based on designing and fabrication of a magnetic gold nanocomposite, Electroanalysis 30 (2018) 2160–2166.

[46] T.T. Baby, S. Ramaprabhu, SiO_2 coated Fe_3O_4 magnetic nanoparticle dispersed multi-walled carbon nanotubes based amperometric glucose biosensor, Talanta 80 (2010) 2016–2022.

[47] J. Li, X. Wei, Y. Yuan, Synthesis of magnetic nanoparticles composed by Prussian blue and glucose oxidase for preparing highly sensitive and selective glucose biosensor, Sens. Actuators B Chem. 139 (2009) 400–406.

[48] X. Chen, J. Zhu, Z. Chen, C. Xu, Y. Wang, C. Yao, A novel bienzyme glucose biosensor based on three-layer Au–Fe_3O_4@ SiO_2 magnetic nanocomposite, Sens. Actuators B Chem. 159 (2011) 220–228.

[49] J. Qiu, H. Peng, R. Liang, Ferrocene-modified Fe_3O_4@ SiO_2 magnetic nanoparticles as building blocks for construction of reagentless enzyme-based biosensors, Electrochem. Commun. 9 (2007) 2734–2738.

[50] M. Martín, P. Salazar, R. Villalonga, S. Campuzano, J.M. Pingarrón, J.L. González-Mora, Preparation of core–shell Fe_3O_4@ poly (dopamine) magnetic nanoparticles for biosensor construction, J. Mater. Chem. B 2 (2014) 739–746.

[51] N. Gan, X. Yang, D. Xie, Y. Wu, W. Wen, A disposable organophosphorus pesticides enzyme biosensor based on magnetic composite nano-particles modified screen printed carbon electrode, Sensors 10 (2010) 625–638.

[52] A. Onay, Ü. Dogan, H. Ciftci, D. Cetin, Z. Suludere, U. Tamer, Amperometric glucose sensor based on the glucose oxidase enzyme immobilized on graphite rod electrode modified with Fe_3O_4-CS-Au magnetic nanoparticles, Ionics 24 (2018) 4015–4022.

[53] C. Hou, D. Zhao, Y. Wang, S. Zhang, S. Li, Preparation of magnetic Fe_3O_4/PPy@ ZIF-8 nanocomposite for glucose oxidase immobilization and used as glucose electrochemical biosensor, J. Electroanal. Chem. 822 (2018) 50–56.
[54] B. Borisova, A. Sánchez, S. Jiménez-Falcao, M. Martín, P. Salazar, C. Parrado, et al., Reduced graphene oxide-carboxymethylcellulose layered with platinum nanoparticles/PAMAM dendrimer/magnetic nanoparticles hybrids. Application to the preparation of enzyme electrochemical biosensors, Sens. Actuators B Chem. 232 (2016) 84–90.
[55] W. da Silva, C.M. Brett, Novel biosensor for acetylcholine based on acetylcholinesterase/poly (neutral red)–deep eutectic solvent/Fe_2O_3 nanoparticle modified electrode, J. Electroanal. Chem. (2020) 114050.
[56] C. Sarika, M.S. Shivakumar, C. Shivakumara, G. Krishnamurthy, B. Narasimha Murthy, I.C. Lekshmi, A novel amperometric catechol biosensor based on α-Fe_2O_3 nanocrystals-modified carbon paste electrode, Artif. Cell Nanomed. Biotechnol. 45 (2017) 625–634.
[57] A. Umar, R. Ahmad, S.W. Hwang, S.H. Kim, A. Al-Hajry, Y.B. Hahn, Development of highly sensitive and selective cholesterol biosensor based on cholesterol oxidase co-immobilized with α-Fe_2O_3 micro-pine shaped hierarchical structures, Electrochim. Acta 135 (2014) 396–403.
[58] V. Sethuraman, P. Muthuraja, J.A. Raj, P. Manisankar, A highly sensitive electrochemical biosensor for catechol using conducting polymer reduced graphene oxide–metal oxide enzyme modified electrode, Biosens. Bioelectron. 84 (2016) 112–119.
[59] J. Singh, M. Srivastava, P. Kalita, B.D. Malhotra, A novel ternary $NiFe_2O_4$/CuO/FeO-chitosan nanocomposite as a cholesterol biosensor, Process. Biochem. 47 (2012) 2189–2198.
[60] F.S. Yardımcı, M. Şenel, A. Baykal, Amperometric hydrogen peroxide biosensor based on cobalt ferrite–chitosan nanocomposite, Mater. Sci. Eng. C 32 (2012) 269–275.
[61] J. Singh, A. Roychoudhury, M. Srivastava, V. Chaudhary, R. Prasanna, D.W. Lee, et al., Highly efficient bienzyme functionalized biocompatible nanostructured nickel ferrite–chitosan nanocomposite platform for biomedical application, J. Phys. Chem. C 117 (2013) 8491–8502.
[62] Y. Li, M. Zhao, J. Chen, S. Fan, J. Liang, L. Ding, et al., Self-assembled $NiFe_2O_4$/carbon nanotubes sponge for enhanced glucose biosensing application, Appl. Surf. Sci. 362 (2016) 115–120.
[63] J. Das, K.B. Cederquist, A.A. Zaragoza, P.E. Lee, E.H. Sargent, S.O. Kelley, An ultrasensitive universal detector based on neutralizer displacement, Nat. Chem. 4 (2012) 642.
[64] J.D. Watson, F.H. Crick, Molecular structure of nucleic acids, Nature 171 (1953) 737–738.
[65] J.D. Watson, F.H. Crick, Genetical implications of the structure of deoxyribonucleic acid, Nature 171 (1953) 964–967.
[66] T. Kurkina, A. Vlandas, A. Ahmad, K. Kern, K. Balasubramanian, Label-free detection of few copies of DNA with carbon nanotube impedance biosensors, Angew. Chem. Int. Ed. 50 (2011) 3710–3714.
[67] H. Liu, A.P. Johnston, A programmable sensor to probe the internalization of proteins and nanoparticles in live cells, Angew. Chem. Int. Ed. 52 (2013) 5744–5748.
[68] H. Wang, R. Yang, L. Yang, W. Tan, Nucleic acid conjugated nanomaterials for enhanced molecular recognition, ACS Nano 3 (2009) 2451–2460.
[69] K. Li, Y. Lai, W. Zhang, L. Jin, Fe_2O_3@ Au core/shell nanoparticle-based electrochemical DNA biosensor for *Escherichia coli* detection, Talanta 84 (2011) 607–613.

[70] A. Benvidi, S. Jahanbani, Self-assembled monolayer of SH-DNA strand on a magnetic bar carbon paste electrode modified with Fe_3O_4@ Ag nanoparticles for detection of breast cancer mutation, J. Electroanal. Chem. 768 (2016) 47–54.
[71] P. Miao, Y. Tang, L. Wang, DNA modified Fe_3O_4@ Au magnetic nanoparticles as selective probes for simultaneous detection of heavy metal ions, ACS Appl. Mater. Interfaces 9 (2017) 3940–3947.
[72] M. Arvand, M. Sanayeei, S. Hemmati, Label-free electrochemical DNA biosensor for guanine and adenine by ds-DNA/poly (L-cysteine)/Fe_3O_4 nanoparticles-graphene oxide nanocomposite modified electrode, Biosens. Bioelectron. 102 (2018) 70–79.
[73] T. Gu, J. Wang, H. Xia, S. Wang, X. Yu, Direct electrochemistry and electrocatalysis of horseradish peroxidase immobilized in a DNA/chitosan-Fe_3O_4 magnetic nanoparticle bio-complex film, Materials 7 (2014) 1069–1083.
[74] Y. Yang, M. Kang, S. Fang, M. Wang, L. He, X. Feng, et al., A feasible C-rich DNA electrochemical biosensor based on Fe_3O_4@3D-GO for sensitive and selective detection of Ag^+, J. Alloy Compd. 652 (2015) 225–233.
[75] F. Sadat Vajedi, H. Dehghani, A high-sensitive electrochemical DNA biosensor based on a novel ZnAl/layered double hydroxide modified cobalt ferrite-graphene oxide nanocomposite electrophoretically deposited onto FTO substrate for electroanalytical studies of etoposide, Talanta 208 (2020) 120444.

Application of magnetic nanomaterials in magnetic field sensors

14

Mazaher Ahmadi, Arash Ghoorchian, Mahdie Kamalabadi, Zahra Amouzegar, Tayyebeh Madrakian and Abbas Afkhami
Faculty of Chemistry, Bu-Ali Sina University, Hamedan, Iran

14.1 Introduction

Magnetic nanomaterials (MNMs) have been utilized as both sorbents and sensors in analytical chemistry [1–8]. Their wide application originates mainly from their superparamagnetism property as well as the high surface-to-volume ratios. Small enough ferromagnetic particles (typically smaller than 20 nm) exhibit a special magnetic property called "superparamagnetism" [9]. Despite bulk materials, superparamagnetic nanoparticles exhibit single magnetic domains that align with the magnetic field quickly under an external magnetic field. Unlike bulk materials, superparamagnetic nanoparticles show no net magnetism by removing the external magnetic field, leading to easy manipulation of these materials using magnetic fields, since the magnetic moment quickly reverses [10]. The superparamagnetism property makes MNMs ideal sorbents since they can adsorb the target analyte with high adsorption capacities and can be easily separated from the medium using an external magnetic field within seconds or a few minutes. By far, MNMs have been utilized as sorbents for food analysis, bioanalysis, environmental monitoring, and forensic chemistry. On the other hand, these magnetic materials can also be utilized as sensing materials in colorimetric, resonance light scattering, electrochemical sensors, and biosensors.

Magnetic field sensors can be divided into two categories: magnetoresistive (MR) sensors and nonmagnetoresistive sensors. Magnetoresistive sensors rely on the dependence of material resistance to its magnetization state that comprises anisotropic magnetoresistance (AMR), giant magnetoresistance (GMR), tunneling magnetoresistance (TMR), and giant magnetoimpedance (GMI) sensors. The nonmagnetoresistive sensors can be utilized under high magnetic fields without being bothered by magnetic saturation. Commonly used nonmagnetoresistive magnetic field sensors include fluxgate sensors, Hall sensors, and superconducting quantum interference devices (SQUIDs) [11–14]. The magnetic field sensors have been used as detectors of MNM labels. It should be noted that designing the magnetic field sensors does not easily expand in the detection of multianalytes since the analytical

Magnetic Nanomaterials in Analytical Chemistry. DOI: https://doi.org/10.1016/B978-0-12-822131-0.00007-8

signal is generated by the magnetic moment. The magnetic moment as a single physical parameter creates this constraint. However, using superparamagnetic nanoparticles with different features (in size or/and material type) can overcome this limitation. Furthermore, superparamagnetic nanoparticles due to their unique non-magnetization curves allow the multianalyte sensing by magnetic field devices.

The application of MNMs in the magnetic field sensors has been less investigated so far. Therefore this chapter provides fundamental knowledge on the magnetic field sensor from the principal to the sensor design. Also, some recent applications of these sensors in analytical chemistry with a focus on bioanalysis are reviewed.

14.2 Magnetic field sensors

14.2.1 Magnetoresistive sensors

The magnetic field sensors can be constructed based on the magnetoresistive effect. The magnetoresistive effect explains the dependence of material's resistance to its magnetization state. Magnetoresistive sensors are simply powered by a constant current. By applying a magnetic field to these sensors, the electrical resistance changes are measured in the presence and absence of the analyte. So, the analytical signal is obtained from small changes of the magnetic field and it depends on the magnetic field along the sensor region. The output voltage of these sensors is a measure of the magnetic field. Therefore their design is low cost and their use is suitable for many applications [11,15].

The magnetization of some materials can occur/change by an externally applied magnetic field. Magnetoresistive sensors utilize the resistance change (ΔR) generated by an external magnetic field (H). The values for the magnetoresistance are given as the ratio of the resistance changes to the resistance ($\Delta R/R$) at which the resistance is set for when the external magnetic field is zero ($H = 0$). This ratio is usually small and therefore DC voltage change is small. As a result, it is necessary to use a bridge circuit or other methods to minimize the voltage in the off state (DC offset) [15]. Four different types of these sensors, which are based on different mechanisms, are described in the following sections.

14.2.1.1 Anisotropic magnetoresistance sensors

In AMR sensors, the resistance is determined by the angle between the direction of the magnetization and current flow. However, the magnetization of the substrate will rotate toward the direction of the magnetic field and the rotation angle is dependent on the magnitude of the external magnetic field. As the angle between the direction of the magnetization and the current flow increases, the resistance of substance alloy decreases. When this angle is in the orthogonal position, the resistance is the lowest. Accordingly, the corresponding resistance changes with the magnetization of the film [15].

Normally, an AMR sensor is constructed with four Permalloy resistors sputter-deposited on a silicon substrate. These four resistors design forms a two-legged current path called a bridge. If the resistance of one leg is different from the other, a voltage is created between them. The long path allows high resistance on the bridge. With this design, the power required by the sensor can be minimized [15]. The stray magnetic field created by MNMs label affects the magnetization of the thin magnetic film. Subsequently, the magnetization of the film affects the corresponding resistance and the analytical signal [12].

Some substances, such as Permalloy, show AMR. Permalloy, the most customary substance for manufacturing the AMR sensors, is a magnetic alloy with a ratio of about 4:1 of nickel and iron, respectively. The relatively large magnetoresistance for this alloy makes it suitable for designing this type of sensors [15]. The high tolerance for the defects in the film makes the AMR sensor cost-effective. Also, the AMR sensor has favorable characteristics, such as relatively appropriate sensitivity, superb flexibility, and great stability. This magnetoresistive sensor can be utilized as a wearable electronic device for navigation, a diagnosis/health monitoring sensor, etc. [12].

14.2.1.2 Giant magnetoresistance sensors

In flat metal structures, magnetoresistance is observed with larger changes, called giant magnetoresistance. Unlike the previous sensor (AMR), which consists of a single-layer film, GMR is based on a multilayered design. The simplest form of this sensor consists of a four-layer structure: two thin ferromagnetic layers (e.g., NiFe or CoFe), a conductive layer (e.g., Cu) that acts as a separator for the two ferromagnetic layers, and an antiferromagnetic layer (e.g., IrMn) to prevent the rotation of one of the ferromagnetic layers and to stabilize the magnetization of the same. Due to the role of the fourth (antiferromagnetic) layer, this composition is sometimes called the three-layer structure. The fixed ferromagnetic layer, called the hard ferromagnet, is between the conductive layer and the antiferromagnetic layer. The unfixed ferromagnetic layer is called the soft ferromagnet. This set is known as spin valve [12,15].

In GMR sensors, electrons pass through the multilayered structure that includes the three main layers of hard, conductive, and soft layers. The magnetization state of two hard and soft layers relative to each other determines the resistance/scattering of electrons in this structure. If the magnetization of these two layers is parallel, the electrons will experience less resistance/scattering to pass from an electronic band structure in one layer into the same electronic band structure in the other layer. Naturally, if the magnetization of these layers is antiparallel, the electrons will have more resistance/scattering to pass through these layers. At room temperature, the resistance difference for both parallel and antiparallel magnetization status can be as high as 12.8%. GMR sensors gain the variation of magnetoresistance of 15% [12,15]. Also, to increase the change in resistance, the main three- or four-layer structure can be repeated. It is necessary that the layers be very thin (a few nanometers thickness) and smooth [15].

14.2.1.3 Tunneling magnetoresistance sensors

TMR sensors, sometimes known as spin-dependent tunneling sensors, were first designed in 1995. TMR sensors have a four-layer structure, similar to GMR sensors, except that the separating layer is an insulator layer that is usually made of MgO or Al_2O_3. The thickness of this layer must be about 1 nm. In these sensors, the insulator layer acts as an obstacle for electrons. It provides resistance against passing the electrons and the conduction is arisen by the electron tunneling phenomenon across the insulator layer. As explained earlier for GMR sensors, one of the ferromagnetic layers is pinned by an antiferromagnetic layer [12,15].

However, TMR sensors have higher magnetoresistance and base impedance values than GMR sensors, and therefore consume less energy. Also, TMR sensors promise higher stability compared to GMR sensors. The performance of TMR may be influenced by some drawbacks such as high coercivity or low linearity [12,15].

14.2.1.4 Giant magnetoimpedance sensors

Historically, GMI has attracted particular interest in various research fields, since Panina and Mohri promulgated their discovery of the GMI effect in Co-based amorphous wires in 1994 [16,17]. If a soft ferromagnetic conductor is exposed to a small AC current and a magnetic field is applied, the conductor experiences a large change in the AC complex impedance, which is known as the GMI effect. The relative impedance change (*Z*) with the applied field (*H*), defined as the GMI effect, is expressed by Eq. (14.1), where H_{max} is the external magnetic field sufficient to saturate the magnetoimpedance [18].

$$\%\frac{\Delta Z}{Z} = \frac{Z(H) - Z(H_{max})}{Z(H_{max})} \times 100 \tag{14.1}$$

Applied DC magnetic field, AC current, and its frequency are directly effective on the magnitude of GMI effect. Depending on the kind of GMI material, the applied DC magnetic field and AC current are different. These two parameters need to be chosen as small as possible. Also, the magnitude and sensitivity of GMI change with the frequency fluctuation of the AC current. For a given material, it is necessary to define a suitable frequency range. A relatively low-frequency ranges lead to a single-peak behavior of a GMI and it is desirable for practical applications. In contrast, at the high-frequency ranges, the double peak behavior of a GMI is observed and it is undesirable. Therefore the appropriate selection of these parameters becomes extremely important to design efficient GMI sensors [19].

The magnetic materials with low resistivity, high magnetic permeability, high saturation magnetization, and a low damping parameter are suitable for the GMI effect. Hence, cobalt-based amorphous and iron-based nanocrystalline materials including thin films, ribbons, and wires are good nominees for the construction of GMI sensors. In terms of cost, iron-based materials are more economical than others [19].

The theoretical models to interpret the basic GMI features for some magnetic conductors are quasistatic, eddy current, domain, electromagnetic, and exchange-conductivity models. The mentioned models cannot explain all of the GMI features over a wide range of the AC current frequency. Therefore it seems necessary to provide other models [19].

14.2.2 Nonmagnetoresistive sensors

The nonmagnetoresistive sensors can be utilized under high magnetic fields without being bothered by magnetic saturation. Therefore the nonmagnetoresistive sensors hold advantages in some special conditions such as strong magnetic field environments or ultralow magnetic signal detection. Several types of these sensors have been developed in different sizes. Among them, the Hall, fluxgate, and SQUID magnetic sensors have obtained some interest in various research fields [12].

14.2.2.1 Fluxgate sensors

Typically, fluxgate sensors consist of two coils and a ferromagnetic material. Two coils (known as a drive or exciting coil and a sense coil) are wrapped around a ferromagnetic material. By passing a sufficiently large sinusoidal current through the drive coil, a field is generated that saturates the magnetic material of the sensor core, periodically in both directions. In this way, the core of the sensor reaches its saturated magnetization state twice per cycle. As the core becomes saturated, its permeability to the external magnetic field decreases. Reducing the current in the drive coil leads to reduce the saturation of the core. In this state (unsaturated core), the external magnetic field is integrated into the core. The alternative changes in the external magnetic field are detected by the sense coil. Usually, the results of these changes are recorded in the form of voltage-output, but some of the fluxgate sensors operate in current-output mode [15,20].

Fluxgate sensors are able to detect a small magnetic field but there is one limitation: large size in the design. According to Faraday's law, an induced electromotive force is proportional to the time derivative of the magnetic field flux. Therefore reasonable amplitude signals of a microfluxgate can only be obtained by complex circuits for signal amplification [12].

14.2.2.2 Hall sensors

The Hall sensors benefit from a physical phenomenon called Hall Effect. Consider a thin gold rectangle perpendicular to a strong magnetic field. A voltage difference occurs across this golden rectangle when an electric current is applied along with it. An electron moving in a magnetic field perpendicular to its direction experiences a force called the Lorentz force, perpendicular to both moving electron direction and magnetic field direction. The response to this force causes a Hall voltage. This phenomenon is discovered by E.H. Hall more than 150 years ago and has been used in many devices to date [15].

Although the Hall Effect on metal conductors is very small, semiconductors have a greater effect. Inexpensive Hall sensors are often made of silicon. Using special semiconductors (III−V) that have higher electron mobility than silicon, more sensitive Hall sensors can be designed [15].

A Hall sensor is capable to develop a wide linear range but has two drawbacks that prevent its broad application: the first is lower sensitivity small magnetic field than magnetoresistive sensors and the second is less sensitive than smaller ones to detect small magnetic beads (MBs) [12]. The latter drawback is due to that a local quasi-uniform magnetic field from the pole of a dipole is captured only by a small Hall sensor with a size similar to MBs. The passing direction of the stray field from the MB is both downward and upward through the large-sized sensor. In this way, the output voltage of the Hall sensor is neutralized. Therefore designing small-sized Hall sensors is necessary to overcome this problem. In general, using smaller nanoparticle beads and taking advantage of their undeniable benefits require a finer design of Hall sensors [12].

14.2.2.3 Superconducting quantum interference device sensors

The basis of the SQUID is the existence of significant interactions between electric currents and magnetic fields and the use of materials that are cooled to below the superconducting transition temperature. These temperature conditions cause the materials used to become superconductors and to lose all resistance against the electricity flow. In such a sensor, the main element is the superconducting ring with point contact junction. The ring is a toroid with a diameter of a few millimeters made of niobium or lead metals. If a line of magnetic flux is created in this ring, the current is induced and remains stable forever, in the absence of a disturbance [15].

However, these sensors have large sizes and special temperature conditions and therefore their utility is not affordable due to the high cost of manufacturing and maintenance. Constructing the necessary equipment to produce low temperatures on a small sensor or chip is difficult and is known as a major drawback of this method [12].

14.3 Application of magnetic nanomaterials in magnetoresistive sensors

In recent years, several magnetic biosensors have been utilized in bioanalysis. These sensors suffer from the large size, poor sensitivity, and the need for a high magnetic field [21]. To overcome these limitations, MR sensors have been proposed and developed because of their superior sensitivity and integration-related advantages [22]. MR sensors, especially AMR, GMR, and TMR sensors, focus on the detection of biomarkers labeled with MNMs [23]. Since GMR and TMR sensors offer advantages in terms of tunable size, excellent compatibility with metal-oxide semiconductor, and high sensitivity, they have been used successfully in many applications [12,22,24].

14.3.1 Application of AMR sensors

AMR sensors have been employed in many biosensing applications because they are simple in technology and exhibit relatively high sensitivity, excellent stability, and flexibility. In fact, AMR sensors are low-cost sensors that are simply fabricated by a magnetic film. The sensing strategy of magnetic biosensors such as the AMR sensor is based on the sensing of the stray field of MNMs that label the analytes. Quynh et al. [25] detected Fe_3O_4 nanoparticle modified with chitosan (50 nm) using a simple device constructed of four AMR sensors in the Wheatstone bridge. Fig. 14.1 shows the schematic illustration of the AMR Wheatstone bridge device. Thin layers of $Fe_{80}Ni_{20}$ alloy as the AMR elements are covered with MNMs by the drop-casting method. A pinned magnetic field is provided using a permanent magnet, resulting in the out-of-plane magnetization of MNMs. An in-plane magnetic field perpendicular to the pinned magnetic field is created by Helmholtz coils because of the sensitivity of sensors toward the in-plane component of the stray field of MNMs. The developed device exhibits the sensitivity up to 2.15 mV Oe^{-1} and the limit of detection (LOD) better than 10^{-6} emu. This low-cost device offers promise for the analysis of biomolecules, provided that the signal-to-noise ratio is increased [21].

Among MNMs, MBs have attracted more interest from the researches in the fabrication of magnetic biosensors. In this context, Hansen et al. presented an AMR sensor using MBs to detect a bacteria DNA target [26]. In this study, DNA

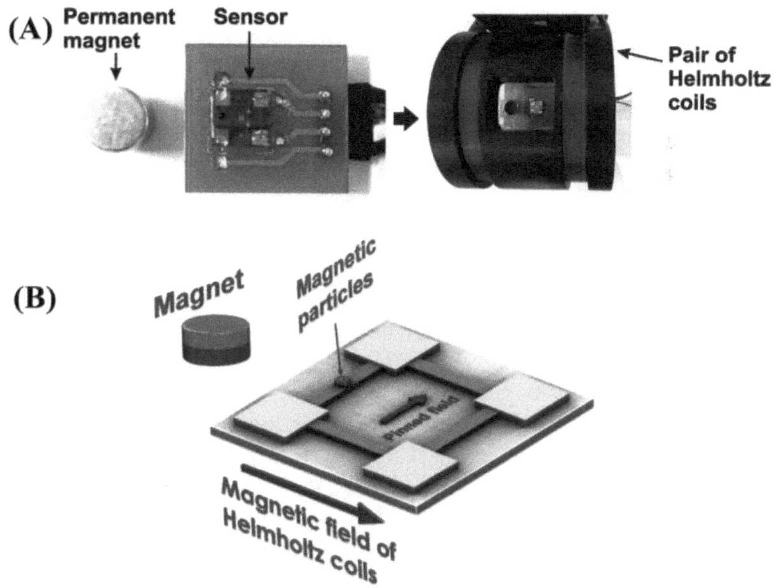

Figure 14.1 The image of (A) and schematic representation (B) of the AMR Wheatstone bridge device.
Source: Reproduced from L. Quynh, B. Tu, D. Dang, D. Viet, L. Hien, D.H. Giang, et al., Detection of magnetic nanoparticles using simple AMR sensors in Wheatstone bridge, J. Sci.: Adv. Mater. Devices 1 (2016) 98–102.

detection probes were immobilized to the surface of a card substrate. Then, biotinylated DNA targets were introduced into the fluid system together with paramagnetic beads labeled with streptavidin. MBs are linked to the sensor surface through the hybridization of biotinylated DNA targets to the DNA detection probes. The DNA-MBs were detected using AMR biosensor based on the Wheatstone bridge configuration. This biosensor showed almost two times higher sensitivity compared to a single AMR sensor [27].

14.3.2 Application of GMR sensors

Baselt et al. presented a GMR sensor for the measuring of magnetic-labeled biomolecules, for the first time [28]. They employed GMR sensors to determine DNA, antibodies, antigens, and receptors and so, proved the capability of GMR sensors as biosensors with single-molecule sensitivity. The research team proposed the use of the Bead Array Counter to hold MBs to a solid substrate. They developed a prototype DNA array composed of 64 GMR sensors. The change in the resistance of the GMR material reaches 15%. Each GMR sensor is a thin-film metal multilayer fabricated by alternatively stacking magnetic and nonmagnetic layers, and can operate independently.

Ferrimagnetic nanoparticles, such as cobalt ferrite ($CoFe_2O_4$) nanoparticles, are one of the MNMs used in biomolecular diagnostics. Zhang and coworkers utilized a GMR biosensor based on $CoFe_2O_4$ to detect viruses with extremely low concentration levels [29].

GMR sensors can simultaneously detect different analytes through the surface modification of various sensors with different capture antibodies. Osterfeld et al. successfully detected multiple biomarkers in serum samples [30]. They extended analytic sensitivity into the low fmol L^{-1} concentrations using magnetic nanotags. Magnetic microsized labels in comparison to MNMs are not desirable for biomolecular assays because they are very bulky, and consequently, they diffuse slowly as well as prevent high-density binding of analytes on the sensor surface. Also, they tend to precipitate due to magnetic interactions [30,31].

The high-moment magnetic nanoparticles are the most promising types of MNMs used in GMR biosensors because they enhance the sensor signal and enable the GMR sensor to be a versatile and robust contender in biosensing applications [32]. Wang et al. used high-magnetic-moment nanoparticles with a size of 50 nm as the biomolecule labels in a GMR sensor array for the simultaneous detection of a series of cardiovascular disease biomarkers in serum samples [31]. This array was designed by the integration of 64 nanosized sensors on a chip (16 mm × 16 mm) in a reaction cell. These GMR sensors can operate individually, although they are simultaneously monitored. The biomarkers are detected in the sandwich assay format. Capture antibodies are bounded to the sensor surface composed of NiFe and CoFe bilayers. The surface of streptavidin-coated MNMs is modified with biotinylated detection antibodies. When MNMs labeled with target antigens are bound to the sensor surface through the biotin–streptavidin interaction, they are magnetized and create the stray magnetic fields that reduce the AC field. The developed sensing method can detect up to 64 proteins simultaneously. The mentioned GMR sensing array exhibited the detection of three portions, including suppression of tumorigenicity 2 (ST2), pregnancy-associated plasma protein-A (PAPP-A), and proprotein convertase subtilisin/Kexin type 9 (PCSK9) with the linear ranges up to four orders of magnitude. LOD of 40 pg mL^{-1} was obtained for ST2.

In another study, a competition bioassay using the GMR sensor was developed to directly detect endoglin, a biomarker of prostate cancer, in human urine samples [33]. In this work, two detection methodologies, including two-layer and three-layer (sandwich) competition detection, were attempted. In the two-layer detection methodology, the sensor surface was modified with capture antibodies. Competition for binding to capture antibodies occurs when the GMR sensor is exposed with a mixture of high-moment MNM-labeled endoglin and -unlabeled endoglin. In the second method, biotin-labeled analyte and -unlabeled analyte compete for capture antibodies on the functionalized sensing surface. Then, streptavidin–FeCo nanoparticle conjugates are employed to bind with MNMs-labeled analyte captured on the biosensor surface. In both methodologies, the GMR biosensor senses the dipole field produced by MNMs captured on the biosensor surface, which is sensitive to the distance between the sensor surface and the MNM label. The signal generated from the three-layer bioassay was greater than that of the two-layer bioassay. The bioassay developed by this team quantified endoglin with LOD of 83 fmol L^{-1}. Also, this research team employed the nanomagnetic competition assay using a GMR sensor for fast determination of interleukin-6, a low-abundance protein, and a potential cancer biomarker, in human serums [34]. The two-layer detection bioassay displayed a concentration range of 0.125–41.5 pmol L^{-1} and higher sensitivity in comparison to the three-layer detection assay for biomarker detection.

14.3.3 Application of TMR sensors

High sensitivity and the ability of the weak magnetic field detection ($<10^{-9}$ to 10^{-10} T) make TMR sensors or spin-valve read sensors excellent candidates for biosensing applications. Wu et al. described a practical method using TMR sensor for the rapid quantification of *Escherichia coli* O157:H7 (*E. coli* O157:H7) bacteria [35]. In the first step, the sensor surface was covered with a 50-nm gold film through the sputtering technique, and then the gold film was functionalized with monoclonal antibodies. Magnetic particles were modified with anti-*E. coli* antibodies and captured to the TMR surface through an antibody-antigen-antibody complex. MNMs change the magnetoresistance of the TMR sensor proportional to *E. coli* concentration. The needed time to detect the minimum concentration (100 CFU mL^{-1}) of bacteria was 5 hours.

Mu et al. developed a rapid method for ricin detection through the combination of the magnetic lateral flow immunochromatography technology and TMR magnetic biosensor for the first time [36]. The surface of carboxyl-modified Fe_3O_4 nanoparticles was modified with antiricin polyclonal antibodies. The MNMs labeled with antibodies were immobilized on the immunochromatographic test strip that is placed under the sensor. The detection of magnetic field intensity generated by MNMs results in the sensing and quantifying of ricin. The detection range of 1–200 μg mL^{-1} and LOD of 1 ng mL^{-1} were achieved with this new method.

14.3.4 Application of GMI sensors

GMI sensor was first coupled with biosensing applications by Kurlyandskaya et al. [37]. They presented a biosensor based on GMI using an amorphous ribbon element, $Co_{67}Fe_4Mo_{1.5}Si_{16.5}B_{11}$, for the detection of a magnetic fluid. The surface of the ribbon is covered with commercial Ferrofluid liquid thin layer in which magnetic nanoparticles of 10 nm in size are embedded. The developed GMI sensor can detect the external fringe magnetic fields produced by MNMs. Hence, GMI sensors based on a sensitive element can be employed as the biosensor in systems with molecular recognition processes comprising MNMs. A four-point measurement system containing a Cu–Pb imprinted circuit was performed to determine magnetoimpedance. The AC drive current in 0.3–10 MHz frequency range flows through sensitive magnetic element along the longitudinal GMI ribbon axis. A pair of Helmholtz coils was used to generate a 150 Oe magnetic field in the parallel direction to the axis of the ribbon. The results of this work showed that applying the current with frequencies of 1–10 Hz could enhance the sensitivity of biosensor as a result of GMI ratio enhancement.

Owing to their fascinating soft magnetic properties, including low coercivity, high permeability, and high saturation magnetization, iron–nickel alloy (NiFe) has been widely used in the fabrication of highly sensitive GMI sensors [38]. Wang et al. designed a GMI biosensor composed of Cr/Cu/NiFe/Cu/NiFe/Al_2O_3/Cr/Au films to determine alpha-fetoprotein (AFP) [39]. Sandwich NiFe/Cu/NiFe films and Au film act as GMI elements and biomolecular immunoplatform, respectively. Al_2O_3 film is served as a protective layer to avoid contamination and damage of immunoplatform. AFP monoclonal antibodies (Ab1) are immobilized on the surface of a self-assembled monolayer deposited on the Au film. A sandwich-assay configuration is formed through capturing of AFP between Ab1 antibodies and the detection monoclonal antibodies labeled with streptavidin–biotin magnetic Dynabeads (Fig. 14.2A). AFP was sensitively quantified in the range of 1–10 ng mL^{-1}. The experimental setup of the biosensor for analyte detection is presented in Fig. 14.2B. To provide a transverse AC magnetic field (H_t), an AC current with a 0.1–5 MHz frequency range is applied to the sensing elements. An external magnetic field (H_e) is provided parallel to the axis of the sensing elements, and the stray magnetic field created by Dynabeads is detected by the sensing element films. Following the same strategy, the researchers as mentioned earlier employed the GMI sensor to quantify the carcinoembryonic antigen in the concentration range of 0.001–10 ng mL^{-1} [40].

14.4 Application of magnetic nanomaterials in nonmagnetoresistive sensors

As one of the most developed non-MR sensors, fluxgate sensors have advantages of reliability, high sensitivity, wide measurement range, and low cost [41]. Hence, they are employed to measure weak magnetic fields. However, the practical

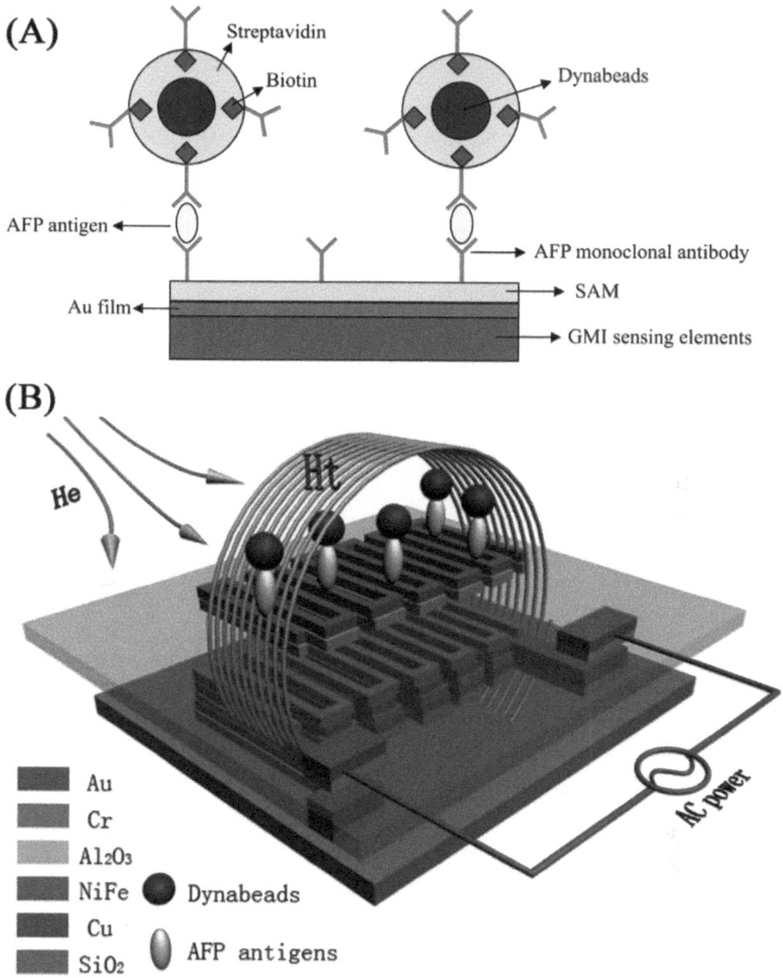

Figure 14.2 (A) Double-body sandwich assay on the GMI gold layer for AFP detection and (B) the experimental setup for the detection of AFP.
Source: Reproduced from T. Wang, Z. Yang, C. Lei, J. Lei, Y. Zhou, An integrated giant magnetoimpedance biosensor for detection of biomarker, Biosens. Bioelectron. 58 (2014) 338–344.

application of present fluxgate sensors for biomagnetic detection is still limited because of their poor long-term stability, large design sizes, and heaviness. Biomagnetic sensors should be portable, user-friendly, and miniaturized. Many researchers have improved the performance of fluxgate sensors. One proper candidate for the fabrication of efficient fluxgate sensors is the employment of microelectromechanical system (MEMS) technology to fabricate microfluxgate sensors [42]. MEMS-based fluxgate sensors have several advantages over the conventional fluxgate sensors: compact size, excellent long-term stability, and low power consumption. For example, a MEMS-based fluxgate sensor to detect *E. coli* as a target bacterium was described by Sun et al. [43]. A testing system was set up to study the developed fluxgate sensor, as indicated in Fig. 14.3A. At first, the effect of Dynabeads concentrations on the magnetic response of the sensor was studied. A magnetic dipole is produced in the Dynabeads by applying a DC magnetic field in the fluxgate sensor. This generated field is opposite to the DC field. An obvious decrease in the DC magnetic field occurred in the presence of MBs. Therefore the changes in the magnetic field are recorded proportionally to the concentration of Dynabeads. A more detailed description of the fluxgate sensor theory was organized

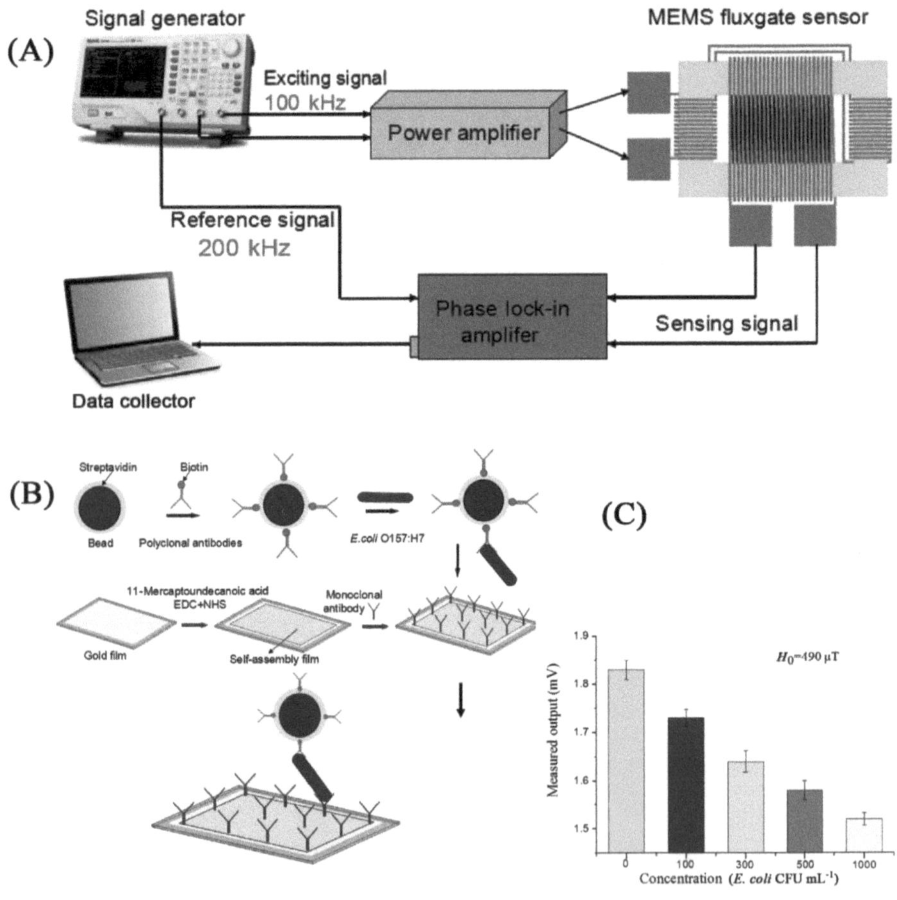

Figure 14.3 (A) The block diagram of the microfluxgate-based biosensing system, (B) schematic illustration of the detection strategy developed for the detection of *E. coli*, and (C) the changes in output signal in terms of *E. coli* concentration.
Source: Reproduced from X.-c. Sun, Z. Yang, C. Lei, Y. Liu, L. Guo, Y. Zhou, An innovative detecting way of *Escherichia coli* O157H:H7 by a micro-fluxgate-based bio-sensing system, Sens. Actuators B: Chem. 221 (2015) 985–992.

by Cao and coworkers [44]. Afterward, to prepare *E. coli* samples, a double-antibody sandwich assay was used. Fig. 14.3B illustrates a schematic overview of the detection strategy developed for the detection of *E. coli*. Upon the addition of *E. coli*, the output signal was decreased in the range of 450–700 μT. The reported immunosensor has a LOD of 100 CFU mL^{-1} for *E. coli*. The maximum sensory response for *E. coli* was obtained at 490 μT. The output signals of the as-prepared fluxgate sensor for various concentrations of *E. coli* for the fixed DC field of 490 T were displayed in Fig. 14.3C. The value of the output signal decreased with the increase in *E. coli* concentration. Later they successfully applied their method for the determination of prostate-specific antigen [45].

Besides, a flexible MEMS-based fluxgate sensor based on magnetic Dynabeads for the quantification of cardiac troponin T (cTnT) was developed by Guo et al. [46]. They employed immunomagnetic Dynabeads to specifically label cTnT through the conjugation of cTnT to anti-cTnT. The MB–cTnT conjugates were immobilized on an 11-MUA (11-mercaptoundecanoic acid) layer and a thiolated protein layer. The immobilized cTnT–Dynabeads conjugates were quantified by detecting the stray field of cTnT-labeled beads. The biosensing system displayed a linear dynamic range from 0.01 to 10 ng mL^{-1}, with a LOD of 0.01 ng mL^{-1}.

Hall sensors have some advantages containing small offset voltage, wide dynamic range, low noise, and linear response [47,48]. Despite all the advantages of Hall sensors mentioned earlier, two main drawbacks limit their application in analytical chemistry. Compared to MR devices, Hall sensors are less sensitive to the relatively small magnetic field. Another considerable limitation is the low sensitivity of larger Hall sensors to detect small MBs. In other words, a small MB can be detected by a relatively small Hall sensor, due to the similarity in their size. Nevertheless, researchers have developed microsized Hall sensors to overcome the aforementioned limitations.

Because of the advantages of Hall sensors, they can serve as high-performance biosensors. Sinha et al., in 2014, reported an aptamer-based Hall sensor to thrombin based on the sandwich-type immunoassay [49]. The sensing surface (with an active area of 50 μm × 50 μm) was constructed by immobilization of thiol-labeled primary aptamer on an Au thin film, and thrombin was introduced to the surface of modified electrode. Afterward, a secondary aptamer was employed to detect thrombin, as indicated in Fig. 14.4. Finally, the biosensor response was recorded using the addition of streptavidin-coated MNMs. The developed biosensor responses to thrombin in the linear range of 86 pmol L^{-1} to 8.6 μmol L^{-1} with a detection limit down to 86 pmol L^{-1} and a sample volume of 2 μL.

In another work, a label-free micro-Hall sensor based on the nonfunctionalized graphene was presented to determine L-histidine and urea as two important biomolecules [50]. The authors suggest that the sensing mechanism of the Hall sensor is based on the interaction between the graphene and produced ions. The symmetry of the electron–hole movement can be broken in the presence of the ions, and consequently, the shape of the Hall resistance profile is changed. The main advantage of the proposed analytical approach relies on the fact that the determination of analytes is performed in a redox-independent process. Moreover, the as-fabricated

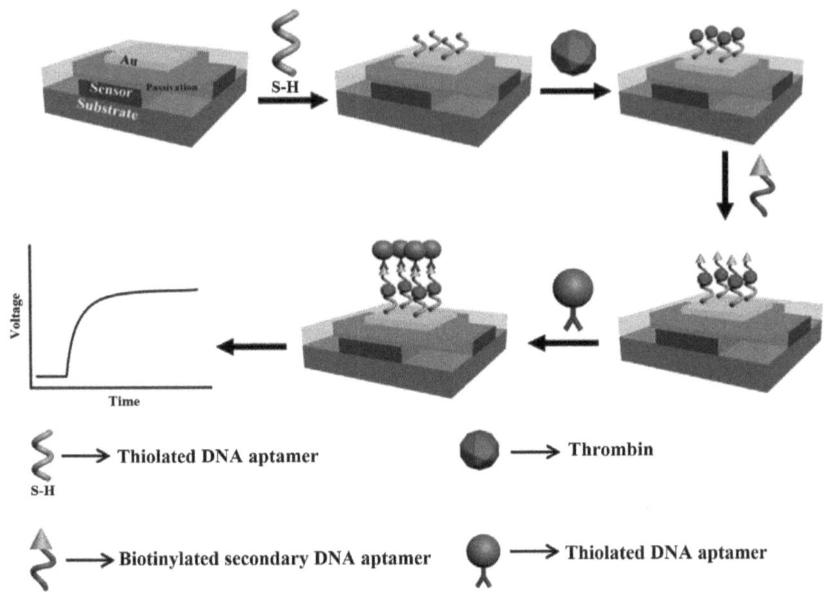

Figure 14.4 Schematic overview of the aptamer-based thrombin detection using the developed Hall sensor.
Source: Reproduced from B. Sinha, T.S. Ramulu, K.W. Kim, R. Venu, J.J. Lee, C.G. Kim, Planar Hall magnetoresistive aptasensor for thrombin detection, Biosens. Bioelectron. 59 (2014) 140–144.

micro-Hall biosensor displayed a higher sensitivity than conventional electrochemical sensors such as potentiometric and amperometric sensors.

Also, some researchers developed the Hall sensors to fabricate gas sensors [51–53]. For example, Gerber et al. [51] reported a gas sensor based on the Hall effect for the sensing of hydrogen using thin Co/Pd alloy film. The chemical interaction between hydrogen molecules and the ferromagnetic surface leads to significant changes in the magnetic properties. The Hall effect sensitivity of the gas sensor toward hydrogen was two orders of magnitude higher than that of the conductance gas sensors.

In addition to biosensors as well as gas sensors, the Hall devices construction can be used for other applications. The photonic spin Hall effect (PSHE) is attractive for many applications [54–56]. In this method, spin-dependent shifts occur when the light changes its direction at an optical interface. A more detailed discussion of the PSHE principle can be found elsewhere [57,58]. Generally, PSHE methods have been applied to measuring the physical parameters of the optical interfaces. In 2019 Liu et al. introduced an analytical application of the PSHE for the detection of the ion concentration in aqueous solution [59]. Under the applied external magnetic field, the change of the optical rotation correlates with the concentration of ions (Na^{+}, Mg^{2+}, and Ca^{2+}) and thus entails precise detection of ion concentration in aqueous medium using the amplified spin Hall shift.

Another type of non-MR sensor is SQUID. They are investigated as biosensors because of their excellent sensitivity toward weak biomagnetic signals. However, the application of SQUIDs in analytical fields is limited because of high costs for manufacturing and maintenance, resulting from low-temperature operating conditions [60]. To overcome this road-block, high transition temperature (high-T_c) has been proposed [61–63]. A high-T_c SQUID sensor was designed by Liao et al. for the determination of AFP, which is an important biomarker of metastatic liver cancer, germ cell tumors, as well as hepatocellular carcinoma [61]. The detection of the target in this non-MR sensor relies on a change in the phase delay (θ) of magnetization relative to applied field strength. The developed SQUID sensor measures the θ angle to achieve the effective relaxation time (τ_{eff}) of biofunctionalized Fe_3O_4 that was conjugated with AFP. Consequently, θ increased by increasing analyte concentration.

14.5 Conclusions

MR sensors, especially AMR, GMR, and TMR sensors, focus on the detection of biomarkers labeled with MNMs. Since GMR and TMR sensors offer advantages in terms of tunable size, excellent compatibility with metal-oxide semiconductor, and high sensitivity, they have been used successfully in many applications. AMR sensors have been employed in many biosensing applications because they are simple in technology and exhibit relatively high sensitivity, excellent stability, and flexibility. AMR sensors are low-cost sensors that are simply fabricated by a magnetic film. The sensing strategy of magnetic biosensors such as the AMR sensor is based on the sensing of the stray field of MNMs that label the analytes. GMR sensors can simultaneously detect different analytes through the surface modification of various sensors with different capture antibodies. The high-moment magnetic nanoparticles are the most promising types of MNMs used in GMR biosensors because they enhance the sensor signal and enable the GMR sensor to be a versatile and robust contender in biosensing applications. High sensitivity and the ability of the weak magnetic field detection make TMR sensors or spin-valve read sensors excellent candidates for biosensing applications.

As one of the most developed non-MR sensors, fluxgate sensors have advantages of reliability, high sensitivity, wide measurement range, and low cost. Hence, they are employed to measure weak magnetic fields. However, the practical application of present fluxgate sensors for biomagnetic detection is still limited because of their poor long-term stability, large design sizes, and heaviness. One proper candidate for the fabrication of efficient fluxgate sensors is the employment of MEMS technology to fabricate microfluxgate sensors. Hall sensors have some advantages containing small offset voltage, wide dynamic range, low noise, and linear response. Despite all the advantages of Hall sensors mentioned earlier, two main drawbacks limit their application in analytical chemistry. Compared to MR devices, Hall sensors are less sensitive to the relatively small magnetic field. Another considerable limitation is the low sensitivity of larger Hall sensors to detect small MBs. In other words, a small MB can be detected by a relatively small Hall sensor, due to the similarity in their size. Nevertheless, researchers have developed microsized Hall sensors to overcome the aforementioned limitations. Another type of non-MR sensor is SQUID. They are investigated as biosensors because of their excellent sensitivity toward weak biomagnetic signals. However, the application of SQUIDs in analytical fields is limited because of high costs for manufacturing and maintenance, resulting from low-temperature operating conditions.

References

[1] L. Vidal, M. Ahmadi, E. Fernández, T. Madrakian, A. Canals, Magnetic headspace adsorptive extraction of chlorobenzenes prior to thermal desorption gas chromatography-mass spectrometry, Anal. Chim. Acta 971 (2017) 40–47.

[2] M. Ahmadi, H. Elmongy, T. Madrakian, M. Abdel-Rehim, Nanomaterials as sorbents for sample preparation in bioanalysis: a review, Anal. Chim. Acta 958 (2017) 1–21.

[3] J.S. Beveridge, J.R. Stephens, M.E. Williams, The use of magnetic nanoparticles in analytical chemistry, Annu. Rev. Anal. Chem. (Palo Alto, Calif.) 4 (2011) 251–273.

[4] M. Faraji, Recent analytical applications of magnetic nanoparticles, Nanochem. Res. 1 (2016) 264–290.

[5] A. Ríos, M. Zougagh, M. Bouri, Magnetic (nano)materials as an useful tool for sample preparation in analytical methods. A review, Anal. Methods 5 (2013) 4558–4573.

[6] A. Afkhami, S. Sayari, F. Soltani-Felehgari, T. Madrakian, $Ni_{0.5}Zn_{0.5}Fe_2O_4$ nanocomposite modified carbon paste electrode for highly sensitive and selective simultaneous electrochemical determination of trace amounts of mercury (II) and cadmium (II), J. Iran. Chem. Soc. 12 (2015) 257–265.

[7] A. Afkhami, H. Khoshsafar, H. Bagheri, T. Madrakian, Preparation of $NiFe_2O_4$/graphene nanocomposite and its application as a modifier for the fabrication of an electrochemical sensor for the simultaneous determination of tramadol and acetaminophen, Anal. Chim. Acta 831 (2014) 50–59.

[8] H. Bagheri, A. Afkhami, P. Hashemi, M. Ghanei, Simultaneous and sensitive determination of melatonin and dopamine with Fe_3O_4 nanoparticle-decorated reduced graphene oxide modified electrode, RSC Adv. 5 (2015) 21659–21669.

[9] Wahajuddin, S. Arora, Superparamagnetic iron oxide nanoparticles: magnetic nanoplatforms as drug carriers, Int. J. Nanomed. 7 (2012) 3445–3471.

[10] M.N. Alves, M. Miró, M.C. Breadmore, M. Macka, Trends in analytical separations of magnetic (nano)particles, TrAC Trends Anal. Chem. 114 (2019) 89–97.

[11] T.A.P. Rocha-Santos, Sensors and biosensors based on magnetic nanoparticles, TrAC Trends Anal. Chem. 62 (2014) 28–36.

[12] H.T. Huang, P. Garu, C.H. Li, W.C. Chang, B.W. Chen, S.Y. Sung, et al., Magnetoresistive biosensors for direct detection of magnetic nanoparticle conjugated biomarkers on a chip, SPIN 09 (2019) 1940002.

[13] L.P. Ichkitidze, S.V. Selishchev, D.V. Telishev, Combined Magnetic Field Sensor with Nanosized Elements, Springer International Publishing, Cham, 2016, pp. 591–601.

[14] S. Tumanski, Thin Film Magnetoresistive Sensors, CRC Press, 2001.

[15] J. Lenz, S. Edelstein, Magnetic sensors and their applications, IEEE Sens. J. 6 (2006) 631–649.

[16] L.V. Panina, K. Mohri, Magneto-impedance effect in amorphous wires, Appl. Phys. Lett. 65 (1994) 1189–1191.
[17] L.V. Panina, K. Mohri, T. Uchiyama, M. Noda, K. Bushida, Giant magneto-impedance in Co-rich amorphous wires and films, IEEE Trans. Magn. 31 (1995) 1249–1260.
[18] A. Mason, S.C. Mukhopadhyay, K.P. Jayasundera, Sensing Technology: Current Status and Future Trends III, Springer International Publishing, 2014.
[19] M.-H. Phan, H.-X. Peng, Giant magnetoimpedance materials: fundamentals and applications, Prog. Mater. Sci. 53 (2008) 323–420.
[20] P. Ripka, Advances in fluxgate sensors, Sens. Actuators A: Phys. 106 (2003) 8–14.
[21] D.L. Graham, H.A. Ferreira, P.P. Freitas, Magnetoresistive-based biosensors and biochips, Trends Biotechnol. 22 (2004) 455–462.
[22] X. Zhang, N. Vernier, Z. Cao, Q. Leng, A. Cao, D. Ravelosona, et al., Magnetoresistive sensors based on the elasticity of domain walls, Nanotechnology 29 (2018) 365502.
[23] H. Lee, T.-H. Shin, J. Cheon, R. Weissleder, Recent developments in magnetic diagnostic systems, Chem. Rev. 115 (2015) 10690–10724.
[24] H. Huang, Z. Wei, J. Liou, W. Zhao, X. Xu, Localization of cells using magnetized patterned thin films, Mater. Sci. Eng.: C. 104 (2019) 109875.
[25] L. Quynh, B. Tu, D. Dang, D. Viet, L. Hien, D.H. Giang, et al., Detection of magnetic nanoparticles using simple AMR sensors in Wheatstone bridge, J. Sci.: Adv. Mater. Devices 1 (2016) 98–102.
[26] M.F. Hansen, G. Rizzi, Exchange-biased AMR bridges for magnetic field sensing and biosensing, IEEE Trans. Magn. 53 (2016) 1–11.
[27] L. Hien, L. Quynh, V. Huyen, B. Tu, N. Hien, D. Phuong, et al., DNA-magnetic bead detection using disposable cards and the anisotropic magnetoresistive sensor, Adv. Nat. Sci.: Nanosci. Nanotechnol. 7 (2016) 045006.
[28] D.R. Baselt, G.U. Lee, M. Natesan, S.W. Metzger, P.E. Sheehan, R.J. Colton, A biosensor based on magnetoresistance technology, Biosens. Bioelectron. 13 (1998) 731–739.
[29] P. Zhang, N. Thiyagarajah, S. Bae, Magnetically labeled GMR biosensor with a single immobilized ferrimagnetic particle agent for the detection of extremely low concentration of biomolecules, IEEE Sens. J. 11 (2010) 1927–1934.
[30] S.J. Osterfeld, H. Yu, R.S. Gaster, S. Caramuta, L. Xu, S.-J. Han, et al., Multiplex protein assays based on real-time magnetic nanotag sensing, Proc. Natl. Acad. Sci. 105 (2008) 20637–20640.
[31] Y. Wang, W. Wang, L. Yu, L. Tu, Y. Feng, T. Klein, et al., Giant magnetoresistive-based biosensing probe station system for multiplex protein assays, Biosens. Bioelectron. 70 (2015) 61–68.
[32] B. Srinivasan, Y. Li, Y. Jing, Y. Xu, X. Yao, C. Xing, et al., A detection system based on giant magnetoresistive sensors and high-moment magnetic nanoparticles demonstrates zeptomole sensitivity: potential for personalized medicine, Angew. Chem. Int. Ed. 48 (2009) 2764–2767.
[33] B. Srinivasan, Y. Li, Y. Jing, C. Xing, J. Slaton, J.-P. Wang, A three-layer competition-based giant magnetoresistive assay for direct quantification of endoglin from human urine, Anal. Chem. 83 (2011) 2996–3002.
[34] Y. Li, B. Srinivasan, Y. Jing, X. Yao, M.A. Hugger, J.-P. Wang, et al., Nanomagnetic competition assay for low-abundance protein biomarker quantification in unprocessed human sera, J. Am. Chem. Soc. 132 (2010) 4388–4392.
[35] Y. Wu, Y. Liu, Q. Zhan, J.P. Liu, R.-W. Li, Rapid detection of *Escherichia coli* O157: H7 using tunneling magnetoresistance biosensor, AIP Adv. 7 (2017) 056658.

[36] X.-H. Mu, H.-F. Liu, Z.-Y. Tong, B. Du, S. Liu, B. Liu, et al., A new rapid detection method for ricin based on tunneling magnetoresistance biosensor, Sens. Actuators B: Chem. 284 (2019) 638–649.

[37] G. Kurlyandskaya, M. Sanchez, B. Hernando, V. Prida, P. Gorria, M. Tejedor, Giant-magnetoimpedance-based sensitive element as a model for biosensors, Appl. Phys. Lett. 82 (2003) 3053–3055.

[38] T. Wang, Y. Zhou, C. Lei, J. Lei, Z. Yang, Development of an ingenious method for determination of Dynabeads protein A based on a giant magnetoimpedance sensor, Sens. Actuators B: Chem. 186 (2013) 727–733.

[39] T. Wang, Z. Yang, C. Lei, J. Lei, Y. Zhou, An integrated giant magnetoimpedance biosensor for detection of biomarker, Biosens. Bioelectron. 58 (2014) 338–344.

[40] T. Wang, L. Guo, C. Lei, Y. Zhou, Ultrasensitive determination of carcinoembryonic antigens using a magnetoimpedance immunosensor, RSC Adv. 5 (2015) 51330–51336.

[41] P. Ripka, New directions in fluxgate sensors, J. Magn. Magn. Mater. 215–216 (2000) 735–739.

[42] J. Lei, T. Wang, C. Lei, Y. Zhou, Detection of dynabeads using a micro-electro-mechanical-systems fluxgate sensor, Appl. Phys. Lett. 102 (2013) 022413.

[43] X.-c Sun, Z. Yang, C. Lei, Y. Liu, L. Guo, Y. Zhou, An innovative detecting way of *Escherichia coli* O157H:H7 by a micro-fluxgate-based bio-sensing system, Sens. Actuators B: Chem. 221 (2015) 985–992.

[44] Y. Cao, D. Cao, Theory of fluxgate sensor: stability condition and critical resistance, Sens. Actuators A: Phys. 233 (2015) 522–531.

[45] X.-c Sun, C. Lei, L. Guo, Y. Zhou, Sandwich immunoassay for the prostate specific antigen using a micro-fluxgate and magnetic bead labels, Microchim. Acta 183 (2016) 2385–2393.

[46] L. Guo, Z. Yang, S. Zhi, Z. Feng, C. Lei, Y. Zhou, Sensitive detection of cardiac troponin T based on superparamagnetic bead-labels using a flexible micro-fluxgate sensor, RSC Adv. 7 (2017) 52327–52336.

[47] L. Ejsing, M.F. Hansen, A.K. Menon, H. Ferreira, D. Graham, P. Freitas, Planar Hall effect sensor for magnetic micro-and nanobead detection, Appl. Phys. Lett. 84 (2004) 4729–4731.

[48] B. Sinha, T. Quang Hung, T. Sri Ramulu, S. Oh, K. Kim, D.-Y. Kim, et al., Planar Hall resistance ring sensor based on NiFe/Cu/IrMn trilayer structure, J. Appl. Phys. 113 (2013) 063903.

[49] B. Sinha, T.S. Ramulu, K.W. Kim, R. Venu, J.J. Lee, C.G. Kim, Planar Hall magneto-resistive aptasensor for thrombin detection, Biosens. Bioelectron. 59 (2014) 140–144.

[50] H. Zhan, J. Cervenka, S. Prawer, D.J. Garrett, Molecular detection by liquid gated Hall effect measurements of graphene, Nanoscale 10 (2018) 930–935.

[51] A. Gerber, G. Kopnov, M. Karpovski, Hall effect spintronics for gas detection, Appl. Phys. Lett. 111 (2017).

[52] J.Y. Lin, Z.X. Chen, X.L. He, W.M. Xie, Detection of H_2S at room temperature using ZnO sensors based on Hall Effect, Int. J. Electrochem. Sci. 12 (2017) 6465–6476.

[53] S.S. Das, G. Kopnov, A. Gerber, Detection of hydrogen by the extraordinary Hall effect in CoPd alloys, J. Appl. Phys. 124 (2018).

[54] X. Qiu, X. Zhou, D. Hu, J. Du, F. Gao, Z. Zhang, et al., Determination of magneto-optical constant of Fe films with weak measurements, Appl. Phys. Lett. 105 (2014) 131111.

[55] J. Zhou, H. Qian, C.-F. Chen, J. Zhao, G. Li, Q. Wu, et al., Optical edge detection based on high-efficiency dielectric metasurface, Proc. Natl. Acad. Sci. 116 (2019) 11137–11140.

[56] T. Zhu, Y. Lou, Y. Zhou, J. Zhang, J. Huang, Y. Li, et al., Generalized spatial differentiation from the spin hall effect of light and its application in image processing of edge detection, Phys. Rev. Appl. 11 (2019) 034043.
[57] J. Hirsch, Spin hall effect, Phys. Rev. Lett. 83 (1999) 1834.
[58] O. Hosten, P. Kwiat, Observation of the spin Hall effect of light via weak measurements, Science 319 (2008) 787–790.
[59] J. Liu, K. Zeng, W. Xu, S. Chen, H. Luo, S. Wen, Ultrasensitive detection of ion concentration based on photonic spin Hall effect, Appl. Phys. Lett. 115 (2019).
[60] A. Mahdi, L. Panina, D. Mapps, Some new horizons in magnetic sensing: high-Tc SQUIDs, GMR and GMI materials, Sens. Actuators A: Phys. 105 (2003) 271–285.
[61] S.-H. Liao, Y.-K. Su, Determining the time-dependent effective relaxation time of biofunctionalized magnetic nanoparticles conjugated with biotargets by using a high-Tc SQUID-based ac susceptometer for a magnetic immunoassay, Sens. Actuators B: Chem. 238 (2017) 66–70.
[62] Y.R. Chemla, H. Grossman, Y. Poon, R. McDermott, R. Stevens, M. Alper, et al., Ultrasensitive magnetic biosensor for homogeneous immunoassay, Proc. Natl. Acad. Sci. 97 (2000) 14268–14272.
[63] S.Y. Yang, J.J. Chieh, W.C. Wang, C.Y. Yu, N.S. Hing, H.E. Horng, et al., Magnetic nanoparticles for high-sensitivity detection on nucleic acids via superconducting-quantum-interference-device-based immunomagnetic reduction assay, J. Magn. Magn. Mater. 323 (2011) 681–685.

Index

Note: Page numbers followed by "*f*" and "*t*" refer to figures and tables, respectively.

Kolibri 360 GmbH
Ferdinand-Jühlke-Straße 7
99095 Erfurt, Deutschland
produktsicherheit@kolibri360.de